Methods in Molecular Biology

Series Editor
John M. Walker
School of Life Sciences
University of Hertfordshire
Hatfield, Hertfordshire, AL10 9AB, UK

For other titles published in this series, go to
www.springer.com/series/7651

Genetic Variation

Methods and Protocols

Edited by

Michael R. Barnes

Medicines Research Centre, GlaxoSmithKline Research & Development Limited, Stevenage, Hertfordshire, UK

Gerome Breen

Division of Psychological Medicine and Social, Genetic and Developmental Psychiatry Centre, Institute of Psychiatry, King's College London, London, UK

Humana Press

Editors
Michael R. Barnes
Medicines Research Centre
GlaxoSmithKline Research &
Development Limited
Stevenage, Hertfordshire SG1 2NY
UK
Michael.R.Barnes@gsk.com

Gerome Breen
Social, Genetic &
Developmental Psychiatry Centre
Institute of Psychiatry
King's College London
London SE5 8AF
UK
gerome.breen@gmail.com
Gerome.Breen@iop.kci.ac.uk

ISSN 1064-3745 e-ISSN 1940-6029
ISBN 978-1-60327-366-4 e-ISBN 978-1-60327-367-1
DOI 10.1007/978-1-60327-367-1
Springer New York Dordrecht Heidelberg London

Library of Congress Control Number: 2009943535

Printed on acid-free paper

Humana press is part of Springer Science+Business Media (www.springer.com)

Preface

"Your genome is an email attachment"

What a difference a few years can make? In 2001, to a global fanfare, the completion of the first draft sequence of the human genome was announced. This had been a Herculean effort, involving thousands of researchers and millions of dollars. Today, a project to re-sequence 1,000 genomes is well underway, and within a year or two, your own "personal genome" is likely to be available for a few thousand pounds, a price that will undoubtedly decrease further. We are fast approaching the day when your genome will be available as an email attachment (about 4 Mb). The key to this feat is the fact that any two human genomes are more than 99% identical, so rather than representing every base, there is really only a requirement to store the 1% of variable sequence judged against a common reference genome. This brings us directly to the focus of this edition of *Methods in Molecular Biology*, Genetic Variation.

The human genome was once the focus of biology, but now individual genome variation is taking the center stage. This new focus on individual variation ultimately democratizes biology, offering individuals insight into their own phenotype. But these advances also raise huge concerns of data misuse, misinterpretation, and misunderstanding. The immediacy of individual genomes also serves to highlight our relative ignorance of human genetic variation, underlining the need for more studies of the nature and impact of genetic variation on human phenotypes.

In March 2009, the US Congress passed the American Recovery and Reinvestment Act, which, among other things, granted the US National Institutes of Health an additional $8.2 billion in funding to disburse over the next 2 years. A substantial amount of this investment is likely to be channelled towards the re-sequencing of thousands of additional human genomes. When combined with the substantial amounts of data that already exist, data availability will no longer be a barrier to the understanding of human genetic variation. Against this background, we feel this edition of *Methods in Molecular Biology* is probably very timely. Although no publication could hope to comprehensively address all forms of human variation, our contributors have tried to provide coverage of most forms. This includes single nucleotide polymorphisms (SNPs), insertions/deletion (indels), copy number variation (CNVs), variable number tandem repeats (VNTRs), mitochondrial variation, mobile elements, and epigenetic variation. In the tradition of the series, we consider both laboratory and in silico methods, in many cases both in the same review. We believe that this underscores the need for increasing interactions between bench scientists and bioinformaticians. Neither breed of scientists can be independently successful in understanding the full impact of variation, but by working together they may have a fighting chance.

Stevenage, Hertfordshire — *Michael R. Barnes*
Denmark Hill, London — *Gerome Breen*

Contents

Contributors

FAHAD R. ALI • *Division of Physiology, School of Biomedical Sciences, University of Liverpool, Liverpool, UK*

MICHAEL R. BARNES • *Medicines Research Centre, GlaxoSmithKline Research & Development Limited, Stevenage, Hertfordshire, UK*

IAN P. BARRETT • *Cancer Bioscience, AstraZeneca, Macclesfield, Cheshire, UK*

MARK A. BATZER • *Department of Biological Sciences, Biological Computation and Visualization Center, Center for BioModular Multi-Scale Systems, Louisiana State University, Baton Rouge, LA, USA*

JOHN K. BLACKWOOD • *Mitochondrial Research Group, Newcastle University, Newcastle upon Tyne, Tyne and Wear, UK*

CHRISTOPH BOCK • *Max-Planck-Institut für Informatik, Saarbrücken, Germany*

GEROME BREEN • *Division of Psychological Medicine and Social, Genetic and Developmental Psychiatry Centre, Institute of Psychiatry, King's College London, London, UK*

HSUEH-WEI CHANG • *Department of Biomedical Science and Environmental Biology, Center of Excellence for Environmental Medicine, Graduate Institute of Natural Products, Kaohsiung Medical University, Kaohsiung, Taiwan*

YU-HUEI CHENG • *Department of Biomedical Science and Environmental Biology, Kaohsiung Medical University, Kaohsiung, Taiwan*

LI-YEH CHUANG • *Department of Biomedical Science and Environmental Biology, Kaohsiung Medical University, Kaohsiung, Taiwan*

BRADLEY P. COE • *Department of Cancer Genetics and Developmental Biology, British Columbia Cancer Research Centre, Vancouver, BC, Canada*

RICHARD CORDAUX • *Laboratoire Ecologie, Evolution et Symbiose, CNRS UMR 6556, Université de Poitiers, Poitiers, France*

UDAY S. EVANI • *Department of Medical and Molecular Genetics, Center for Computational Biology and Bioinformatics, Indiana University School of Medicine, Indianapolis, IN, USA*

ELIN GRUNDBERG • *Department of Human Genetics, McGill University and Génome Québec Innovation Centre, McGill University, Montréal, QC, Canada*

DE-LEUNG GU • *Department of Biomedical Science and Environmental Biology, Kaohsiung Medical University, Kaohsiung, Taiwan*

KATE HADDLEY • *Division of Physiology, Division of Human Anatomy & Cell Biology, School of Biomedical Sciences, University of Liverpool, Liverpool, UK*

KONSTANTIN HALACHEV • *Max-Planck-Institut für Informatik, Saarbrücken, Germany*

KYUDONG HAN • *Department of Biological Sciences, Biological Computation and Visualization Center, Louisiana State University, Baton Rouge, LA, USA*

HURNG-WERN HUANG • *Department of Biomedical Science and Environmental Biology, Kaohsiung Medical University, Kaohsiung, Taiwan*
EWEN F. KIRKNESS • *The J. Craig Venter Institute, Rockville, MD, USA*
MIRIAM K. KONKEL • *Department of Biological Sciences, Biological Computation and Visualization Center, Center for BioModular Multi-Scale Systems, Louisiana State University, Baton Rouge, LA, USA*
KIM J. KRISHNAN • *Mitochondrial Research Group, Newcastle University, Newcastle upon Tyne, Tyne and Wear, UK*
VIDHYA G. KRISHNAN • *Department of Medical and Molecular Genetics, Center for Computational Biology and Bioinformatics, Indiana University School of Medicine, Indianapolis, IN, USA*
TONY KWAN • *Department of Human Genetics, McGill University and Génome Québec Innovation Centre, McGill University, Montréal, QC, Canada*
WAN L. LAM • *Department of Cancer Genetics and Developmental Biology, British Columbia Cancer Research Centre, Vancouver, BC, Canada; Department of Pathology and Laboratory Medicine, Interdisciplinary Oncology Program, University of British Columbia, Vancouver, BC, Canada*
THOMAS LENGAUER • *Max-Planck-Institut für Informatik, Saarbrücken, Germany*
BO ESKEROD MADSEN • *AgroTech, Institute for Agri Technology and Food Innovation, Aarhus N, Denmark*
SEAN D. MOONEY • *Department of Medical and Molecular Genetics, Center for Computational Biology and Bioinformatics, Indiana University School of Medicine, Indianapolis, IN, USA*
ANTON NEKRUTENKO • *Center for Comparative Genomics and Bioinformatics, Huck Institutes for Life Sciences, Penn State University, University Park, PA, USA*
PAULINE C. NG • *The J. Craig Venter Institute, Rockville, MD, USA*
TOMI M. PASTINEN • *Department of Human Genetics, McGill University and Génome Québec Innovation Centre, McGill University, Montréal, QC, Canada*
INTI PEDROSO • *NIHR Biomedical Research Centre for Mental Health, South London and Maudsley NHS Foundation Trust and Institute of Psychiatry and MRC Social Genetic and Developmental Psychiatry Centre, Institute of Psychiatry, King's College London, London, UK*
JOHN P. QUINN • *Division of Human Anatomy & Cell Biology, School of Biomedical Sciences, University of Liverpool, Liverpool, UK*
DAVID A. RAY • *Department of Biology, West Virginia University, Morgantown, WV, USA*
AMY K. REEVE • *Mitochondrial Research Group, Newcastle University, Newcastle upon Tyne, Tyne and Wear, UK*
CLAUDIA RUIVENKAMP • *Center for Human and Clinical Genetics, Leiden University Medical Center (LUMC), Leiden, The Netherlands*
JACQUELINE SCHOUMANS • *Department of Molecular Medicine & Surgery, Karolinska Institute, Stockholm, Sweden*
SHURJO K. SEN • *Department of Biological Sciences, Biological Computation and Visualization Center, Center for BioModular Multi-Scale Systems, Louisiana State University, Baton Rouge, LA, USA*

James Taylor • *Departments of Biology and Mathematics & Computer Science, Emory University, Atlanta, GA, USA*

Robert W. Taylor • *Mitochondrial Research Group, Newcastle University, Newcastle upon Tyne, Tyne and Wear, UK*

Kelsie L. Thu • *Department of Cancer Genetics and Developmental Biology, British Columbia Cancer Research Centre, Vancouver, BC, Canada; Interdisciplinary Oncology Program, University of British Columbia, Vancouver, BC, Canada*

Douglass M. Turnbull • *Mitochondrial Research Group, Newcastle University, Newcastle upon Tyne, Tyne and Wear, UK*

Palle Villesen • *Bioinformatics Research Center (BiRC), University of Aarhus, Aarhus C, Denmark*

Greg Von Kuster • *Center for Comparative Genomics and Bioinformatics, Huck Institutes for Life Sciences, Penn State University, University Park, PA, USA*

Emily A. Vucic • *Department of Cancer Genetics and Developmental Biology, British Columbia Cancer Research Centre, Vancouver, BC, Canada; Department of Pathology and Laboratory Medicine, University of British Columbia, Vancouver, BC, Canada*

Jerilyn A. Walker • *Department of Biological Sciences, Biological Computation and Visualization Center, Louisiana State University, Baton Rouge, LA, USA*

Michael E. Weale • *Department of Medical and Molecular Genetics, King's College London, Guy's Hospital, London, UK*

Ariane C. Williams • *Department of Cancer Genetics and Developmental Biology, British Columbia Cancer Research Centre, Vancouver, BC, Canada*

Carsten Wiuf • *Bioinformatics Research Center (BiRC), University of Aarhus, Aarhus C, Denmark*

Peter M. Woollard • *Computational Biology, Quantitative Sciences, GlaxoSmithKline Pharmaceuticals, Stevenage, Hertfordshire, UK*

Cheng-Hong Yang • *Department of Biomedical Science and Environmental Biology, Kaohsiung Medical University, Kaohsiung, Taiwan*

Chapter 1

Genetic Variation Analysis for Biomedical Researchers: A Primer

Michael R. Barnes

Abstract

Biomedical researchers studying gene function should consider the impact of variation, even if genetics is not the primary objective of an investigation. Information on genetic variation can provide a valuable insight into the functional range and critical regions of a gene, protein or regulatory element. Genetic variants may be diverse in nature, ranging from single nucleotide variants, tandem repeats, small insertions or deletions to large copy number variants. Until recently, information on genetic variation was quite limited, but now a range of large scale surveys of variation have made plentiful data on common variation and a picture is beginning to emerge from the driving forces in human evolution and population diversification. Next-generation sequencing technologies are moving knowledge into a new phase focused on the individual genome and complete disclosure of individual variation, including the rarest of variants. The consequences of these advances in medicine are unresolved, but it is clear that biomedical researchers cannot afford to ignore this information. This review presents a broad overview of the *in silico* methods that will allow a researcher to quickly review known variation in a gene of interest, providing some pointers for further investigation.

Key words: SNP, CNV, VNTR, INDEL, Polymorphism, Genome, Bioinformatics, Variation, Mutation

1. Introduction

1.1. Genetic Variants: From Phenotypic Determinants to Commodity

Genetic variation is a key biological determinant underpinning evolution and defining the heritable basis of phenotype. How a researcher might want to deal with genetic data really depends on the viewpoint of the researcher. The viewpoint of the biomedical researcher tends to be gene- or phenotype-centric. Gene function cannot be fully understood without awareness of the potential variability within a gene. This means that biomedical researchers, studying genes need to know what variants exist and what impact

Michael R. Barnes and Gerome Breen (eds.), *Genetic Variation: Methods and Protocols*, Methods in Molecular Biology, vol. 628, DOI 10.1007/978-1-60327-367-1_1, © Springer Science+Business Media, LLC 2010

these variants might have on gene function and consequently phenotype. The viewpoint of the geneticist tends to be variant centric. To geneticists, genetic variation is essentially a commodity used as a marker of phenotype to pinpoint the specific variants that determine phenotype. These are quite differing objectives, which mean that although both groups of researchers need to access the same comprehensive datasets of variation, the kinds of operations they need to perform may be distinct. This review is intended for the biomedical researcher, who would like to know how genetic variation could impact their gene(s) of interest. It will present an overview of some of the different forms of genetic variation that should be considered, highlighting the key databases from which this data can be accessed and manipulated. This is intended as a basic review with no prior assumption of knowledge of the field. For simplicity, this review will also largely avoid coverage of the key genetic concept of linkage disequilibrium and the work done by the HapMap consortium to study haplotype structure (see (1) for a review of this area). Instead the focus here is on the variants, those with some pre-existing knowledge of the field, particularly geneticists are directed to the more detailed reviews in this volume.

1.2. SNPs are Not the Only Variants

As this manuscript will evidence, the single nucleotide polymorphism (SNP) will dominate in any review of genetic variation data. To some extent, the preoccupation with SNPs is driven by technology. Put simply, SNPs are easy and cheap to assay, and so they have become the tool of choice for studying human genetics. SNPs are the commodity of genetic research. However, SNPs are also the commonest form of human variation, occurring approximately every kilobase when two chromosomes are compared (2). This makes SNPs the most likely candidates for the determination of phenotype. Other forms of variation do not come close to the SNP in terms of frequency; as a result, this review will without apology focus most attention on SNP variation. However, other forms of variation such as tandem repeats, small insertions or deletions and large copy number variants also exist and should not be overlooked. Fortunately, data on these forms of variation is also improving, particularly in genome browsers, so these will also be reviewed. As next-generation sequencing technologies begin to focus on re-sequencing of individual genomes, the understanding of all forms of variation will increase, and rarer forms of variation are likely to receive much more attention. The consequences of this shift in focus for medicine remains to be seen, but it is clear that biomedical researchers cannot afford to ignore this information.

1.3. Types of Genetic Variation

Genetic variation takes many forms, but all forms originate from just two types of mutation event. The simplest type of variation results from a simple base substitution. This type of mutation event accounts for the commonest form of variation, the single nucleotide polymorphism (SNP), but also rare mutations which may show Mendelian inheritance in families. Most of the other types of variation result directly or indirectly from the insertion or deletion of a section of DNA. At the simplest level, this can result in the insertion or deletion of one or more nucleotides, the so called insertion/deletion (Indel) polymorphisms. The most common insertion/deletion events occur in repetitive sequence elements, where repeated nucleotide patterns, so called "variable number tandem repeat polymorphisms" (VNTRs), expand or contract as a result of insertion or deletion events. VNTRs are further sub-divided on the basis of the size of the repeating unit; minisatellites are composed of repeat units ranging from ten to several hundred base pairs. Simple tandem repeats (STRs or microsatellites) are composed of 2–6 bp repeat units. Insertion/deletion events involving large regions ranging from a few kilobases to several megabases are known as copy number variations (CNVs). These events may occur as a result of recombination between flanking repetitive elements. CNVs were once thought to be very rare, restricted to severe genomic syndromes; however, the initial sequencing of the human genome and more recently, studies of samples ascertained for the HapMap project have provided evidence that CNVs are actually commoner than previously expected with most individuals carrying substantial deletion or duplications of DNA possibly with little phenotypic impact (3).

1.4. How Much Variation Exists in the Average Individual?

The quantity of genetic variation in the human genome is something that until relatively recently we have only been able to make an educated guess to estimate. Empirical studies quite quickly identified that, on average, comparison of chromosomes between any two individuals will generally reveal common SNPs (>20% minor allele frequency) at 0.3–1 kb average intervals, which scales up to 5–10 million SNPs across the genome (2). Large scale SNP discovery projects such as the HapMap (4), have proved these early estimates to be remarkably accurate. The number of potentially polymorphic VNTRs, can now be determined from the complete human genome sequence, there are >600,000 simple repeats (see Note 1). All could potentially be mutable in some individuals, but those with greater than 8–12 repeats are most likely to be polymorphic (5). Other forms of variation such as small insertion deletions have been more technologically difficult to quantify, although they are likely to fall somewhere between SNPs and VNTRs in numbers. Large CNVs have for a long time remained the most unquantifiable form of variation in the genome. Until recently, quantification of CNVs was only possible by intensive cytogenetic methods (6). CNVs cannot be reliably identified from assembled genome sequence. In fact they are implicitly an obstacle to genome assembly, as large duplications are often incorrectly collapsed into a single assembly. The breakthrough for CNV analysis came with the development of more sensitive methods for SNP analysis, which actually allowed CNV calling on the basis of signal intensity in a given region (7). This allowed Redon et al. (3) to screen for CNVs in 270 individuals from four populations with ancestry in Europe, Africa or Asia (the HapMap collection). A total of 1,447 CNVs were identified, covering 360 megabases (12% of the genome). More recent studies have downgraded this estimate somewhat, suggesting more complexity, but the figures are similar, with 1,320 CNVs existing at population frequencies of greater than 1% (8).

1.5. Polymorphism or Mutation?

Before getting into the specifics of the analysis of variation, it is worth clarifying some of the terminologies that are applied to genetic variants. In the simplest sense, the terminology for a variant is defined by allele frequency. A variant, occurring in a population at a frequency of >1% is generally termed a polymorphism. When a variant occurs at <1%, it is considered to be a mutation. However, this definition looks increasingly artificial, particularly with wider availability of next-generation sequencing data, which is allowing re-sequencing of 1,000s of individuals enabling the identification of many rare variants (9). Instead "mutations" occurring in more than one individual but at <1% in general populations might more appropriately be termed rare variants. The term "mutation" is widely used in a Mendelian disease context to describe germline variants identified in related individuals that segregate with a disease phenotype. However, the case for the description of Mendelian variants as "mutations" is not very well defined. A grey area exists which argues against rigid division of germline variation data. Some autosomal recessive Mendelian mutations have been linked to complex disease susceptibility in a heterozygote form and indeed are relatively widely spread in populations. For example, homozygote mutations in the cystathione beta synthase gene cause homocystinuria, a rare disorder inducing multiple strokes at an early age. The heterozygotes do not share this severe disorder, but do have an increased lifetime risk of stroke (10). In Caucasians, the population frequency of homozygote homocystinuria mutations is only 1/126,000, but in the same population, heterozygote frequency is relatively high at 1/177. Many other examples exist of Mendelian mutations that actually exist at appreciable heterozygote levels in general populations, particularly isolated populations, e.g. mutations in the breast cancer susceptibility gene, BRCA1 have been found in 1–2% of Jewish populations (11) and mutations in the CFTR gene, cause cystic fibrosis – the most common autosomal recessive disease in the Caucasian population, with a carrier frequency of around 2% (12).

1.6. Somatic Mutations

Perhaps the most appropriate use of the term "Mutation" is to describe somatic variants that arise in cancer cells cannot be inherited. This review will not consider somatic mutation data; however, it is important to be aware if a mutation is somatic or germline in origin and two should not be confused. COSMIC is a database of somatic mutations in cancer which can be consulted for more information (13).

1.7. The Natural History of Genetic Variants

The presence of heterozygous Mendelian "mutations" in general populations illustrates the point that it may not always be helpful to consider polymorphism and mutation data separately. Actually as Mendelian mutations are increasingly "re-discovered" in large scale population screens for variation such as the 1,000 genomes project (www.1000genomes.org/), mutations are increasingly finding their way into dbSNP (Table 1). This is being reflected in OMIM (Table 1; (14)), one of the best repositories of Mendelian mutation data, which now links to dbSNP where appropriate records exist.

As discussed previously, both SNPs and point mutations arise by the same mechanism; selection is the force that influences their spread in populations. Miller and Kwok (15) presented a detailed review of the "life cycle" of a single nucleotide variation (Fig. 1). They defined SNP and mutation evolution in four phases:

1. Appearance of a new variant allele by mutation
2. Survival of the allele through early generations against the odds
3. Increase of the allele to a substantial population frequency
4. Fixation of the allele in populations.

Each of these stages goes to the heart of the differences and similarities between SNPs and mutations. Both arise by the same mechanism; nucleotide substitution is DNA sequence context dependent – substitution rates are influenced by 5′ and 3′ nucleotides. This effect is most dramatic for CT and GA transitions; these CpG dinucleotides are methylated and tend to deaminate to either a TpG or CpA dinucleotide (16). This makes these dinucleotides the most likely locations for point mutation in the human genome, with G>A or C>T transitions accounting for 25% of all SNPs and mutations in the human genome (15). In itself this molecular mechanism accounts for the deficiency of CG dinucleotides in the human genome. The creation of new CG dinucleotides is not an adequate counter balance against this effect, due to the lower frequency of tranversions back to CpG. While SNPs and mutations both arise in the same way, their survival in populations is likely to be quite different. Most newly arisen mutations are likely to be lost in early generations by random sampling of the gene pool alone. For example, if a heterozygous individual

Table 1
Tools for genetic characterisation

Central databases (SNPs and mutations)	
dbSNP	http://www.ncbi.nlm.nih.gov/SNP/
HapMap	http://www.hapmap.org
The SNP consortium (TSC)	http://snp.cshl.org/
Mutation databases	
OMIM	http://www.ncbi.nlm.nih.gov/Omim/
HGMD	http://www.hgmd.org
Locus specific mutations	http://www.hgvs.org/dblist/glsdb.html
CNV databases	
Db of genomic variants	http://projects.tcag.ca/variation/
Structural variation db	http://Humanparalogy.gs.washington.edu/
Decipher	https://decipher.sanger.ac.uk/
VNTR databases	
Tandem repeats database	https://tandem.bu.edu/cgi-bin/trdb/trdb.exe
uniSTS	http://www.ncbi.nlm.nih.gov/sites/entrez?db=unists
Linkage disequilibrium visualisation and analysis	
SNAP	http://www.broad.mit.edu/mpg/snap/
HaploView	http://www.broad.mit.edu/mpg/haploview/
Gene orientated SNP and mutation visualisation	
LocusLink	http://www.ncbi.nlm.nih.gov/LocusLink/
HUGE navigator	http://www.hugenavigator.net/
SNPper	http://bio.chip.org:8080/bio/snpper-enter
BrainArray	http://brainarray.mbni.med.umich.edu/Brainarray/
Biomart	http://www.ensembl.org/biomart/martview?
Genome orientated SNP and mutation visualisation	
Ensembl	http://www.ensembl.org
UCSC	http://genome.ucsc.edu/index.html
Map viewer	http://www.ncbi.nlm.nih.gov/projects/mapview/
1,000 genomes project	http://www.1000genomes.org/
NHGRI catalog of GWAS	http://www.genome.gov/gwastudies/

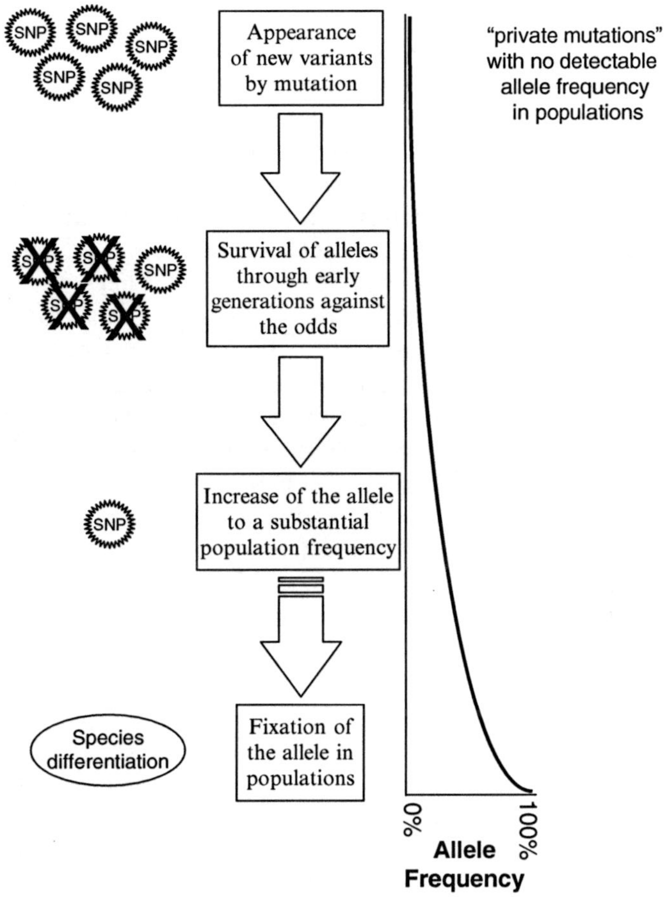

Fig. 1. The life cycle of SNPs and mutations. SNP and mutation evolution occurs in four main phases; (1) Appearance of a new variant allele by mutation; (2) Survival of the allele through early generations against the odds; (3) Increase of the allele to a substantial population frequency; (4) Fixation of the allele in populations

for a selectively neutral mutation has two offspring, there is a 0.75 probability that the mutation will be found in at least one child. If each generation has two children, the probability of loss of the new mutation is $1-(0.75)^g$, where g = generations. To give a worked example, this relates to a 94% probability of loss of a mutation or SNP in 10 generations (approximately 200 years). Where a heterozygous mutation has an early onset deleterious effect, natural selection is likely to further increase the rate of loss of the allele from populations. The same pressures do not apply to late onset diseases, perhaps explaining the proliferation of such diseases in humans.

If an SNP or mutation survives early generations and increases in frequency sufficiently to become homozygous in some individuals, the risk of loss of the allele will be reduced. At this stage, the frequency of the allele in a population is likely to vary, with higher frequency alleles being consistently favoured, especially when populations are subject to severe bottlenecks in size. Reich et al. (17) studied linkage disequilibrium data between SNPs to present convincing evidence for such a bottleneck in recent Northern European population history. In the face of these fluctuations of allele frequency, an SNP or mutation will cease to exist in populations, either by disappearing or by reaching a 100% allele frequency, in which case the variant becomes an allele that helps to define a species. Interestingly, SNPs have been shown to be shared between closely related species e.g. Rhesus and Cynomologus Macaques (18). However, there is less evidence of conservation of SNPs between more distantly related species. Hodgkinson et al. (19) were able to identify over 11,000 SNPs in the same location between humans and chimpanzees. However, they concluded that these were unlikely to herald from a common ancestor, based on a repeat of their analysis using human and macaque SNPs, which showed a higher occurrence of shared SNPs than would be expected considering the distance of a shared common ancestor. Overall, the pattern of coincident SNPs was also inconsistent with ancestral polymorphism. These lines of evidence all suggest that the lifetime of an SNP is considerably shorter than the divergence of species. Miller et al. (20) estimated that the average period from original mutation to species fixation of an allele was 284, 000 years.

1.8. Recognising "Private Mutations"

Although these considerations of evolution of genetic variation might appear highly academic in nature, they do have an immediate relevance to the data currently being generated by genome re-sequencing efforts. If a variant is seen in only one individual, it should be considered to be a "private mutation" – often referred to as a "private SNP" although this description is rather an oxymoron. Once a variant is seen for the second time, then arguably it is no longer a private mutation, but instead it may be a population polymorphism. Awareness of this definition is important, this means that every single variant seen in an individual genome could potentially be a population based polymorphism, but only when it is seen twice. Until then, the individual bearing the "private mutation" might be an interesting model of phenotype, but arguably, considering the massive odds against fixation of a mutation in a population, the variant bears little relevance to phenotype on a population level.

1.9. Genetic Variation Databases

For the reasons discussed earlier, among genetic variation databases, the SNP is far and away the most highly documented genetic variant. Interest in the SNP as the driving force of new genetic technologies led to the early development of a predominant SNP database – dbSNP at the NCBI (21) (Table 1). Other forms of variation have their own databases, although none is as well organized and fast growing as dbSNP (see Note 2). Some of the key databases are briefly reviewed subsequently. Table 1 lists a selection of the best databases.

1.9.1. SNP and Indel Databases

The National Center for Biotechnology Information (NCBI) established the dbSNP database in September 1998 as a central repository for both SNPs and short Indel polymorphisms. In June 2009, human build 130 of dbSNP contained 79 million SNPs (see Note 3). These SNPs cluster into a non-redundant set of 17.8 million SNPs, known as Reference SNPs (RefSNPs). 7.3 M SNPs are located in gene regions. 6.6 M SNPs have frequency information and so are considered to be validated. As dbSNP is part of the NCBI-Entrez suite of tools, SNP records are highly integrated with other information, particularly gene and genomic information. The individual SNP reports in dbSNP are generally very good, with reports on SNP functional impact and population frequency. It is possible to query dbSNP directly, but the search interface can be a little confusing at times. Sometimes it is easier to query indirectly using other tools, such as the UCSC genome browser and Ensembl (Table 1). Examples of both types of query are given in Subheading 3. Most of the variation data in dbSNP is likely to be functionally neutral, resolving the functional variants is essentially the objective of genome wide association studies (GWAS) that are currently being used to study complex disease genetics. Many of the dbSNP variants that have already been associated with complex phenotypes have been recorded and are searchable in the NHGRI Catalogue of GWAS (Table 1).

1.9.2. Mendelian Mutation Databases

A large number of Mendelian disease mutations have been identified over the past 50 years. These have helped to define many key biological mechanisms, including gene regulatory motifs and protein–protein interaction. Many highly specialised locus specific databases (LSDBs) have been established to collate this data (Table 1). This review could not hope to cover all these databases; however, OMIM is one database that serves as a good starting point for most searches. Online Mendelian Inheritance in Man (OMIM) is an online catalogue of human genes and their associated mutations, based on the long running catalogue Mendelian Inheritance in Man (MIM), started in 1967 by Victor McKusick at Johns Hopkins (14). OMIM is an excellent resource for getting a quick background biology on genes and diseases, it includes information on the most common and clinically significant mutations and also polymorphisms in genes. Despite the name, OMIM also covers complex diseases to varying degrees of detail. OMIM is curated by a dedicated but small group of curators, but the limits of a manual curation process mean that entries may not be current or comprehensive. With this caveat aside OMIM is a very valuable database, which usually presents a very accurate digest of the literature (it would be difficult to do this in such a focused way automatically). A major added bonus of OMIM is that it is very well integrated with the NCBI database family, most recently with dbSNP, this makes movement from a disease to a gene to a locus and vice versa fairly effortless.

1.9.3. VNTR and Microsatellite Databases

Highly polymorphic microsatellites have been shown to have considerable utility as markers in genetic studies; however, much evidence also exists to demonstrate that tandem repeats can also exert a direct functional effect when located in or near gene coding or regulatory regions. Thus, VNTRs in themselves can be candidates for disease causing genetic variants. The most well-characterised of these are the triplet repeat expansion diseases (22). Tandem repeats have also been associated with complex diseases, for example different alleles of a 14-mer VNTR in the insulin gene promoter region, have been associated with different levels of insulin secretion. Different alleles of this VNTR have been robustly linked with type I diabetes (23).

In comparison to the hundreds of thousands of VNTR polymorphisms in the genome, only ~18,000 VNTRs have been genetically characterised. Several highly characterised subsets of these markers have been arranged into well defined linkage marker panels by the Marshfield Institute and Genethon. Almost all genetically characterised VNTRs are stored centrally in the NCBI uniSTS database (Table 1). Potentially polymorphic novel VNTRs can be identified from genomic sequence using the tandem repeat finder tool (24) (http://tandem.bu.edu/trf/trf.html). A complete analysis of the human genome sequence using tandem repeat finder is presented in the UCSC human genome browser in the "simple repeats" track (Table 1).

1.9.4. Copy Number Variation Databases

Technologies enabling the recognition of CNVs and other structural variants are evolving rapidly and so too are databases to enable the documentation and analysis of these variants. The Database of Genomic Variants (DGV) (Table 1) was established to catalogue genomic variation from human control samples. Considering this, the level of copy number variation is remarkable. Indeed, it is probably worth bearing in mind the caveat that although these individuals are deemed to be "healthy controls," the amount of phenotypic documentation is limited. A control subject for a cancer study, for example, may not have been assessed with respect to blood pressure, psychiatric disorder, etc. (the same probably applies to any control subject). Moving from controls to subjects with defined phenotypes, the DECIPHER database, contains data on chromosomal imbalance and other structural variants (Table 1). These are mostly highly penetrant variants that cause overt phenotypes such as dysmorphic syndromes and cognitive impairment. As understanding of structural variation advances, the overlap of content in "control" and "disease" databases such as DGV and DECIPHER, respectively, is likely to increase, just as we have seen growing overlaps between SNP and mutation databases.

2. Materials

All the tools described here are freely available internet web tools which would run on any PC, Mac or Unix workstation with web access. For more sophisticated analysis of large datasets on a genomic scale, see (25). A list of genomic tools and databases mentioned and used in this review is given in Table 1.

3. Methods

3.1. Using dbSNP to Identify Known SNPs in a Gene of Interest

Although there are many tools for identification of SNPs in genes, almost all use data from dbSNP, but if the dbSNP database version is not given by the tool, it is not always possible to determine if the data is fully up to date (see Note 4). To ensure that all available data is obtained, this example queries dbSNP directly. Some later examples will illustrate indirect methods of querying dbSNP data.

1. Navigate to the NCBI dbSNP resource (see Table 1).
2. Select "Gene" from the "Search Entrez" pull down menu. Enter the gene symbol of interest, e.g. CCR5. Click on the gene symbol for the species you are interested in, e.g. Human. The Entrez Gene summary for your gene of interest will be displayed.
3. On the right hand side, there is a list of Links, press the "SNP:GeneView" link. This returns a report which by default lists all SNPs in the coding region of the gene.
4. To view all SNPs in the gene, including those in introns and promoter regions, press the "in gene region" button near the top of the report and press refresh. This returns all SNPs in the gene.
5. To view all SNPs in the gene with known allele frequency information, press the "has frequency" button and press refresh. This returns a report on all SNPs in the gene with known allele frequency information (Fig. 2.)
6. From the results, a range of validated SNPs are evident in the CCR5 gene. There is also an Indel polymorphism (rs333). Links to further analysis and more information are given

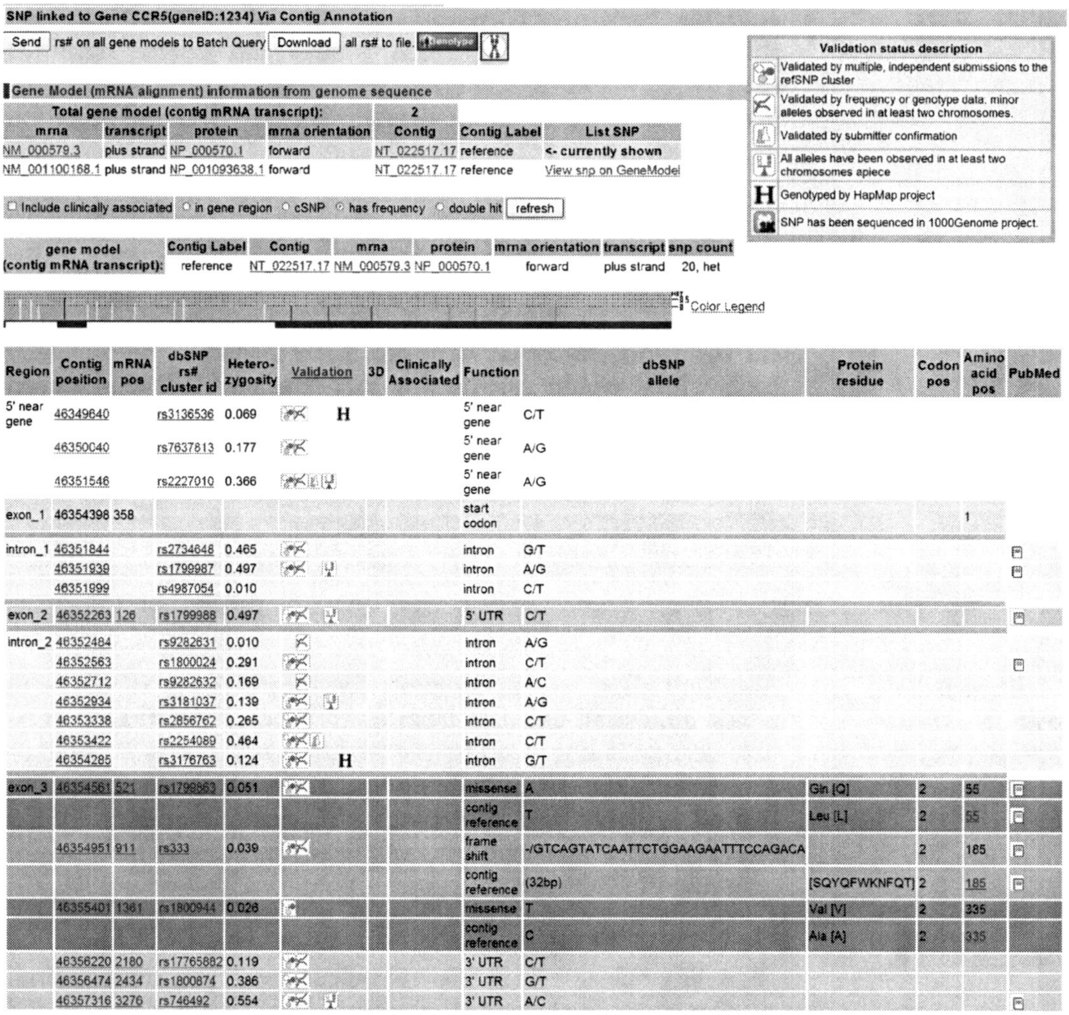

Fig. 2. The dbSNP gene view

where available, including a link to the CN3D database (26), which places SNPs in a structural context. Links are also sometimes given if the SNP is clinically associated (see Note 5). The validation status of the SNP and an indication of data availability from the HapMap and 1,000 genomes project is indicated by a small icon, the key for which is shown in Fig. 2.

3.2. Obtaining a Genomic Overview of Known Variants Across a Gene

Although tools like dbSNP offer detailed reports of variants across a gene, the results only encompass SNP and Indel polymorphisms. The UCSC genome browser (Table 1; (27)) can be used to quickly generate a comprehensive overview of all types of variation across a gene locus.

1. Navigate to the UCSC genome browser (see Table 1). Enter the gene symbol "HCRT" into the query window
2. A number of UCSC mapped genes are returned. Select the top hit for HCRT from the UCSC genes results (see Note 6). This should return a genomic view of the HCRT gene. In order to view the wider locus around the gene, zoom out by pressing the "zoom out 3×" button.
3. A view of the HCRT gene locus is returned. The browser displays a number of default tracks with information of interest (see Note 7). In this example to focus on genetic variation, press the "hide all" button below the genome view
4. To view the gene and major types of variation, Use the configuration menu at the bottom of the screen to make data of interest visible. In the Genes and gene prediction menu, select "pack" on the "UCSC genes" track. In the "Variation and Repeats" menu, select "pack" for "SNPs (129)", "DGV Struct Var" and "Simple Repeats". These represent SNPs, CNVS and VNTRs, respectively. Press the refresh button.
5. The tool should display all known SNPs, CNVs and Simple repeats across the HCRT locus (Fig. 3).
6. Information about the variants can be reviewed by clicking on the variant. In the case of HCRT, there are 22 SNPs across the gene region, but only one SNP is located in the coding region. SNPs causing a non-synonymous change are indicated in red, synonymous SNPs are indicated in green. SNPs located in the untranslated regions of the gene are indicated in blue. There are also four simple repeats flanking the gene. Clicking on the repeats shows that there is a potentially polymorphic tetrameric CCTT repeat that is repeated perfectly 12 times. There are no CNVs in the immediate HCRT region, where available information on CNVs is linked from the database of genomic variants (see Table 1).

3.3. Annotating Known Variants Across a Gene at the Sequence Level

Sometimes it may be useful to annotate SNPs across a nucleotide sequence (e.g. for publication or primer design). The following example goes through the sequence export and annotation process that is available from the UCSC genome browser (Table 1).

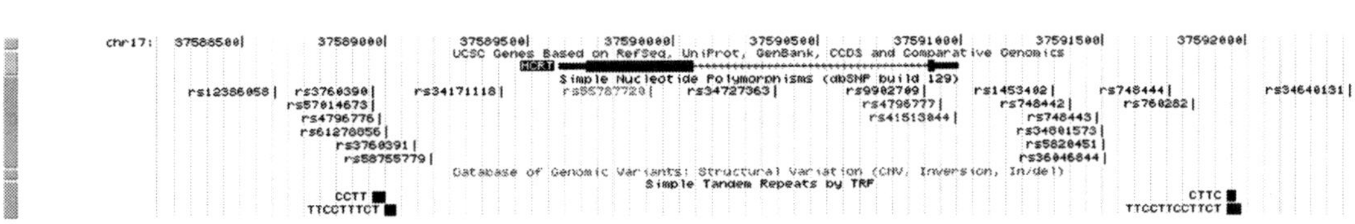

Fig. 3. A UCSC genome browser view of common variation across the HCRT gene

In this case, we will use the HCRT locus queried in the previous example.

1. Navigate to the UCSC genome browser (see Table 1) and retrieve the HCRT locus as described in steps 1 and 2 in the previous example.
2. Click on the "DNA" link in the top menu bar. A DNA export form is returned. Press the "Extended case/colour options" button. This returns a complex form that allows the user to annotate the currently selected UCSC tracks on the locus queried.
3. Mark up the tracks of interest using the colour, toggle and underline features (see Fig. 4). HCRT is located on the reverse strand, so tick the "reverse complement" box at the top of the form. Press the submit button.
4. The reverse complemented genomic sequence across the HCRT locus is returned with fully annotated exons and known variants (Fig. 4).

3.4. Using Biomart to Identify SNPs in a Given List of Genes

Most of the previous examples have focused on the analysis of individual genes. In some cases, it may be necessary to retrieve SNP information for multiple genes on different chromosomes. Although this might appear to be a simple query, there are not many simple tools available to carry out a query of this nature against the most current version of dbSNP. The best available is probably the Ensembl Biomart tool (Table 1).

1. Gene IDs of interest may come from multiple sources, but Ensembl IDs are needed to query Biomart. Convert Gene IDs (e.g. HUGO IDs) to Ensembl IDs at the following URL (http://idconverter.bioinfo.cnio.es/).
2. Navigate to the Ensembl Biomart interface (Table 1). Select the "Ensembl Variation" database. Choose the "Homo Sapiens Variation" dataset. Click "Filters" on the left hand menu. Open the "Gene Associated Variation Filters" menu. Tick the Ensembl Gene IDs box and paste the Ensembl Gene IDs from step 1. Press the "Results" button in the top left part of the screen.
3. The query should retrieve all SNPs that are mapped to the queried genes. The default results are very simple, to add annotation of SNP location and other information, Click "Attributes" on the left hand menu. Open the menus of gene and variant associated information and tick the boxes of the desired information. Again press the "Results" button in the top left part of the screen.

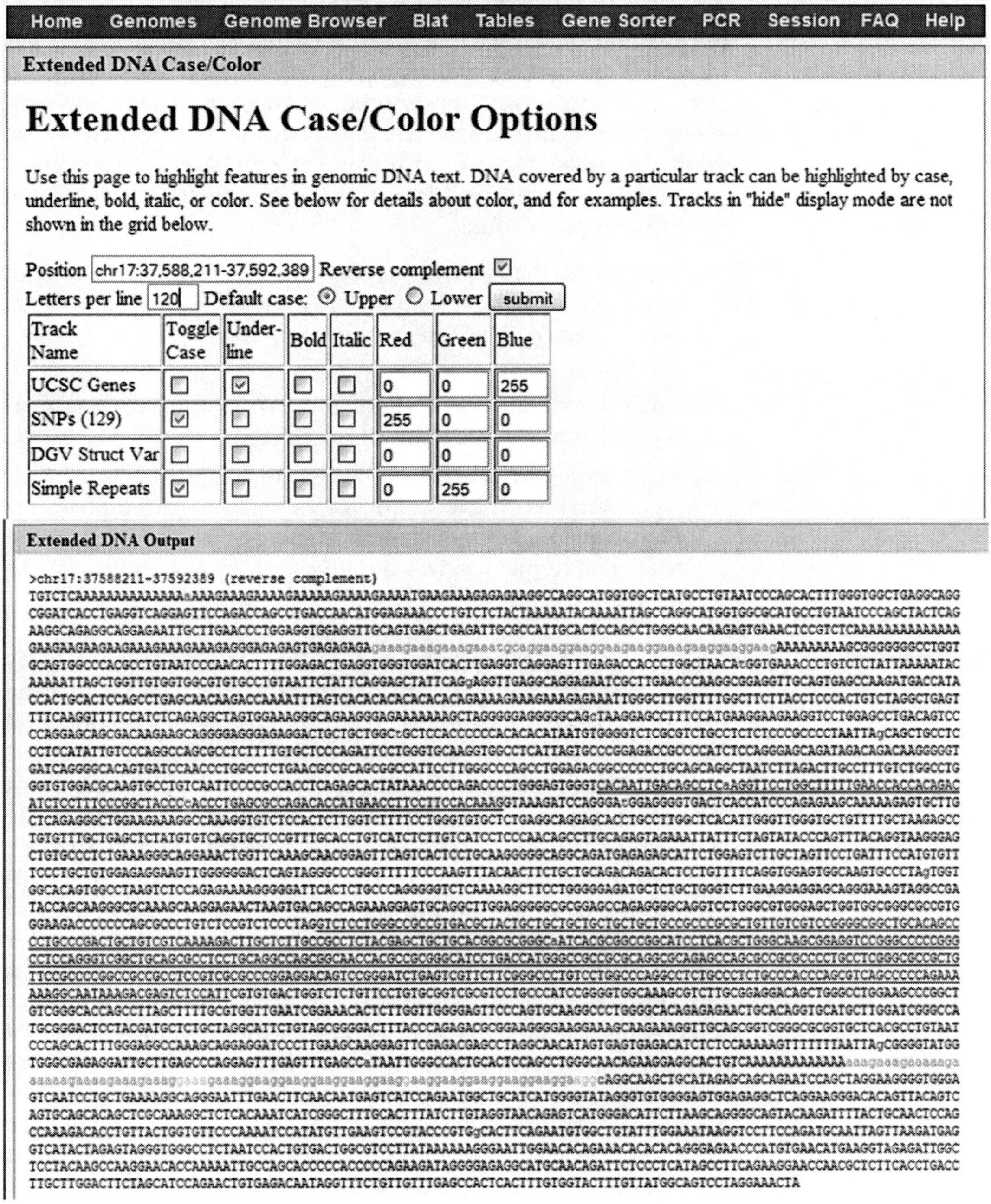

Fig. 4. Use of the UCSC DNA export feature to annotate variants on a DNA sequence

4. This query updates the information for all the SNPs. It is now possible to view and export the information in a variety of formats. For example, to view the data in Microsoft excel format, export results to file and select "XLS" format. Press go and then open the file in Microsoft excel.

3.5. Gaining an Overview of the Population Diversity of Genetic Variants

As projects such as the HapMap (4) have read out, the amount of information on the allele frequency of genetic variation data has increased dramatically. Much of this information has focused on the four population samples used by the HapMap (Caucasian, Hong Kong Chinese, Japanese and Nigerian Yoruba). However, as the HapMap has moved into its third phase, data has been generated on a wider range of populations, including data from 11 different populations.

1. Navigate to the HapMap website (Table 1). On the left menu, select the HapMap Genome Browser (including phase 1, 2, and 3 data). This takes the user to a generic genome browser.
2. Enter your gene or SNP of interest in the "landmark or region" box and press the search button.
3. The query returns a view of the gene locus and HapMap genotyped variants. Allele frequency of variants in the phase III HapMap populations is indicated in a tiny barchart next to the SNP. Full frequency details can be seen by passing the mouse over the variant (Fig. 5)

3.6. Conclusion

This review has covered a great deal of ground in a relatively small space, but the sheer complexity of genetics means that the material covered here is just a small start to help biomedical researchers to broaden their consideration of genetic variation. For brevity, this review has skimmed over some very important areas that should be given further consideration. Perhaps the most important are the relationships between variants that are revealed by the analysis of linkage disequilibrium (LD) and haplotype structure (1). Put simply, genetic variants do not present independently in genomes, they are connected in a way that reflects their shared ancestry. Taking LD into consideration, it becomes clear that to effectively consider the impact of variants in genes, it may be necessary to consider the combined impact of variants that are completely correlated by LD. This review has also avoided any detailed consideration of the methods for evaluating the potential impact of a variant on a gene or regulatory element. This is covered in some detail by Mooney et al. (28) in this volume. Putting aside the shortcomings of this review, hopefully it illustrates, that visualisation and analysis of variation data is quite achievable using publicly available web resources. With accurate and comprehensive information on variation in hand, the next steps towards the better understanding of phenotype should naturally follow.

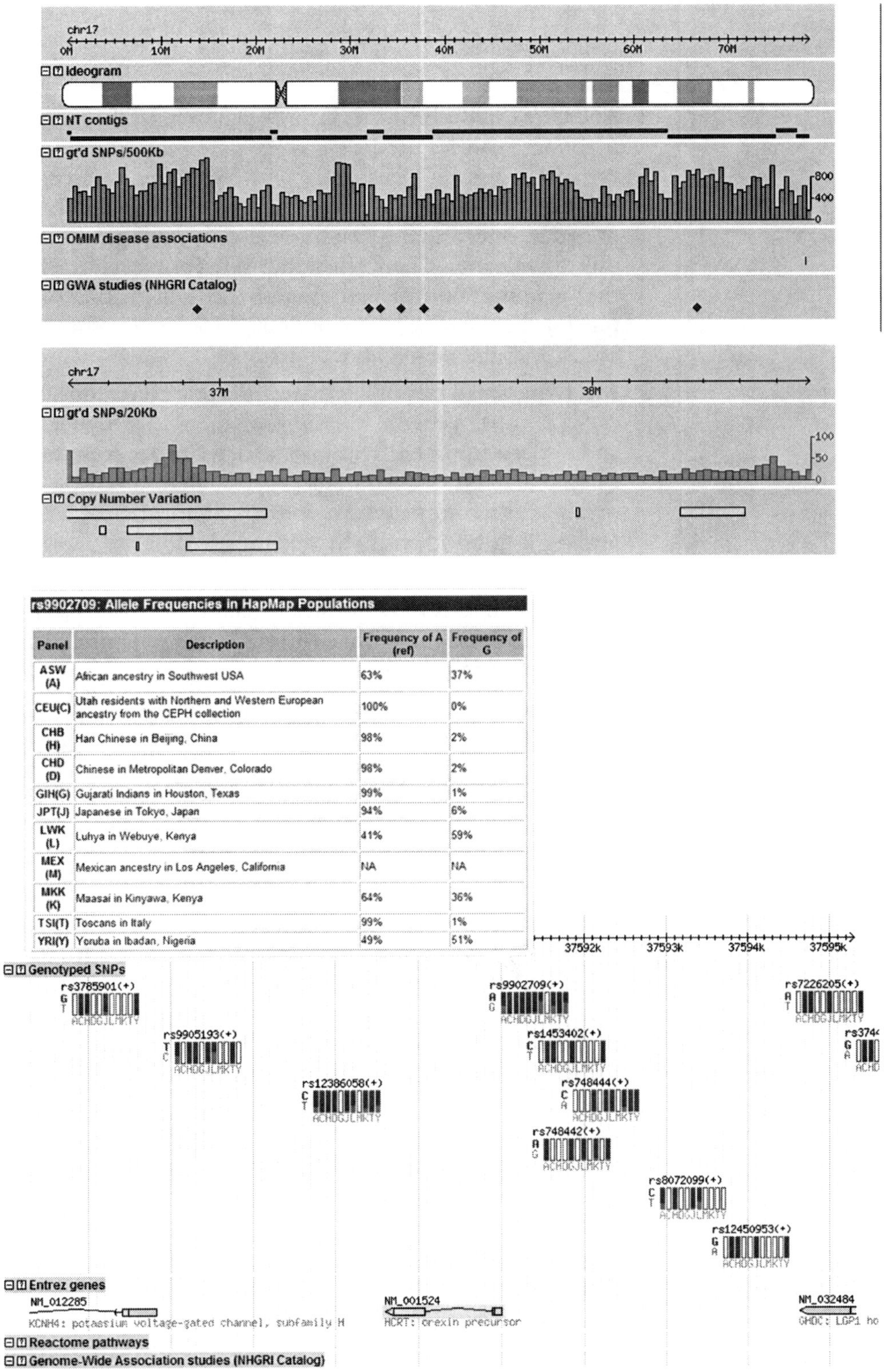

Panel	Description	Frequency of A (ref)	Frequency of G
ASW (A)	African ancestry in Southwest USA	63%	37%
CEU(C)	Utah residents with Northern and Western European ancestry from the CEPH collection	100%	0%
CHB (H)	Han Chinese in Beijing, China	98%	2%
CHD (D)	Chinese in Metropolitan Denver, Colorado	98%	2%
GIH(G)	Gujarati Indians in Houston, Texas	99%	1%
JPT(J)	Japanese in Tokyo, Japan	94%	6%
LWK (L)	Luhya in Webuye, Kenya	41%	59%
MEX (M)	Mexican ancestry in Los Angeles, California	NA	NA
MKK (K)	Maasai in Kinyawa, Kenya	64%	36%
TSI(T)	Toscans in Italy	99%	1%
YRI(Y)	Yoruba in Ibadan, Nigeria	49%	51%

Fig. 5. A HapMap view of SNP frequency

4. Notes

1. The UCSC table browser (http://genome.ucsc.edu/cgi-bin/hgTables) can be used to quickly determine the number of variants in a given locus or the entire genome. As an example, the number of simple repeats in the genome can be determined by selecting the "variation and repeats" group from the pull down menu. Then, select the "simple repeats" track. Click the "genome" button for the region and then click the summary button. In this case, there are 633,715 simple repeats in the NCBI36 build of the human genome.
2. It is rather a sad fact of life that with a few exceptions, notably dbSNP, most genetic variation databases are under-funded and under-resourced. This means that they are commonly out of date and on occasion inaccessible. Before relying on any database as a comprehensive source of information, it is worth getting a good idea of the update schedule of the database and the date of the last update.
3. It is possible to review the latest statistics of the dbSNP database on the summary page (http://www.ncbi.nlm.nih.gov/SNP/snp_summary.cgi).
4. When querying SNP data in tools other than dbSNP, it is important to determine the version of dbSNP being used by the tool. Many tools offer better interfaces to dbSNP data than dbSNP but do not contain the most current data. Tools that reliably contain current dbSNP data include Ensembl, UCSC and Biomart. Other tools should be treated with caution.
5. As Fig. 2 illustrates, dbSNP and OMIM do have reciprocal links to clinical associations; however, the linking is somewhat erratic. The indel represented by rs333, is the CCR5 delta 32 allele, which confers resistance to HIV infection (29). Neither this is indicated in the dbSNP report, nor the RSid is linked in the OMIM report. This shows that some care is needed in the interpretation of data from both resources.
6. The search window of the UCSC genome browser offers a very flexible query interface. Users can directly enter a genome position, an accession number, SNP ID, gene ID or keywords. Depending on the query used, the results returned may be a little confusing. Usually, the desired target of the query is reported at the top of the list, but sometimes it may not be, so it is worth inspecting closely the results, does the detail the gene of interest, are there multiple hits, is result homology 100%, etc.

7. Selection and configuration of track information in the UCSC genome browser: Over 100 tracks of information are available to view in the UCSC human genome browser. These tracks contain highly specific information across many fields. However, for general applications, 20–30 tracks are likely to see the most regular use. More importantly, selection of more than 10–15 tracks is likely to slow the browser down considerably, so it is worth turning off tracks which are not being used. In order to determine the best track for the job, it is worth reading the track documentation to check the provenance and age of the data.

References

1. Barnes, M.R. (2006) Navigating the HapMap. *Brief. Bioinform.*, 7, 211–224.
2. Altshuler, D., Pollara, V.J., Cowles, C.R., Van Etten, W.J., Baldwin, J., Linton, L. and Lander, E.S. (2000) An SNP map of the human genome generated by reduced representation shotgun sequencing. *Nature*, **407**, 513–516.
3. Redon, R., Ishikawa, S., Fitch, K.R., Feuk, L., Perry, G.H., Andrews, T.D., *et al.* (2006) Global variation in copy number in the human genome. *Nature*, **444**, 444–454.
4. The International HapMap Consortium (2005) A haplotype map of the human genome. *Nature*, **437**, 1299–1320.
5. Gelfand, Y., Rodriguez, A. and Benson, G. (2007) TRDB – the Tandem Repeats Database. *Nucleic Acids Res.*, **35**, D80–D87.
6. Gratacos, M., Nadal, M., Martin-Santos, R., Pujana, M.A., Gago, J., Peral, B., *et al.* (2001) A polymorphic genomic duplication on human chromosome 15 is a susceptibility factor for panic and phobic disorders. *Cell*, **106**, 367–379.
7. Komura, D., Shen, F., Ishikawa, S., Fitch, K.R., Chen, W., Zhang, J., *et al.* (2006) Genome-wide detection of human copy number variations using high-density DNA oligonucleotide arrays. *Genome Res.*, **16**, 1575–1584.
8. McCarroll, S.A., Kuruvilla, F.G., Korn, J.M., Cawley, S., Nemesh, J., Wysoker, A., *et al.* (2008) Integrated detection and population-genetic analysis of SNPs and copy number variation. *Nat.Genet.*, **40**, 1166–1174.
9. Bentley, D.R. (2006) Whole-genome re-sequencing. *Curr. Opin. Genet. Dev.*, **16**, 545–552.
10. Kluijtmans, L.A., van den Heuvel, L.P., Boers, G.H., Frosst, P., Stevens, E.M., van Oost, B.A., *et al.* (1996) Molecular genetic analysis in mild hyperhomocysteinemia: a common mutation in the methylenetetrahydrofolate reductase gene is a genetic risk factor for cardiovascular disease. *Am. J. Hum. Genet.*, **58**, 35–41.
11. Bahar, A.Y., Taylor, P.J., Andrews, L., Proos, A., Burnett, L., Tucker, K., *et al.* (2001) The frequency of founder mutations in the BRCA1, BRCA2, and APC genes in Australian Ashkenazi Jews: implications for the generality of U.S. population data. *Cancer*, **92**, 440–445.
12. Roque, M., Godoy, C.P., Castellanos, M., Pusiol, E. and Mayorga, L.S. (2001) Population screening of F508del (DeltaF508), the most frequent mutation in the CFTR gene associated with cystic fibrosis in Argentina. *Hum. Mutat.*, **18**, 167.
13. Forbes, S.A., Bhamra, G., Bamford, S., Dawson, E., Kok, C., Clements, J., *et al.* (2008) The catalogue of somatic mutations in cancer (COSMIC). *Curr. Protoc. Hum. Genet.*, **Chapter 10**, Unit.
14. Amberger, J., Bocchini, C.A., Scott, A.F. and Hamosh, A. (2009) McKusick's online Mendelian inheritance in man (OMIM). *Nucleic Acids Res.*, **37**, D793–D796.
15. Miller, R.D. and Kwok, P.Y. (2001) The birth and death of human single-nucleotide polymorphisms: new experimental evidence and implications for human history and medicine. *Hum. Mol. Genet.*, **10**, 2195–2198.
16. Cooper, D.N. and Youssoufian, H. (1988) The CpG dinucleotide and human genetic disease. *Hum. Genet.*, **78**, 151–155.
17. Reich, D.E., Cargill, M., Bolk, S., Ireland, J., Sabeti, P.C., Richter, D.J., *et al.* (2001) Linkage disequilibrium in the human genome. *Nature*, **411**, 199–204.
18. Street, S.L., Kyes, R.C., Grant, R. and Ferguson, B. (2007) Single nucleotide polymorphisms

(SNPs) are highly conserved in rhesus (Macaca mulatta) and cynomolgus (Macaca fascicularis) macaques. *BMC Genomics*, **8**, 480.

19. Hodgkinson, A., Ladoukakis, E. and Eyre-Walker, A. (2009) Cryptic variation in the human mutation rate. *PLoS Biol.*, 7, e1000027.
20. Miller, R.D., Taillon-Miller, P. and Kwok, P.Y. (2001) Regions of low single-nucleotide polymorphism incidence in human and orangutan xq: deserts and recent coalescences. *Genomics*, **71**, 78–88.
21. Sherry, S.T., Ward, M.H., Kholodov, M., Baker, J., Phan, L., Smigielski, E.M. and Sirotkin, K. (2001) dbSNP: the NCBI database of genetic variation. *Nucleic Acids Res.*, **29**, 308–311.
22. Orr, H.T. and Zoghbi, H.Y. (2007) Trinucleotide repeat disorders. *Annu. Rev. Neurosci.*, **30**, 575–621.
23. Lucassen, A.M., Julier, C., Beressi, J.P., Boitard, C., Froguel, P., Lathrop, M. and Bell, J.I. (1993) Susceptibility to insulin dependent diabetes mellitus maps to a 4.1 kb segment of DNA spanning the insulin gene and associated VNTR. *Nat. Genet.*, **4**, 305–310.
24. Benson, G. (1999) Tandem repeats finder: a program to analyze DNA sequences. *Nucleic Acids Res.*, **27**, 573–580.
25. Woollard, P. (2009) Asking complex questions of the genome without programming. *Methods Mol. Biol.*
26. Porter, S.G., Day, J., McCarty, R.E., Shearn, A., Shingles, R., Fletcher, L., Murphy, S. and Pearlman, R. (2007) Exploring DNA structure with Cn3D. *CBE Life Sci. Educ.*, **6**, 65–73.
27. Kuhn, R.M., Karolchik, D., Zweig, A.S., Wang, T., Smith, K.E., Rosenbloom, K.R., *et al.* (2009) The UCSC Genome Browser Database: update 2009. *Nucleic Acids Res.*, **37**, D755–D761.
28. Mooney, S., Krishnan, V. and Evani, U.S. (2009) Bioinformatic tools for identifying disease gene and SNP candidates. *Methods Mol. Biol.*
29. Samson, M., Libert, F., Doranz, B.J., Rucker, J., Liesnard, C., Farber, C.M.S. *et al.* (1996) Resistance to HIV-1 infection in caucasian individuals bearing mutant alleles of the CCR-5 chemokine receptor gene. *Nature*, **382**, 722–725.

Chapter 2

Exploring the Landscape of the Genome

Michael R. Barnes

Abstract

Genome browsers are powerful tools for biologists – offering fundamental information on genes, regulatory elements, genomic variants, genome structure, and evolution. The comprehensive range of information presented in tools such as the UCSC genome browser and Ensembl enables integrated queries of data that are otherwise reserved to the most skilled computational biologists. However, for the non-specialist user, the juxtaposition of so many different forms of data in one small space can be an information overload. Getting the most out of these tools requires some understanding of the key concepts and caveats of genome visualization and annotation. Genome analysis can be carried out at different levels of detail – at a macro level; it improves understanding of issues like genome structure and species evolution. While at a micro level, genome annotation can help to describe the full complexity of gene regulation, variation, and transcript diversity. Once demystified, it is clear that genome browsers are more than the sum of their parts – they are the most comprehensive portals available for browsing and analysis of biological data.

Key words: Genome, Bioinformatics, Variation, Gene, Regulation, FTO, Evolution

1. Introduction

To understand genes and their role in the biology and the genetics of an organism, it is necessary to understand genome sequences. A good familiarity with the landscape and mechanics of the genome can really help in the study of biology. Genomes are pertinent to the study of many different types of data, for example, in the case of genetic variation, a single sequence variant could impact function at many levels, including gene function, gene regulation, splicing, genomic stability or epigenetic modification, or indeed all, or some of these in combination. With this in mind, this review will focus on the study of genetic variation in a genomic context purely as an illustration of the range of analysis that is possible using genomic information and the tools that are used to

Michael R. Barnes and Gerome Breen (eds.), *Genetic Variation: Methods and Protocols*, Methods in Molecular Biology, vol. 628, DOI 10.1007/978-1-60327-367-1_2, © Springer Science+Business Media, LLC 2010

analyze genomic data. These principles can be generalized to any form of data that can be mapped to a genome.

Although there are many ways to access genomic information in an integrated manner, there are two primary tools that are the acknowledged leaders in the field, the UCSC human genome browser (1) and ENSEMBL (2). Although both tools have many similarities, each contains distinct information and data interpretation, and so it usually pays to consult both viewers, if only for a second opinion (both viewers provide reciprocal links). The UCSC genome browser has one great advantage over Ensembl for macro scale genome analysis as it allows detailed visualization across regions greater than 1 Mb or even whole chromosomes. This really makes the UCSC browser an exceptional tool for integrated genomic analysis, and so most examples given below focus on this tool, but almost all the examples are possible to complete using either tool.

2. Materials

All the tools described here are freely available internet web tools, which would run on any PC, Mac or Unix workstation with web access. For more sophisticated analysis of large datasets on a genomic scale, see (3). A list of genomic tools and databases mentioned and used in this review is given in Table 1.

3. Methods

3.1. Representing User Data in the UCSC Genome Browser

The UCSC genome browser allows the user to easily represent data in a genomic context with the *custom track* or *genome graph* tools. While both these mechanisms can be used to represent quite complex data, at the most basic level, a lot can be achieved with a very simple tab delimited format. So for example, the *genome graph* function can be used to represent the results of a genetic association analysis across a region by simply uploading a list of SNP ids (which are mapped to the genome by the browser) and –log p values in a tab delimited format. An example of this format is given below:

```
SNP ID        Log p value
RS1477196     0.824065115
RS1121980     6.490366087
RS7193144     7.841293827
RS16945088    0.737468229
RS8050136     7.698837565
RS9926289     6.682249971
```

```
SNP ID        Log p value
RS9939609     7.28029591
RS9930506     5.856133069
RS1115005     1.094787167
RS11075994    0.225325403
```

Table 1
Tools for genomic characterisation

Tool	URL
Genome visualization	
UCSC Genome Browser	genome.ucsc.edu
ENSEMBL	www.ensembl.org
NCBI MapViewer	www.ncbi.nlm.nih.gov/mapview/map_search.cgi/
LD and haplotype data	
HapMap website	www.hapmap.org
HapMap Genome Browser	www.hapmap.org/cgi-perl/gbrowse/gbrowse/
SNAP	www.broad.mit.edu/mpg/snap/
Structural genome analysis	
Db of Genomic Variants	projects.tcag.ca/variation/
Structural Variation db	humanparalogy.gs.washington.edu/
Building biological rationale	
GNF SymAtlas	symatlas.gnf.org/SymAtlas/
HUGE Navigator	www.hugenavigator.net/
Stanford SOURCE	source.stanford.edu
STITCH (Pathways)	stitch.embl.de/
UniProt	www.uniprot.org

The *Genome Graphs* input form can be accessed from the left hand menu on the UCSC home page (Table 1). Once the desired genome has been selected, press the upload button, enter the details of the data and paste the text into the genome graph data

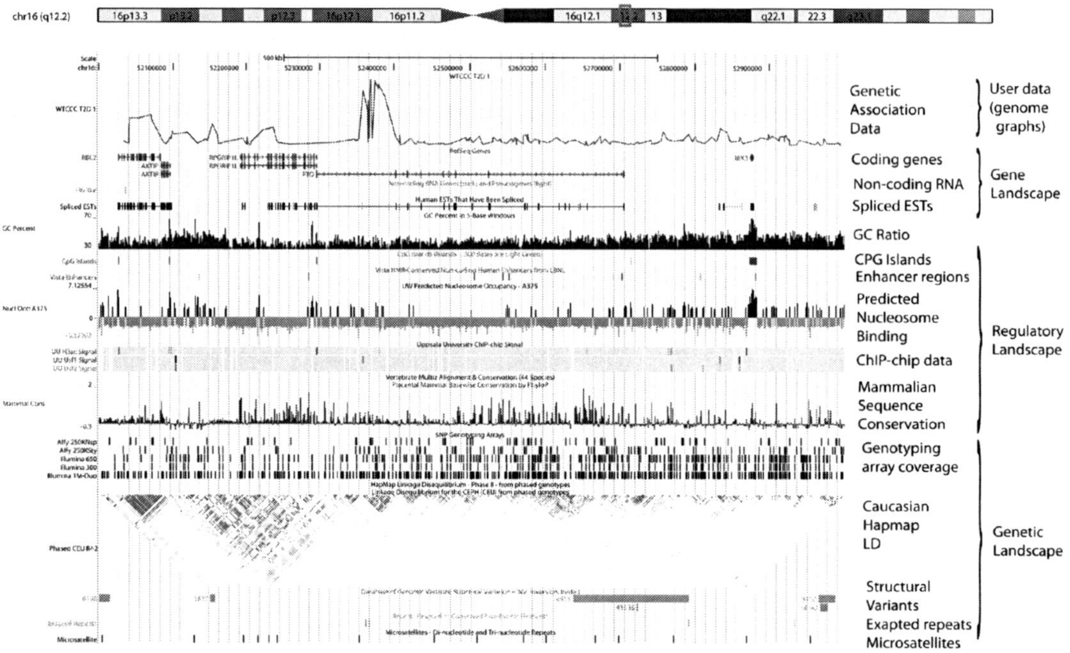

Fig. 1. The Genomic Macro-Environment. A view of the 1 Mb region (52–53 Mb) around the FTO gene, generated with the UCSC human genome browser. User submitted genetic association data is displayed using the UCSC Genome Graph and Custom Track functions. Genetic association with type II diabetes is plotted across the region (–log *p* values) and shows that the association is restricted to the FTO gene. The macro HapMap LD structure across the region also supports this. Descriptive information for each UCSC dataset can be accessed by pressing the grey button in the UCSC browser to the left of each track

window. After loading the data, a chromosome ideogram is returned. Select your graph from the pull down menu and the –log *p* values are annotated as a graph across the Chr. 16 ideogram. If the "browse regions2 button" is followed, then the data is displayed in the genome view as shown in Fig. 1. For more information about the Genome Graph function, see the UCSC help documentation (see Note 1).

Similarly, the UCSC *Custom Track* feature can be used to annotate a list of SNPs or any other genomic features by providing chromosome number, start and end genome coordinates and optionally a name. For example to view all the SNPs that are in LD with an associated SNP across a genomic region (see Note 2), the following format can be used:

```
Chr        Chr start    Chr end     Name
Track name=LD_SNPs description="SNPs_in_LD_with_
  associated_SNP"
chr22      20100000     20100001    RS5346536
chr22      20100011     20100012    RS346976
chr22      20100215     20100216    RS2658758
```

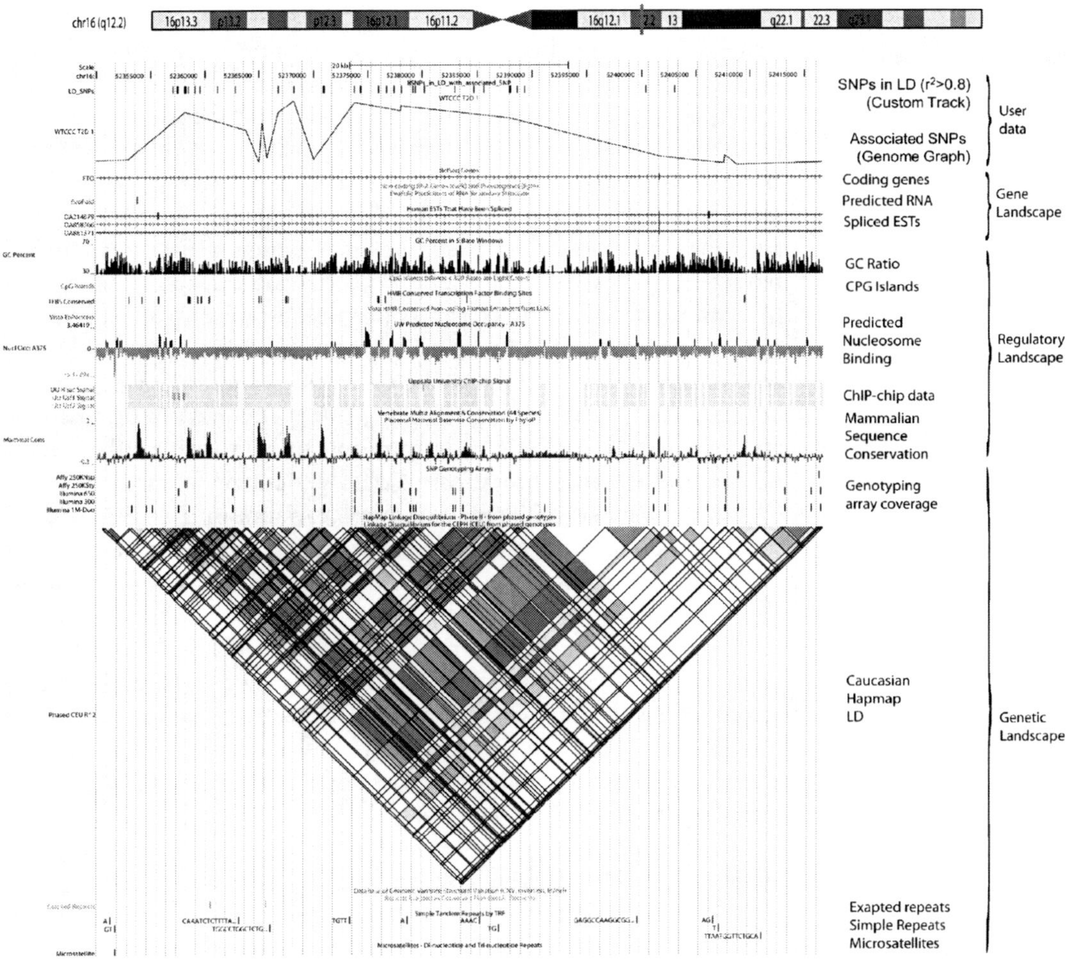

Fig. 2. The Genomic Micro-Environment. A closer view of the 100 Kb associated region in intron 1 of the FTO gene, generated with the UCSC Human genome browser. User submitted genetic association data is displayed using the UCSC Genome Graph and Custom Track functions. This view shows more detail of known regulatory elements across the region and allows the user to identify variants in these regions

After pasting this text into the browser window, the SNPs of interest are annotated in the genome view (see Fig. 2). For more information about custom track formats, see the UCSC help documentation (see Note 3).

3.2. Evaluating a Genetic Association at a Genomic Level

The custom track and genome graph facilities make the UCSC genome browser a powerful tool for evaluating the genomic context of a genetic association. A new generation of genome-wide association studies (GWAS) is revolutionizing our understanding of human disease genetics (4), and so it seems fitting to use the output of such a study as an example. To demonstrate the process in Figs. 1 and 2, a genome graph is used to plot type II diabetes

(T2D) association data across the fat mass-and obesity-associated gene (FTO) (5). This region of chromosome 16 has been reproducibly associated with fat mass and body mass index (BMI), risk of obesity, and adiposity, however no clear molecular mechanism for the action of FTO in obesity has been determined (see Note 4). By placing the association data into a genomic context, a clearer picture of the nature of this association emerges. As the amount of information available in a genome browser can be bewildering, it is often beneficial to consider a genomic region in both terms of the macro and micro-environment. The genomic *macro-environment* (Fig. 1) informs on the overall physical structure of the region, including the GC ratio, long range LD, recombination rate, structural variants and overall gene content. Gaining an understanding of the wider region can help with further study design and the interpretation of data generated from the region, e.g. the presence of large structural variants in a region would need to be factored into primer design or the interpretation of expression or association data. The genomic *micro-environment* (Fig. 2) encompasses all the features present in the immediate region around the most strongly associated SNPs (and SNPs in LD with these SNPs). These features can usually be defined at a sequence level and are immediately relevant to the regulation or function of a gene. In the sections below and in Figs. 1 and 2, the Type II diabetes association across the FTO gene is considered at both levels.

3.3. Evaluating Genomic Information: The Genomic Macro-Environment

In Fig. 1, a UCSC genome browser view of the 1 Mb region (52–53 Mb) around the FTO gene is presented with a number of tracks that highlight the properties of the macro-environment of the genetic association with type II diabetes. Firstly, the genome graph function is used to plot the –log p value of the T2D GWAS. A custom track is also used to annotate all SNPs in LD ($r^2 > 0.5$) (see Note 2) with the most strongly associated SNP in the region (RS7193144). Descriptive information for each UCSC dataset can be accessed by pressing the grey button in the UCSC browser to the left of each track. A great deal of configurable extra information is also available but not shown here for brevity (see Note 5).

In order to evaluate the initial association, the UCSC has a track which shows the SNP coverage of the major genotyping panels. The WTCCC diabetes study was completed with the affymetrix 500 K platform, which is actually composed of two chips each with 250K SNPs, these chips are presented in the "Affy 250KNsp" and "Affy 250KSty" tracks. This shows relatively good coverage across the entire region with the notable exception of a large gap in coverage immediately to the 5′ (left) of the association signal. This lack of coverage should be considered when the association is evaluated – as there is no marker coverage over the first exon or the promoter region of FTO – further follow-up studies of this association would need to provide better coverage of this region. Incidentally, coverage by the Illumina 550 K genotyping panel is also displayed and this seems to provide adequate physical coverage of the entire region.

Marker coverage across a region also needs to be considered in terms of capture of variation by LD (so called SNP tagging) rather than physical spacing alone. There have been several good comparisons of the capture of variation by commercial marker panels (6,7). Tools on the HapMap website (Table 1) also present comprehensive web-based views of LD and haplotype structure; however, they offer limited genomic information. For the purposes of an initial evaluation of the LD around a genetic association, the UCSC genome browser excels on a number of levels. Both Ensembl and the UCSC genome browser offer an integrated view of HapMap LD data, however, the UCSC browser allows visualization of LD across regions of greater than 1 Mb or even whole chromosomes. This is demonstrated in Fig. 1, where the Macro LD structure across the 1 Mb region containing the FTO gene is shown. This clearly delineates LD into Blocks of high LD punctuated by recombination hotspots, which are also displayed, based on calculations from HapMap data. From this data, it looks likely that the FTO association is restricted to an LD block in intron 1 of the gene (see Note 6). Zooming in closer to a 100 Kb view in Fig. 2, this correlation is even clearer and is also backed up by the map locations of the SNPs that are known to be in LD with the most highly associated SNP. All this information points strongly to the involvement of a genetic variant that is present in a restricted region shown in Fig. 2.

3.4. The Genomic Micro-Environment: The Nuts and Bolts of Gene Function

After defining a locus of interest, one of the key questions to ask is – what genes are located in the locus? A genome viewer is the best tool to ask this question, if known genes are all that is required then the answer is routine, but if a comprehensive answer is needed – all known and novel genes, and all transcript variants of these genes – then analysis is non-trivial. The UCSC human genome browser and Ensembl both run the human genome sequence through sophisticated gene prediction and sequence mapping pipelines (1,2). Genome viewers offer a comprehensive view of supporting evidence for genes, such as ESTs, CPG islands and both predicted and experimentally determined regulatory regions. Homology with the constantly expanding collection of genomes from other species is also presented. At the time of writing (March 09), 43 vertebrate genomes were mapped against human sequence in the UCSC genome browser. It is important to be aware of the provenance of the data presented – in effect genome annotations can be viewed as a hierarchy of evidence, with known genes at the top, hypothetical genes, spliced ESTs, sequence conservation and finally unspliced ESTs at the bottom.

Ideally, most genes should be evidenced by several of these features, e.g. a spliced EST supported by vertebrate sequence conservation is fairly reliable supporting evidence for a novel gene. Improvement on the quality of annotation provided by Ensembl and the UCSC requires an in-depth understanding of the intricacies of gene prediction, which are not within the scope of this review. Instead, it is probably best to focus on the available data to build gene models based on existing annotation. Once all the genes in the locus have been identified, the next logical step would be to investigate each gene for involvement in the phenotype being studied (see Note 7). Searching the literature can give some clues about gene function and the likelihood of involvement in a specific phenotype.

Returning to the FTO case study, let us review the genetic association signal (Fig. 2). This should be considered to encompass the associated SNPs plotted in the genome graph and also the SNPs in LD with the associated SNPs – so called proxy SNPs. Comparison of the location of the associated SNPs and the proxy SNPs against genes, ESTs and non-coding RNA, appears to restrict the association to intron 1, part of intron 2 and possibly exon 2 of the FTO gene. Reviewing the known gene information, all the associated SNPs and proxy SNPs are intronic. It is also worth reviewing EST data for evidence of novel splice variants. In this case, there is an EST (DA214879) that may represent a novel FTO exon leading to a novel splice variant. Again there are no SNPs in this EST. The magnitude and repeated replication of the association signal in the FTO region, suggest the involvement of a common variant (8). As there are no exonic variants showing association or LD with associated SNPs, it seems reasonable to assume that the causal variant(s) are likely to be intronic or alternatively there may be as yet un-characterized variants, for example copy-number variants or repeat sequences. Genome browsers are ideal tools to enable further exploration of some of these possible hypotheses to explain the functional nature of the FTO association, helping to formulate lab testable hypotheses.

3.5. The Regulatory and Epigenetic Landscape

If the FTO causative alleles are most likely to be restricted to introns, then it is clearly important to evaluate the regulatory landscape of the associated region. The traditional view of gene regulation usually focuses on the promoter region of a gene. However, regulatory sequences can be located throughout a gene, in the 5′ regions, the introns, exons, splice boundaries and 3′ untranslated region (Fig. 3). Regulatory elements may also take many forms, including highly specific transcription factor binding sites, or extended enhancer regions controlling tissue-specific expression or alternative splicing (9).

A key field which is helping to shed light on the basis of gene regulation is the emergent science of Epigenomics – the study of

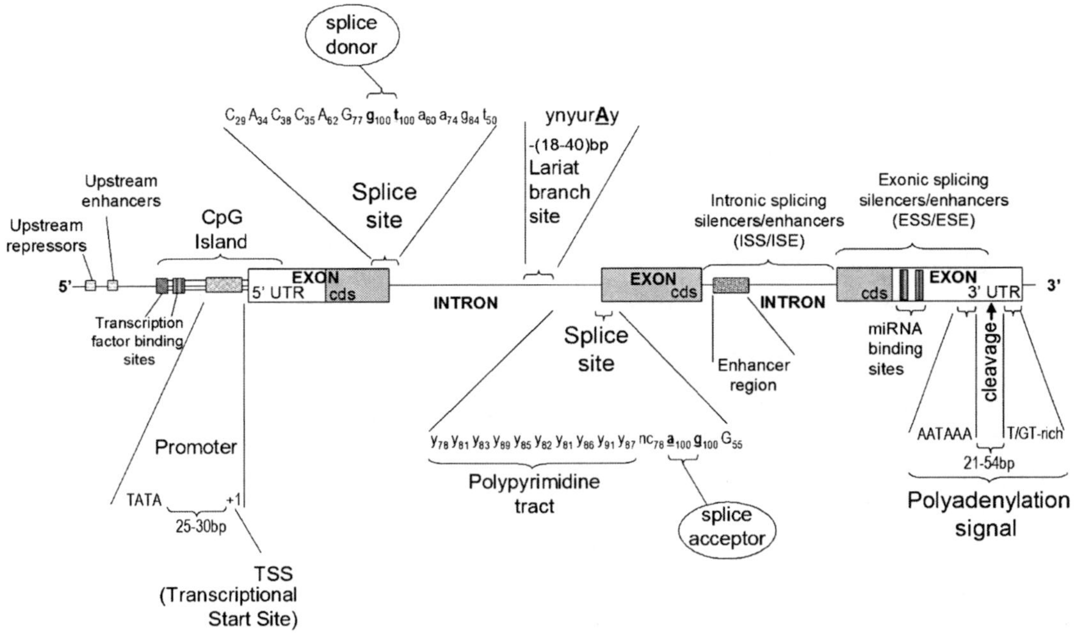

Fig. 3. The anatomy of a gene. This figure illustrates some of the key regulatory regions, which control the transcription, splicing, and post-transcriptional processing of genes and transcripts. Polymorphisms in these regions should be investigated for functional effects

epigenetic modification on a genome-wide scale. Epigenetics is concerned with the study of heritable changes other than those in the DNA sequence and encompasses two major modifications of DNA or chromatin: DNA methylation, the covalent modification of cytosine, and post-translational modification of histones including methylation, acetylation, phosphorylation and sumoylation (10). In terms of function, epigenetic modifications act to regulate gene expression and stabilize adjustments of gene dosage, as seen in X inactivation, gene silencing and genomic imprinting.

3.6. Epigenetic Insight into Gene Regulation

At the most basic level, the sequence composition of a specific region of DNA can give some clues about its regulatory potential. Inside the nucleus, DNA is wrapped into a complex molecular structure called chromatin, which is composed of a fundamental unit of approximately 150 bp of DNA organized around an eight-histone protein complex known as the nucleosome. The local organization of nucleosomes defines the accessibility of DNA to protein binding and hence the regulatory potential of a region. An excellent example of the role of the nucleosome in gene regulation was reviewed by Costello and Vertino (11) based on the work of Futscher et al. (12). This example is based on studies of the tissue-specific regulation of SERPINB5 which is controlled at the level of the nucleosome by the methylation and acetylation state of the promoter region of the gene. This regulatory mechanism

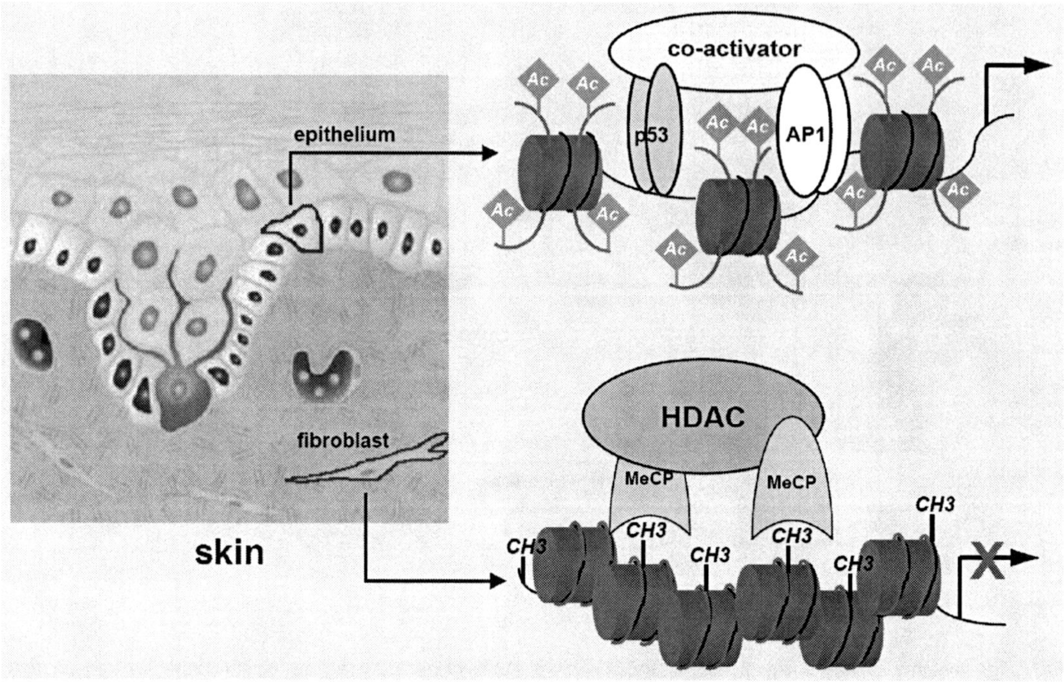

Fig. 4. Epigenetic control of SERPINB5 tissue-specific expression. Expression is mediated by methylation leading to the opening and closing of chromatin structure. (reproduced from (11) with permission from Nature publishers)

is equally applicable to regulatory elements in introns. Figure 4 shows a model of the tissue-specific control of SERPINB5 expression by methylation leading to the opening and closing of chromatin structure. The SERPINB5 promoter is unmethylated in skin epithelial cells allowing the sequence specific occupation by the transcription factors AP1 and p53. In addition, the histones bound in that region are acetylated (Ac), limiting histone–histone interactions and opening up the chromatin structure to allow binding by other transcription factors required for SERPINB5 expression. By contrast, in skin fibroblasts, the promoter is completely methylated (CH3), this is associated with hypoacetylated histones and adopts a tighter inaccessible state that is transcriptionally inactive. DNA Methylation is a key to this model as it allows the binding of methyl CpG-binding proteins (MeCP), which mediate histone deactylase (HDAC) and chromatin remodeling complexes to direct the compression of the chromatin structure into the transcriptionally inactive state. In this model, methylation is a primary impediment to SERPINB5 expression and thus determines the cell type-specificity. This is a good example, where the consideration of epigenetics could help genetic analysis. SNPs in CpG sites could lead to loss or gain of cytosine–guanine dinucleotide (CpG) methylation sites – and hence an indirect impact on regulation at nearby sites. Rakyan et al. (13) suggested that CpG polymorphisms might affect the overall methylation profile of a locus and, consequently, promoter activity and gene expression. Alternatively, a non-CpG SNP located within an epigenetically sensitive regulatory element could also influence the epigenetic makeup of that region. Therefore, mutations in regulatory sequences could influence epigenetic profiles, resulting in altered phenotypes.

Moving back to the FTO case study, several UCSC tracks included in Figs. 1 and 2 give some indication of the epigenetic environment and hence the regulatory potential of the associated region. Examining Fig. 1, G/C nucleotide ratio is plotted across the region. Extended GC rich regions of the genome, known as CpG islands, are also shown. These usually correlate with gene promoter regions – this region is no exception, and it is possible to see a clear correlation between CpG islands and the start of genes in the region. As Fig. 1 shows, the GC ratio of a DNA region is also somewhat predictive of nucleosome occupancy, but GC ratio alone is a crude measure, so the UCSC browser also has a track with predicted nucleosome occupancy scores produced by a cell-line trained model (14). Aside from the predicted data, the UCSC browser also presents several valuable epigenomic data sets. These include a number of ChIP on chip data sets, representing laboratory observed nucleotide binding by specific transcription factors. In Fig. 1, data is displayed for three transcription factors generated by the University of Uppsala (15). It is notable that binding sites for M3ac and Usfl are present in the FTO associated region. Data is also presented on known enhancer elements in the Vista enhancers track (16). This is a fascinating genome-wide set of enhancer regions that show super-conservation (>99% conservation) over 100–250 bp in human, mouse and rat. Pennacchio et al. (16), showed that when inserted upstream of a lacZ construct these enhancer regions drove highly tissue- specific expression. Enhancers in grey showed no activity in constructs, while enhancers in black drive tissue-specific expression. By clicking on each of the enhancer elements, it is possible to view the in situ expression information for each enhancer. From Fig. 1, it is clear that none of these enhancers fall in the region of association, however there are a remarkably large number of active (black) enhancers across the larger region. Three are located within the FTO gene. The closest to the association is in intron 7 of the FTO gene. Interestingly, this ultra-conserved enhancer element was shown to drive hindbrain specific expression in mouse embryos (see Note 8). Regions arising from the embryonic hindbrain in adults are known as a key region for mediation of appetite and satiety. In genome-wide terms, these enhancers are quite rare and so although there is no direct evidence that the association extends across this region, further investigation would clearly be sensible.

For example, the association might be linked to a structural variant which could extend across the enhancer element.

Perhaps the most fundamental source of information which can be used to infer genome function is conservation. In Fig. 1, mammalian conservation determined from alignment of 43 different species is plotted across the genome. Sequence conservation is a universal measure of preserved function caused by evolutionary constraint. Conservation is usually highest in coding exons of genes; however, high levels of conservation are also seen in promoter regions and other regulatory regions, like the enhancer regions discussed above. A quick scan across the sequence conservation in Fig. 1 reveals high conservation across exons, but there are also conserved sequences across the entire region. A review of the conservation across the associated region in Fig. 2 shows several intronic regions that appear to be more highly conserved than exon 2. These are clearly of interest and might be considered for further in silico and laboratory investigation.

3.7. The Variant Landscape

Once all the genes and regulatory features in the region have been identified, the next step is to determine how variants in the region might impact function, explaining the association. As SNP genotyping is the technology of choice for most genetic association studies, accordingly a large amount of the information presented at the UCSC relates to known SNPs and HapMap LD data. However, SNPs are not the only form of variation and a great deal of information is also available relating to non-SNP variants, such as structural variants, microsatellites and other repeat sequences. One track is available which maps all published structural variants in the *Database of Genomic Variants* (17). Until recently, structural variants in the human genome were rarely reported, but several studies help us to appreciate the contribution that copy-number variants (CNVs) may be making to clinical phenotypes (18,19). Identifying and evaluating the impact of a CNV is quite a complex process, and determining the true impact of CNVs is likely to be a big challenge for genetics in the coming years (20). In the case of the FTO case study, examining the wider FTO region, several structural variants are present, although none appear to be located in the region of association. Some information is also presented on other types of repeat sequences, in the context of the FTO association. One of the most interesting is the exapted repeat track. This track displays conserved non-exonic elements that have been deposited by mobile elements, these regions were identified during a genome-wide survey (21) with the expectation that regions of this type may act as distal transcriptional regulators for nearby genes. A previous case study experimentally verified an exapted mobile element acting as a distal enhancer (22). It is tempting to speculate that exapted repeats in the FTO locus may also play some sort of enhancer role.

3.8. Dealing with "Personal Genome" Data

One of the weaknesses of the sequence based view of the genome is that a single sequence does not effectively represent the full dynamic range of variation that may be seen within and between populations. The human genome sequence represented in genome browsers like the UCSC and Ensembl is actually a composite sequence generated from several individuals. With the rise of next-generation sequencing technologies (23), there are now several projects completed or underway that are resequencing individual genomes (24). The most high profile "individual genome" sequences have been those generated for James Watson (25) and Craig Venter (26). These projects are now being overshadowed by the "1000 genomes" project which seeks to re-sequence the genomes of 1,000 individuals around the world (http://www.1000genomes.org/). The Ensembl and UCSC genome browsers have already developed views of individually sequenced genomes. In the case of Ensembl, a *Resequencing Alignment View* is available which presents the sequences of James Watson, Craig Venter and four other anonymous individuals across a user defined genomic region (http://www.ensembl.org/Homo_sapiens/sequencealignview?). In Fig. 5, a small region of the FTO gene is shown with an SNP highlighted in grey. Intriguingly, this shows a tri-allelic SNP position that is not represented in dbSNP, the human genome reference sequence (REF:36) shows a T allele shared with two of the anonymous individuals. The other two anonymous individuals have an A allele, while Craig Venter carries a C allele and James Watson has the A/C ambiguity call, M, showing a heterozygote A/C call at this base. As more individual sequence data becomes available, this type of view may become an increasingly important consideration in the study of any genomic region.

3.9. Using UCSC Custom Tracks and Table Browser to Intersect Genomic Features and Identify Potentially Functional Variants

In addition to visualization, the UCSC browser is also a powerful tool for large scale analysis of the genomic context of a given list of genome features, such as SNPs. A causal SNP is unlikely to be tested directly in a genome scan, but in principle it may be in LD with markers that have been genotyped (This is the principle underlying association analysis). After creation of a custom track containing the SNPs of interest (see above), the SNPs can be queried using the UCSC *Table browser* (27). Table Browser, which is accessed from the "*Tables*" link in the main browser, is an excellent tool that effectively allows the user to perform complex queries between data sets, including custom tracks loaded by the user. For example, it is possible to identify all SNPs (your custom track) that overlap with conserved transcription factor binding sites (TFBS). To do this, take the following steps:

1. Entering the Table Browser and select the "Custom Tracks" from the pull down "group" menu

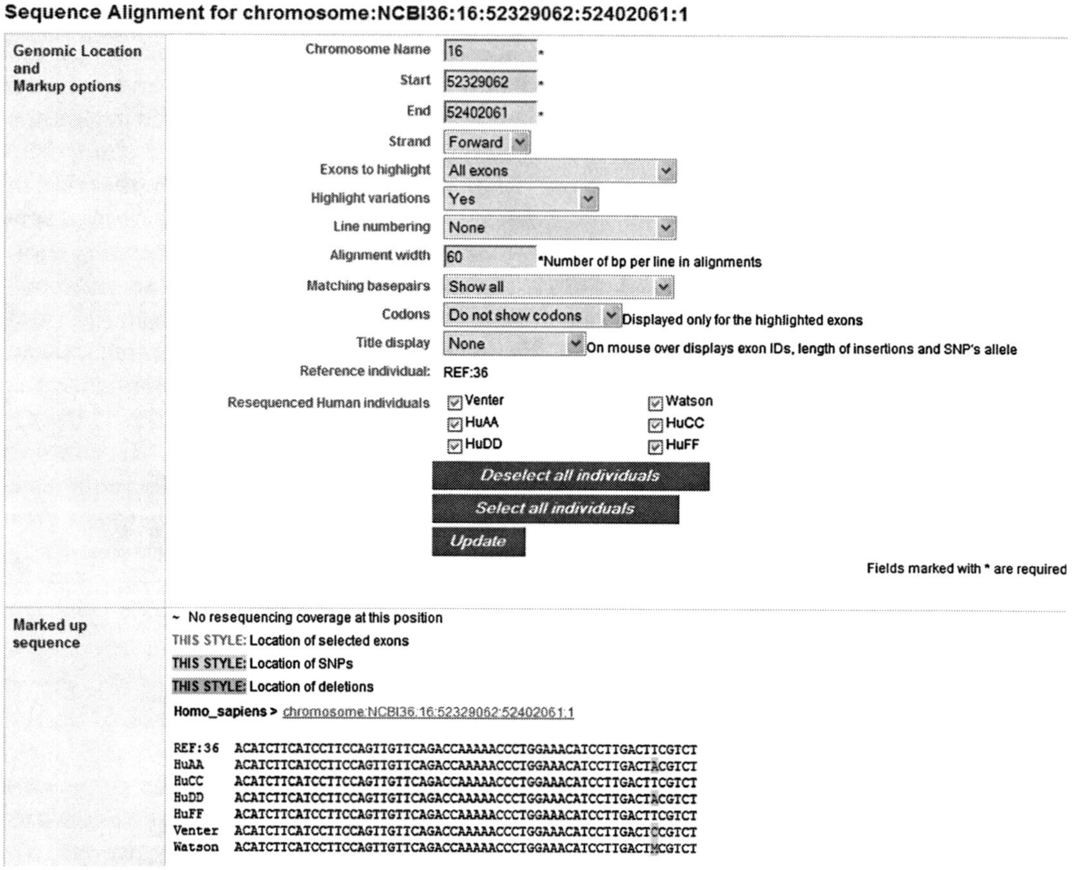

Fig. 5. Individual genome sequence data. An Ensembl *Resequencing Alignment View* of six individual genome sequences, including sequences from James Watson, Craig Venter and four other anonymous individuals across a user defined genomic region in the FTO gene. (http://www.ensembl.org/Homo_sapiens/sequencealignview?). SNPs are highlighted in grey, in this case, a tri-allelic SNP position is shown that is not represented in dbSNP

2. Select your custom track of choice from the "track" menu
3. Press the [create] intersection button
4. Select the group and track you are interested in, e.g. The "Regulation" group and the "TFBS Conserved" track. Press [submit]
5. To view a summary of overlaps, press the [summary/statistics] button
6. To view SNPs overlapping TFBS sites, press the [get output] button.

This basic process can be used for very large complex queries, making the UCSC table browser one of the most useful tools available for biologists and geneticists. It is possible to take this type of analysis to an even higher level of sophistication using UCSC data focused workflow tools such as Galaxy (3).

3.10. Conclusion

In the case study used in this review, the association seen between Type II diabetes and the FTO locus has been evaluated at a molecular level. Analysis of the associated locus in the full context of the data annotated by tools like the UCSC genome browser, supported several hypotheses which might explain the association. LD appeared to restrict the association to Intron 1 of the FTO gene, suggesting a possible regulatory element. Review of the data across the region identified an associated variant in an exapted repeat sequence, which is known to show regulatory function. This might warrant further investigation. An ultra-conserved element, directing specific hindbrain expression was also identified neighboring the associated markers, this may also be worth further investigation. As this example illustrates, mastering the in silico data to build a biological rationale around an association is not a trivial process, but it is achievable using publicly available web resources. Ultimately, good in silico analysis may help to align an association to a molecular mechanism, but as a general rule, it will raise more questions than it answers, returning the focus to the experimentalist.

4. Notes

1. The UCSC genome graphs help documentation: (http://genome.ucsc.edu/goldenPath/help/hgGenomeHelp.html).
2. *Retrieving a set of SNPs in Linkage Disequilibrium (LD)*: The SNP Annotation and Proxy Search tool, *SNAP* (Table 1) is a useful tool for identifying SNPs in LD with an SNP of interest. The output of the tool can be rapidly converted into a custom track using a text editor. The r^2 LD threshold is set by default to 0.8, this can be modified to increase or reduce stringency of LD.
3. The UCSC Custom track help documentation: http://genome.ucsc.edu/goldenPath/help/customTrack.html
4. The FTO region case study directly addresses one of the most challenging problems for complex disease genetics. Although an SNP association may be localized to a particular gene, association mapping also needs to take LD into account. An SNP showing association may be in strong LD with an ungenotyped marker nearby or in some cases at a considerable distance from the associated marker. This means that genetic associations need to evaluate the LD across a region, and each marker in LD with the associated SNP needs to be evaluated as a candidate for the molecular basis of the association. Genome browsers are supremely effective tools to assist this search.

5. *Selection and configuration of track information in the UCSC genome browser*: Over 100 tracks of information are available to view in the UCSC human genome browser. These tracks contain highly specific information across many fields. However, for general applications, 20–30 tracks are likely to see the most regular use. More importantly, selection of more than 10–15 tracks is likely to slow the browser down considerably, so it is worth turning off tracks which are not being used. In order to determine the best track for the job, it is worth reading the track documentation to check the provenance and age of the data.
6. *A Caveat to consider when dealing with LD "blocks"*: Although the traditional triangular block structure of an LD plot (Fig. 1) is a useful and intuitive guide to the extent of LD across a region, it is important to be aware that LD may extend across greater distances than the block structure suggests. This may be due to many factors, including the presence of longer rare haplotypes in the population or differences between LD structure in the study population and the HapMap population. Consequently, LD blocks should be taken as guides only and further analysis of the extent of LD should always be carried out.
7. *Building Biological Rationale around genes*: It is important to preface the consideration of biological rationale for genes in phenotypes or diseases, with an acknowledgement that a convincing rationale can be made for almost any gene in almost any phenotype if enough sources of information are mined. However, there are some simple principles and tools (listed in Table 1) that may help to identify genes with good links to a specific phenotype. Firstly, is the gene expressed in the relevant tissue? This can be reviewed with the *SymAtlas* tool. Secondly, is the gene linked to the phenotype in the literature? *Huge Navigator* is a good tool enabling rapid review of the literature around a gene. Finally, does the gene fall into a pathway or interact with other genes with a known involvement in the phenotype? In this case, the EMBL *STITCH* tool is a good place to start. Once these areas have been considered and wishful thinking has been purged, then further investigation can be planned.
8. *Vista Enhancers*: Three ultra-conserved enhancer regions which have been demonstrated to drive tissue specific within the FTO gene, intron 7 (http://enhancer.lbl.gov/cgi-bin/imagedb.pl?form=presentation&show=1&experiment_id=element_155).

References

1. Kuhn, R.M., Karolchik, D., Zweig, A.S., Wang, T., Smith, K.E., Rosenbloom, K.R., *et al.* (2009) The UCSC Genome Browser Database: update 2009. *Nucleic Acids Res.*, **37**, D755–D761.
2. Hubbard, T.J., Aken, B.L., Ayling, S., Ballester, B., Beal, K., Bragin, E., *et al.* (2009) Ensembl 2009. *Nucleic Acids Res.*, **37**, D690–D697.
3. Woollard, P. (2010) Asking complex questions of the genome without programming. *Methods Mol. Biol*, 39–52.
4. Wellcome Trust Case Control Consortium (2007) Genome-wide association study of 14,000 cases of seven common diseases and 3,000 shared controls. *Nature*, **447**, 661–678.
5. Frayling, T.M., Timpson, N.J., Weedon, M.N., Zeggini, E., Freathy, R.M., Lindgren, C.M., *et al.* (2007) A common variant in the FTO gene is associated with body mass index and predisposes to childhood and adult obesity. *Science*, **316**, 889–894.
6. de Bakker, P.I., Yelensky, R., Pe'er, I., Gabriel, S.B., Daly, M.J. and Altshuler, D. (2005) Efficiency and power in genetic association studies. *Nat. Genet.*, **37**, 1217–1223.
7. Anderson, C.A., Pettersson, F.H., Barrett, J.C., Zhuang, J.J., Ragoussis, J., Cardon, L.R. and Morris, A.P. (2008) Evaluating the effects of imputation on the power, coverage, and cost efficiency of genome-wide SNP platforms. *Am. J. Hum. Genet.*, **83**, 112–119.
8. Dina, C. (2008) New insights into the genetics of body weight. *Curr. Opin. Clin. Nutr. Metab. Care*, **11**, 378–384.
9. Sandelin, A. (2008) Prediction of regulatory elements. *Methods Mol. Biol.*, **453**, 233–244.
10. Callinan, P.A. and Feinberg, A.P. (2006) The emerging science of epigenomics. *Hum. Mol. Genet.*, **15 Spec No 1**, R95–R101.
11. Costello, J.F. and Vertino, P.M. (2002) Methylation matters: a new spin on maspin. *Nat. Genet.*, **31**, 123–124.
12. Futscher, B.W., Oshiro, M.M., Wozniak, R.J., Holtan, N., Hanigan, C.L., Duan, H. and Domann, F.E. (2002) Role for DNA methylation in the control of cell type specific maspin expression. *Nat. Genet.*, **31**, 175–179.
13. Rakyan, V.K., Hildmann, T., Novik, K.L., Lewin, J., Tost, J., Cox, A.V., *et al.* (2004) DNA methylation profiling of the human major histocompatibility complex: a pilot study for the human epigenome project. *PLoS Biol.*, **2**, e405.
14. Ozsolak, F., Song, J.S., Liu, X.S. and Fisher, D.E. (2007) High-throughput mapping of the chromatin structure of human promoters. *Nat. Biotechnol.*, **25**, 244–248.
15. Rada-Iglesias, A., Ameur, A., Kapranov, P., Enroth, S., Komorowski, J., Gingeras, T.R. and Wadelius, C. (2008) Whole-genome maps of USF1 and USF2 binding and histone H3 acetylation reveal new aspects of promoter structure and candidate genes for common human disorders. *Genome Res.*, **18**, 380–392.
16. Pennacchio, L.A., Ahituv, N., Moses, A.M., Prabhakar, S., Nobrega, M.A., Shoukry, M., *et al.* (2006) In vivo enhancer analysis of human conserved non-coding sequences. *Nature*, **444**, 499–502.
17. Zhang, J., Feuk, L., Duggan, G.E., Khaja, R. and Scherer, S.W. (2006) Development of bioinformatics resources for display and analysis of copy number and other structural variants in the human genome. *Cytogenet. Genome Res.*, **115**, 205–214.
18. Redon, R., Ishikawa, S., Fitch, K.R., Feuk, L., Perry, G.H., Andrews, T.D., *et al.* (2006) Global variation in copy number in the human genome. *Nature*, **444**, 444–454.
19. Cooper, G.M., Zerr, T., Kidd, J.M., Eichler, E.E. and Nickerson, D.A. (2008) Systematic assessment of copy number variant detection via genome-wide SNP genotyping. *Nat. Genet.*, **40**, 1199–1203.
20. McCarroll, S.A. (2008) Extending genome-wide association studies to copy-number variation. *Hum. Mol. Genet.*, **17**, R135–R142.
21. Lowe, C.B., Bejerano, G. and Haussler, D. (2007) Thousands of human mobile element fragments undergo strong purifying selection near developmental genes. *Proc. Natl. Acad. Sci. U.S.A.*, **104**, 8005–8010.
22. Bejerano, G., Lowe, C.B., Ahituv, N., King, B., Siepel, A., Salama, S.R., Rubin, E.M., Kent, W.J. and Haussler, D. (2006) A distal enhancer and an ultraconserved exon are derived from a novel retroposon. *Nature*, **441**, 87–90.
23. Mardis, E.R. (2008) Next-generation DNA sequencing methods. *Annu. Rev. Genomics Hum. Genet.*, **9**, 387–402.
24. Wang, J., Wang, W., Li, R., Li, Y., Tian, G., Goodman, L., *et al.* (2008) The diploid genome sequence of an Asian individual. *Nature*, **456**, 60–65.

25. Wheeler, D.A., Srinivasan, M., Egholm, M., Shen, Y., Chen, L., McGuire, A., *et al.* (2008) The complete genome of an individual by massively parallel DNA sequencing. *Nature*, **452**, 872–876.

26. Levy, S., Sutton, G., Ng, P.C., Feuk, L., Halpern, A.L., Walenz, B.P., *et al.* (2007) The diploid genome sequence of an individual human. *PLoS Biol.*, **5**, e254.

27. Karolchik, D., Hinrichs, A.S., Furey, T.S., Roskin, K.M., Sugnet, C.W., Haussler, D. and Kent, W.J. (2004) The UCSC Table Browser data retrieval tool. *Nucleic Acids Res.*, **32**, D493–D496.

Chapter 3

Asking Complex Questions of the Genome Without Programming

Peter M. Woollard

Abstract

Increasingly, vast amounts of genomics and genetic data are available. Although much of the data is largely accessible to relatively simple web queries, in some cases, more complex queries are required. This paper reviews the hierarchy of tools for querying genetic and genomic data. For querying multiple genes, variants or regions ENSEMBL BioMart and the UCSC Table Browser offer flexible interfaces. For more complex queries, GALAXY is a sophisticated tool for building workflows over existing internet resources. For the most challenging genome scale queries, programmatic access may be required through a defined application programming interface (API) – such as the one provided by Ensembl. All these tools allow one to rapidly ask many questions that were difficult to answer a few years ago, but choosing the appropriate tool for the job is critical.

Key words: Genome, Genetics, SNP, Bioinformatics, Workflow, Pipeline, API

1. Introduction

Biology is an information-driven science. This is self evident in the scale of biological data resources built to support genome projects, transcriptomics, whole genome scans etc. The increasing quantity of the available data means, that there are often many challenges in getting the information you want (1). The productionised science approach over the past few decades has provided biological knowledge and technological infrastructure that has dramatically increased the diversity, coverage and often the quality of genomic information. Fortunately, the importance of data standards underpinning this data has been recognised at an early stage (2), and with the range and depth of ontologies being developed by the community, this all helps to bring meaning to data.

Michael R. Barnes and Gerome Breen (eds.), *Genetic Variation: Methods and Protocols*, Methods in Molecular Biology, vol. 628, DOI 10.1007/978-1-60327-367-1_3,

Long gone are the days you could maintain key information in spreadsheets and it is increasingly difficult even in a well resourced organisation to maintain comprehensive, integrated data systems of genomic and related data. One of the key reasons for this is that genomic data is rarely static, there are frequent updates, and informatics systems need to integrate updates and diverse data sources together in a meaningful way. We are now increasingly reliant on querying data at the data source sites or key data integration centres e.g. ENSEMBL, UCSC, Mouse Genome Informatics, NCBI (Table 1). This trend is set to continue, indeed the biomedical community is following a similar route to the trailblazers of "big science" – the physics community, by an increased reliance on shared super computer centres (1).

1.1. Tools for Querying Genomic Data

For most scientists, access to genomics data does not require a super computer, it usually means using the web query graphical user interfaces (GUIs) e.g. ENSEMBL and UCSC genome browsers (Table 1). These are well designed and allow you to ask relatively simple questions across impressively comprehensive arrays of data sources. Genome browsers are generally designed to query by a single gene, SNP or genomic region – allowing you to visualise and focus in on the relevant data, such as SNPs, transcripts, promoter regions, etc.

If you wish to query with multiple genes or genomic regions, then it is possible to use web applications like BioMart/EnsMart (3) and UCSC Table Browser (4). BioMart does allow you to run

Table 1
Selected list of tools for querying genomic data

Tools	URL
Ensembl Genome Browser	http://www.ensembl.org
UCCS Genome Browser	http://genome.ucsc.edu/
NCBI Mapview	http://www.ncbi.nlm.nih.gov/mapview/
Mouse Genome Informatics (Jackson Labs)	http://www.informatics.jax.org/genes.shtml
Galaxy	http://main.g2.bx.psu.edu/
BioMart/ENSMART	http://www.biomart.org
Taverna	http://taverna.sourceforge.net/
Ensembl API	http://www.ensembl.org/info/docs/api/index.html
NCBI API (eutils)	http://eutils.ncbi.nlm.nih.gov/entrez/query/static/eutils_help.html

quite complex queries, and it conscientiously leads you through the data, building up a query step by step, but to some, this largely linear and somewhat prescriptive approach may feel a little inflexible. The UCSC Table Browser is more flexible allowing the user to intersect any pair of UCSC data tracks, including custom tracks added by the user. Intersection of data is a key concept that is likely to be important to anyone who wishes to study the impact of variation on the genome. So for example, using tools like the UCSC table browser, it is possible to intersect variants such as SNPs or CNVs with functional elements, such as genes, promoters or regulatory regions.

There are still many queries that you will want to ask that you cannot yet do or that you will want to repeat frequently, for this reason, several big genomic data centres now provide programmatic access e.g. the ENSEMBL Perl API (application programming interface), NCBI e-utils (5) and UCSC mySQL (4). There are also statistical packages available which are able to run queries using an API in R-bioconductor (6). It is not difficult to learn how to run queries programmatically, but we all face time challenges and have different aptitudes. Also if you want to do a quick investigation, an API query is often overkill. Figure 1 gives an overview of the different query tools that should be considered depending on the complexity of the query.

The rest of this paper works through some simple examples of the use of some of these tools, mainly focusing on the practical use of Galaxy (7) as an example of how to get complex questions answered non-programmatically. Galaxy is compared and contrasted with some of the other available options. The reason for the focus on Galaxy is that it is freely available, simple to use and is primarily focused on genetic and genomic data.

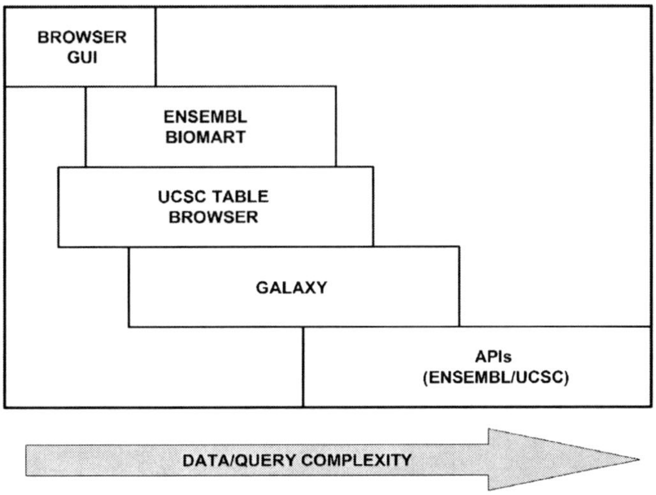

Fig. 1. A simple overview of the scope of the various query tools discussed in this paper

1.1.1. A Word of Caution

Before we proceed, it is important to be aware that any query of a genomic data set is prone to errors. These may be present in the core data or they may be introduced during subsequent data integration, so it is important to always sanity check results. When queries are particularly large, errors can also be caused because of sheer volume of data, leading to truncation of files during download, or due to lack of disk space. It is also important to be aware of different data standards. Finally, a great deal of automation is needed to integrate data. Automated data is error prone, e.g. gene build pipelines are often corrupted by rogue mRNA/CDNA evidence. The good news is that human curation efforts like Havana (8) and reporting of bugs by users are gradually improving the accuracy of genomic data.

1.2. Workflows

A workflow in bioinformatics is typically a series of computational or data manipulation steps (9). Generally, each component step is relatively simple, e.g. all genes on human chromosome 6. With most workflows, it is facile to join together lots of different types of genomic queries and manipulation steps, which becomes something that is complex but readily understood.

There have been many attempts to write bioinformatics workflows with varying degrees of success e.g. Taverna (10) in the public domain, Infosense (11) and Accelrys PipelinePilot (12) commercially. These also offer lots of control and power. Typically, they allow you to fetch data from the genomic and protein worlds, and run a wide variety of data and sequence analysis often using EMBOSS tools. Most of these are great when you are doing the same types of processing repeatedly, but often less good when you want to interactively explore questions.

1.2.1. Galaxy (Example Workflow Tool)

Galaxy provides an easy way to access genetic and genomic data without the need to program. The site also provides many excellent screencasts as introductions and worked examples, so one is soon productive.

Galaxy is a web based interface predominantly to data underlying the UCSC genome browser, but also other sources including BioMart. The key feature it provides over and above the UCSC Genome Browser (4), ENSEMBL Genome Browser (13), NCBI Map Viewer (14) or BioMart/EnsMart (3) is that it allows one to interactively and intuitively join queries together. This means that without programming, one can do complex queries and rapidly get the answers you wish to. It does also mean that each individual query component can be simple without the need to delve into detailed options (although these are available if you want them).

The developers state that it is designed for two different audiences:

1. Experimental biologists: "*I really have no time to program but I want to do whole-genome analyses to find targets for experimental validation*".
2. Computational biologists: "*I develop algorithms but have no time to develop interfaces*".

Galaxy has a similar ethos to the unix/linux operating systems where there are lots of relatively simply understood components, which are combined to produce something very powerful e.g. several get_data components, fetch_sequences, cut (to extract just certain columns), draw histograms etc.

An initial query can be used to extract a starting dataset from the UCSC Table Browser for example, from here a tab delimited file is created with chromosomal coordinates as the output. The subsequent query components work on each of these output files to generate new output files which the next component works on etc. Crucially, the components are biology aware e.g. having extracted a set of genes to get the 5′ prime flank of a gene to any user selected number of basepairs upstream. This gives most of the advantages of the querying provided by the superb existing interfaces to UCSC and ENSEMBL, coupled with the interactive querying that Galaxy offers.

The output of each query is recorded in the history in the right hand window. The output components can be viewed, deleted etc. The session history can easily be saved and reused at a future time.

There is a workflow editor capability in development; this allows the user to create a workflow from scratch or from a previous query history. The workflow can then be reused with different data set queries. However when reviewed (August 2008), this interface was found to have number of bugs, making it difficult to use. It is anticipated that these bugs will be fixed in the near future and generally the potential looks excellent and the mini workflows created do work nicely.

The Galaxy code is all open source and users are actively encouraged to develop their own algorithms.

2. Materials

The web querying was carried out using a PC with Windows XP, Firefox 3.01 and Internet Explorer 6. The Perl coding was carried on a linux workstation using the ENSEMBL API installed locally. The ENSEMBL mySQL db was queried directly at ensembl.org.

3. Methods

3.1. Get Yourself Set Up to Use Galaxy

1. Go to the Galaxy website http://main.g2.bx.psu.edu/
2. If you do not have an account, from the top tool bar select **Account→Create** and enter your details. For me, the password arrived just minutes later. The site states that you do not need to have an account, but having an account allows the user to save data and workflows.
3. Now select **Account→Login** and login with your details.

3.1.1. Getting all the SNPs Associated with Promoter Regions of Genes

We are going to get a set of genes and find the promoter regions. We will then identify which SNPs are located in these promoter regions. In the UCSC table browser, you can intersect any track with any other track, e.g. you could look for the intersection of SNPs with genes or transcription factor binding sites (TFBS). You can also look for intersection with your own custom tracks, e.g. SNPs used in a study.

Get all genes and promoters (first data set)

1. **Tools→Get Data→UCSC Main table browser** (see Fig. 2)
   (a) Group = **Get all genes Genes and Gene Prediction Tracks**
   (b) Track = **UCSC Genes**

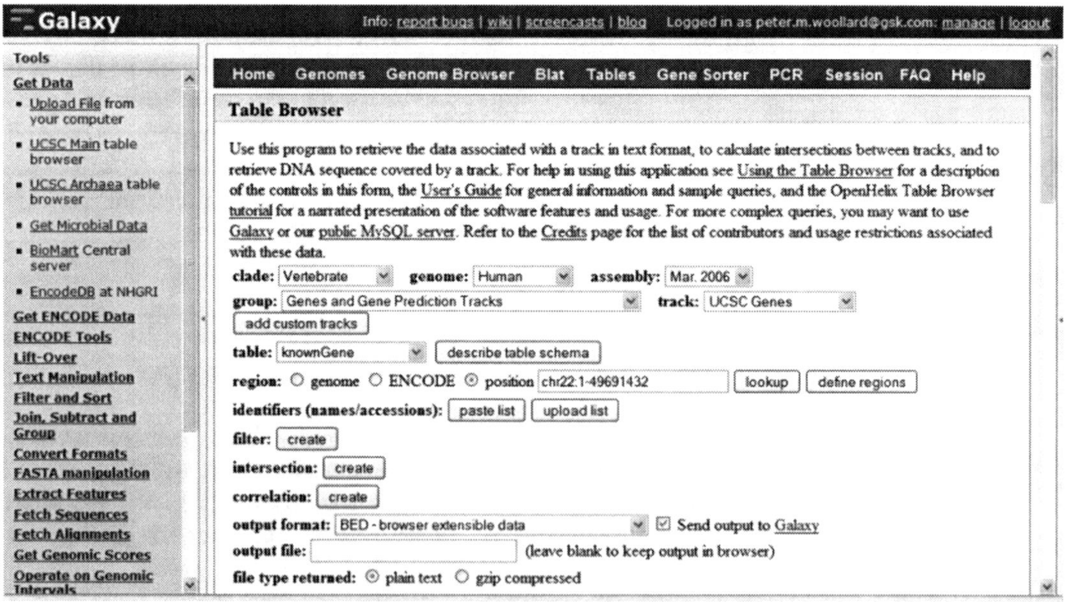

Fig. 2. Choosing the genes using *Get Data* in Galaxy using UCSC Table Browser

   (c) Region = **genome**
   (d) Click on **get output**
   (e) Send Query to Galaxy (making sure "**whole gene**" is selected)

2. **Tools → Operate on Genomic Intervals → Get Flanks Upstream 500 bp flanking.** (this should cover each gene's promoter)
3. **Tools → Text Manipulation → cut** (c1, c2, c3, c4, c6)
4. Click on the **pencil** of the cut output and then change the data type to **interval** (so that it knows the right column types)
   (a) You will also need to assign the column names
5. N.B. for any of the generated data sets, you can click on the pencil and change the display name to be memorable.
6. Next get all the SNPs (second data set)
   (a) **Tools → Get Data → UCSC Main table browser** (see Fig. 3)
   (b) Group = **Variation and Repeats Track**
   (c) Track = **SNPs (129)** and remember to select whole genes

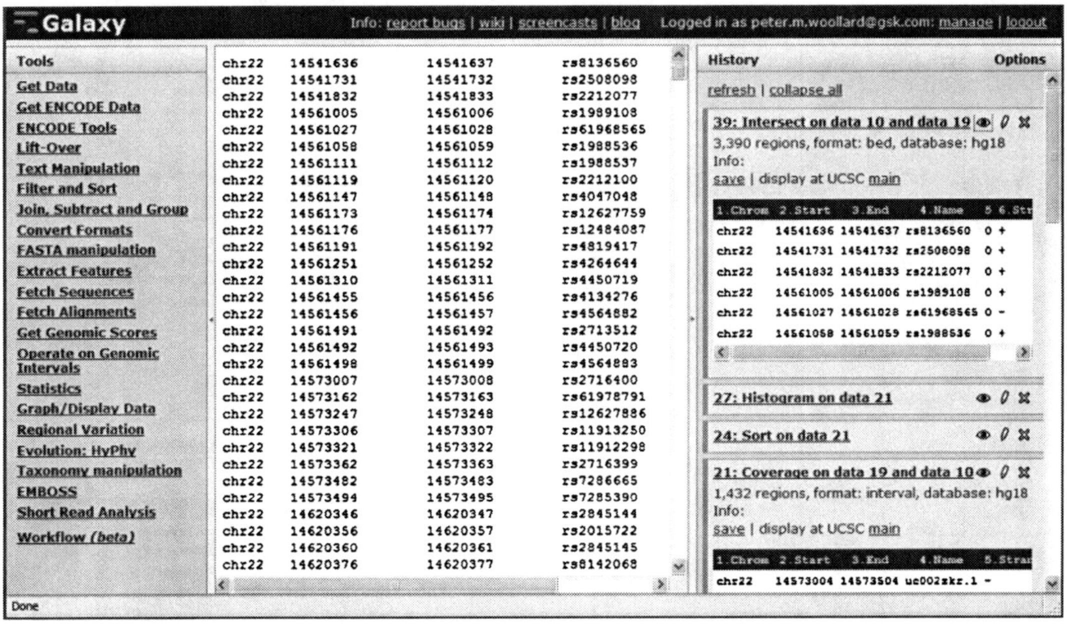

Fig. 3. Galaxy: showing the result of intersecting SNPs with Gene Promoters. Note the mini-tables in the history, with the full data visible in the central panel

7. Combining the two data sets to get the coverage
   (a) **Tools→Operate on Genomic Intervals→Coverage**
   (b) **Graphs/Display Data→Filter**_and_**Sort→sort** (c6 is column 6)
   (c) **Histogram**
8. Combining the two data sets to get the list of SNPs that appear in the promoters
   (a) **Tools→Operate on Genomic Intervals→Intersect** (does an intersect on the chromosomal positions)
   (b) Choose SNPs vs the promoters, so that way you get to see the SNP Ids (Intersect of Promoters and SNPs data sets)
   (c) Then save the resulting output file to your local computer

3.1.2. Using an Automated Workflow

Galaxy has an automated workflow capability in development; this allows the user to create a workflow from scratch or from previous query history. This allows reuse of workflows with different data set queries, and it then automatically runs through the pipeline. The Current prototype still has some bugs, but the potential looks excellent and the mini workflows worked nicely, see Fig. 4.

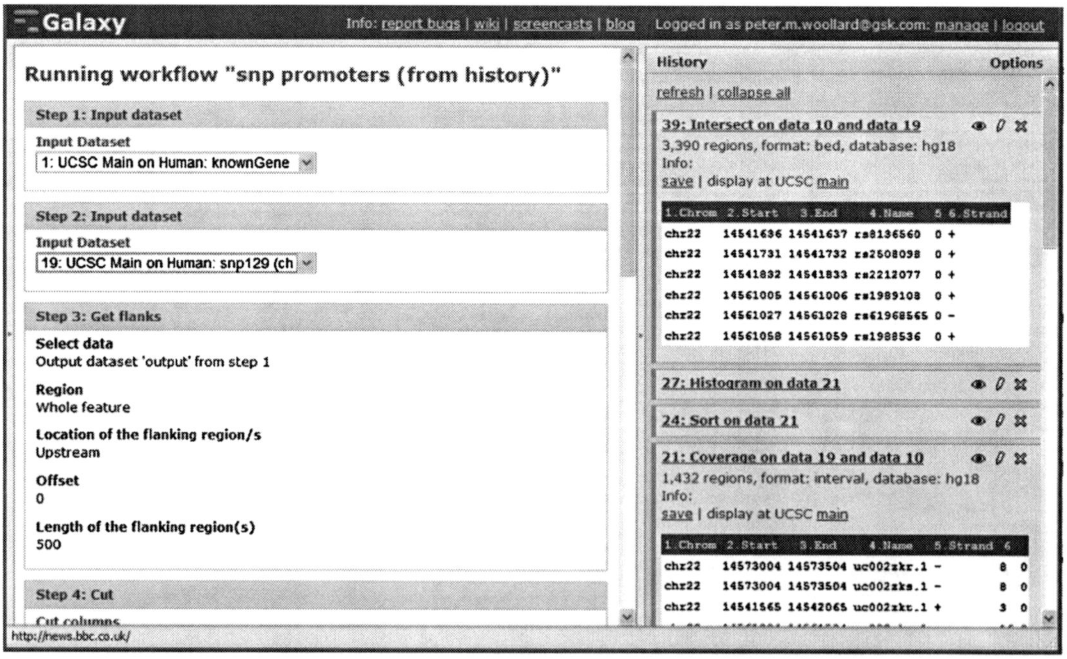

Fig. 4. Using a workflow. The user can choose genes and SNPs to input into the workflow. The latter need not have been a collection of SNPs, it could be any genomic entity with known coordinates or sequence

3.2. Querying Using BioMart/EnsMart

If you wish to find all the SNPs in the protein coding genes known to be involved in the immune system (GO:0002376 immune system process), you can do this easily in

BioMart. Note that this example is making use of the power of ontologies.

1. Go to the Biomart web page: http://www.biomart.org
   (a) Choose Database: **ENSEMBL50** Dataset: **Homo sapiens genes(NCBI50).**
2. Click on **Filters,** so we can refine our query.
3. Open the **Gene Ontology** section and in the **Biological process:** enter **GO:0002376** and check the box on the left.
4. Open the **Gene** section and select Gene type: **protein_coding.**
5. Now we need to choose which columns we want to see in the output, so click on **Attributes.**
6. Open **SNPs** and then check the following boxes **Chromosome Name, Gene Start (bp), Gene End (bp) Strand** and **Associated Gene.**
7. Open **GENE ASSOCIATED SNPS** and check the following boxes for SNP Attributes: **Reference ID, Gene Location and Effect, Synonymous status** and **Location in Gene.**
8. Click on **Results** and you now see a preview (Fig. 5). If you wish to change your **Filters** or **Attributes** (output columns), just click on the filters or attributes on the left panel.

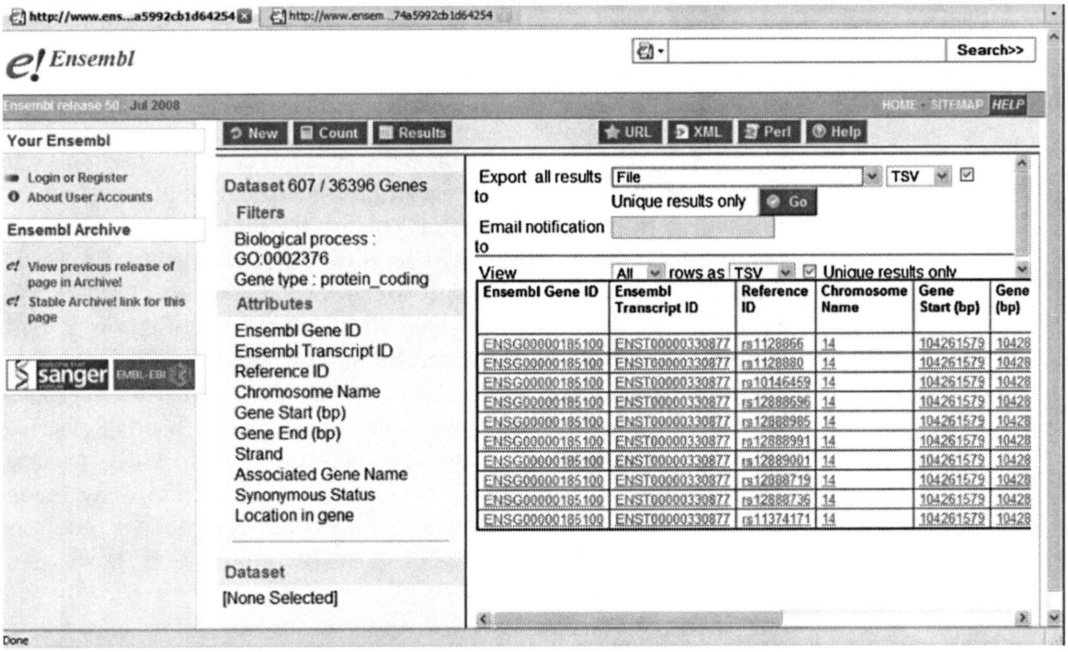

Fig. 5. Biomart showing the results of an ENSEMBL query to get SNPs for particular genes

9. If you click on **Count**, you can see that you have selected just 607 out of 36,396 genes.
10. To save a file with all the results, **Export all results → file** and **TSV** (=tab separated values). Now check the box and press **Go**.
11. A file then soon downloads. You will need to delete the HTML code at the top and bottom of the file, e.g. using a text editor like WordPad before opening in Excel.

Incidentally, there is an API to the biomart and a powerful feature of the Biomart GUI is that if you click on **Perl** on the top panel on your results panel, it shows you the Perl API Biomart code to have produced this query. There is a similar option for those of you who prefer web services, especially if you prefer writing code in Java or Python.

3.3. Querying Using the Ensembl Perl API

Although this review demonstrates the ease of querying genomic data without programming, programmatic queries should still be considered an option for the more advanced user. Direct queries using an API (application programming interface) have a number of queries over workflow tools, they can be more complex and less constrained by interface, and they are faster and can also be automated. Figure 6 shows a fairly simple script to find all SNPs and their frequencies. Starting points could be a slice of a chromosome or a list of genes. There is good documentation and even courses that provide further information about using Perl to access an API (15).

4. Discussion

Before selecting a tool and formulating a query of genomic data, it is worth thinking about what you are trying to achieve. Table 2 summarises the features of the different tools reviewed here. There may be better solutions for specific questions. The UCSC table browser, Biomart or even the genome browsers may well get you what you need quickly or at least allow you to see what types of information are available. If one is particularly interested in mouse genomics and mouse knockouts with human disease relevant phenotypes, then the Jackson Lab's MGI Mouse Genome Browser (Table 1) will be better suited. Thinking more laterally much can be achieved using other approaches, such as literature searching/text mining, tools for which have also improved considerably (16).

Using interactive workflow applications like Galaxy allows one to rapidly get up to speed and start asking interesting questions.

```
#!/GWD/bioinfo/apps/bin/Perl -w

use strict;
use Bio::EnsEMBL::Utils::ConfigRegistry;
use Bio::EnsEMBL::DBSQL::DBAdaptor;
use Bio::EnsEMBL::Variation::DBSQL::DBAdaptor;

use Data::Dumper;

my $registry = 'Bio::EnsEMBL::Registry';
$registry->load_registry_from_db(
    -host => 'ensembldb.ensembl.org',
    -user => 'anonymous' #,-verbose => 1
);
my $reg = $registry;

my $species = 'Human';
my $variation_adaptor =       $reg->get_adaptor($species, "variation", "variation");
my $va = $variation_adaptor;
my $vf_adaptor = $reg->get_adaptor($species, "variation",'VariationFeature'); #get adaptor to
VariationFeature object

my $gene_adaptor = $registry->get_adaptor( $species, 'core', 'Gene' );
my $slice_adaptor = $reg->get_adaptor($species,'core','slice'); #get the database adaptor for
Slice objects
#my $slice = $slice_adaptor->fetch_by_region('chromosome','22'); #get chromosome 22

#doing it by gene
my $geneName = 'GLI2';
my (@genes) = @{$gene_adaptor->fetch_all_by_external_name($geneName)};
my $gene = $genes[0];
print "GENE name=",$gene->external_name , " stable_id=",$gene->stable_id(), "\n";

my $slice = $slice_adaptor->fetch_by_gene_stable_id($gene->stable_id);

my $vfs = $vf_adaptor->fetch_all_by_Slice($slice); #return ALL variations defined in $slice
my $variationTotal = scalar(@{$vfs});
my $variationCount =1;

foreach my $vf (@{$vfs}){
  print $variationCount++,"/$variationTotal", "\tVariation: ", $vf->variation_name, " with
alleles ", $vf->allele_string, " in chromosome ", $slice->seq_region_name, " and position ", $vf-
>start,"-",$vf->end,"\n";
  my $v = $vf->variation();
  my @alleles = @{$v->get_all_Alleles()};
  foreach my $a (@alleles)
    {
      my $p = $a->population();
      if (!defined $p) { next; };
      print "\t",join("\t",,$v->name,$a->allele(), $a->frequency(),$p->name), "\n";
    }
  if($variationCount>20) { last};
  }
1;
```

Fig. 6. A simple Perl script to query the Ensembl API

Galaxy is user friendly enough to make it easy to experiment on, explore and investigate.

The more established workflow applications such as Taverna are better for rapid replication of workflows, where query inputs are of a defined type. This has much more of an inherent learning curve than web based applications such as Galaxy. Arguably, this makes web based workflow tools a little more suited to dynamic exploration.

Programmatic solutions like R-bioconductor, the ENSEMBL API or direct mySQL queries are unarguably the "Rolls-Royce"

Table 2
Feature summary of tools for genome queries

Platform	Genomes	Single gene/chr region	Multiple genes/chr regions	Learning curve	Querying	Visualisation	Example queries
UCSC Genome Browser Ensembl Genome Browser	Multi Multi	Yes	No	Easy	Single gene/ region	Graphical	ADORA1 gene and want to see all the genomic information (tracks).
MGI (Jax) Mouse Genome Browser	Mouse	Yes	No	Easy	Single gene/ region Many options	Graphical	Chr = 1 cM = 10.0–40.0 Phenotypes/Diseases: contains cardiovascular
NCBI Map Viewer	Multi	Yes	No	Easy	Single gene/region Many options (in advanced search)	Graphical	Have genetic mapping data, and want candidate disease genes in that region
BioMart/ EnsMart	Multi	Yes	Yes	Easy	Options e.g. Filtering Leads you through choices	Tables	For a set of Hugo gene names from a study, retrieve all the SNPs in protein encoding genes with a particular ontology property.
UCSC Table Browser	Multi	Yes	Yes	Easy	Intersection of two queries Options e.g. Filtering	Tables	For a set of mouse chromosomal positions, get me all the genes Intersect this with those with Allen Brain expression information.
Galaxy	Multi	Yes	Yes	Easy	Multiple queries	Graphical/tables	Get me all SNPs that are found in a defined promoter region of my choosing in human GPCRs.

General Bioformatics Workflows e.g. Taverna	Multi	Yes	Yes	Medium	Lots of flexibility	Graphical/many	For a set of genes extract the protein translations look for a particular interpro domain and then align those that contain proteins.
Ensembl API R/Bio-Condutor methods	Multi	Yes	Yes	Medium (if you can program)	Lots of flexibility	Text	For a human chromomal region get me the mouse orthologues as proteins, give me the gene alignment percentage identities- Add in the human SNPs with a frequency in specific hapmap population > 0.6
mysql queries (e.g. UCSC & ENSEMBL)	Multi	Yes	Yes	Hard (e.g. need to know the schemas)	Massive flexibility	Text	Limited only mainly by your imagination, data availability and schema design.

solution for data mining, where one wants lots of flexibility. But if your needs are nearer to a "Volkswagen", then tools like the UCSC table browser, BioMart and Galaxy are also reliable and flexible. Ultimately, these are everyday tools. When the needs are exceptional, then programmatic access is probably the method of choice – then it is time to rely on goodwill and a few favours from Perl-savvy colleagues, but by using Galaxy, one may have solved half the problem already.

References

1. Stein LD (2008) Towards a cyberinfrastructure for the biological sciences: progress, visions and challenges. *Nat Genet.* 9(9):678–688.
2. Smith B, Ashburner M, Rosse C, et al. (2007) The OBO Foundry: coordinated evolution of ontologies to support biomedical data integration, 1. *Nat Biotechnol.* 25(11):1251–1255.
3. Kasprzyk A, Keefe D, Smedley D, et al. (2004) EnsMart: A generic system for fast and flexible access to biological data. *Genome Res.* 14:160–169.
4. Karolchik D, Kuhn, RM, Baertsch R, et al. (2008) The UCSC Genome Browser Database: 2008 update. *Nucleic Acids Res.* 36:D773–D779.
5. http://eutils.ncbi.nlm.nih.gov/entrez/query/static/eutils_help.html.
6. Durinck S, Moreau Y, Kasprzyk A, et al. (2005) BioMart and Bioconductor: a powerful link between biological databases and microarray data analysis. *Bioinformatics* 21(16):3439–3440.
7. Giardine B, Riemer C, Hardison RC, et al. (2005) Galaxy: a platform for interactive large-scale genome analysis. *Genome Res.* 15(10):1451–1455.
8. Harrow J, Denoeud F, Frankish A, et al. (2006) GENCODE: producing a reference annotation for ENCODE. *Genome Biol.* Suppl. 1:S4.1–S4.9.
9. http://en.wikipedia.org/wiki/Bioinformatics_workflow_management_systems.
10. Oinn T, Addis M, Ferris J, et al. (2004) Taverna: a tool for the composition and enactment of bioinformatics workflows. *Bioinformatics* 20(17):3045–3054.
11. Inforsense http://www.inforsense.com/.
12. Accelrys SciTegic Pipeline Pilot http://accelrys.com/products/scitegic/.
13. Birney E, Andrews TD, Bevan P, et al. (2004) An overview of Ensembl. *Genome Res.* 14(5):925–928.
14. Wheeler DL, Barrett T, Benson DA, et al. (2008) Database resources of the National Center for Biotechnology Information. *Nucleic Acids Res.* 36(Database issue):D13–D21.
15. Stabenau A, McVicker G, Melsopp C, et al. (2004) The Ensembl core software libraries. *Genome Res.* 14(5):929–933.
16. Cohen KB, Hunter L (2008) Getting started in text mining. *PLoS Comput Biol.* 4(1):e20.

Chapter 4

Laboratory Methods for the Detection of Chromosomal Abnormalities

Jacqueline Schoumans and Claudia Ruivenkamp

Abstract

Constitutional chromosomal aberrations are inborn changes with or without phenotypic consequences. Conventional chromosome analysis has been for a long time the method of choice for identification of such abnormalities. However, over the past decades, several molecular cytogenetic techniques have successfully been introduced into the genetic diagnostic laboratories to increase the detection sensitivity and to outline chromosome rearrangements in more detail. Each method has its strength and limitation, therefore often several techniques are needed to detect and unravel the complexity of chromosome abnormalities. This chapter focuses on the most commonly used methods in the diagnostic setting for detection and characterization of constitutional chromosome abnormalities.

Key words: Chromosomal aberration, Cytogenetic techniques, Diagnostic, Chromosome rearrangements, Aneuploidy, Segmental aneusomies, Translocations, Inversions, G-band, FISH, Karyotyping, MLPA, QF-PCR, Uniparental disomy, Array CGH

1. Introduction

Constitutional chromosomal aberrations are inborn changes that are either inherited from a parent or have occurred as a de novo mutation in one of the gametes that form the zygote. Numerical aberrations comprise aneuploidy, e.g. trisomy or monosomy, and ploidy changes, e.g. triploidy. Structural rearrangements affect the normal structure of one or several chromosomes i.e. deletions, translocations, and inversions. Cytogenetic imbalances are present in 50–60% of first trimester miscarriages and 0.7–0.9% of newborn children. Most of these imbalances are numerical; however, 3% are due to structural changes (1). Individuals with an unbalanced chromosomal rearrangement usually present with symptoms like mental retardation, dysmorphic features, and

Michael R. Barnes and Gerome Breen (eds.), *Genetic Variation: Methods and Protocols*, Methods in Molecular Biology, vol. 628, DOI 10.1007/978-1-60327-367-1_4, © Springer Science+Business Media, LLC 2010

malformations of the internal organs. Structural chromosome abnormalities have been estimated to be present in approximately 0.76% of newborn children when using conventional chromosome analysis (2). Some of these chromosome abnormalities will give rise to an abnormal phenotype in the child, while some will be carriers of a balanced structural rearrangement without phenotypic consequences. However, healthy carries have an increased risk of having offspring with an unbalanced aberration resulting in an abnormal phenotype. Approximately 1/500 individuals is a carrier of a balanced chromosome rearrangement.

A considerable number of cases with mental retardation can be explained by the presence of constitutional chromosome abnormalities. They often cause specific and complex phenotypes resulting from an imbalance in the normal dosage of genes located in a particular chromosomal segment. Chromosome aneuploidies, large segmental aneusomies, translocations, and inversions can be detected by conventional chromosome analysis. Nevertheless, this method has limited resolution and is unreliable for the detection of subtle copy number changes (deletions and duplications) and complex chromosome rearrangements. With the sequence of the human genome assessable, new reliable technologies have rapidly been developed over the past decade, resulting in molecular and cytogenetic methods that are offered in the diagnostic setting for the detection of subtle chromosome imbalances and for a more accurate characterization of complex chromosome rearrangements.

This chapter describes molecular and cytogenetic methods (chromosome analysis by GTG-banding, fluorescence in situ hybridization (FISH), Spectral Karyotyping (SKY)/multicolor FISH (M-FISH), Multiplex Ligation-dependent Probe Amplification (MLPA), and copy number analysis by array) used for outlining chromosomal abnormalities in the clinical diagnostics laboratories. Each method has its advantages and its limitations, but the methods are often a complement to each other when characterizing chromosome aberrations in detail. The usefulness of the different methods in order to unravel different types of chromosome abnormalities are listed in Tables 1 and 2.

2. Materials

2.1. Chromosome Analysis by G-bands using Trypsine and Giemsa (GTG-banding)

1. Fresh blood (~5–10 mL) should be drawn into sodium heparin tubes and kept at room temperature; (the blood should not be frozen). The blood may also be stored overnight at 4°C if necessary (see Note 1).

Table 1
Useful methods for whole genome screening to detect constitutional chromosome abnormalities

	Method			
Abnormality	Karyotyping by GTG banding	SKY/ MFISH	Array CGH	SNP array
Balanced translocation	+	++	–	–
Unbalanced translocation	+	+	++	++
Balanced inversion	+	–	–	–
Insertion	+	++	–	–
Complex rearrangement	±	++	±	±
Deletion	±	±	++	++
Duplication	±	±	++	++
Triplication	±	±	++	++
Trisomy	+	+	+	+
Triploidy	+	+	–	±[b]
Monosomy	+	+	+	+
Uniparental disomy	–	–	–	+[a]
Methylation defect	–	–	–	–
Copy neutral LOH	–	–	–	+[a]

[a]Only detection of isodisomy, if parents are included also heterodisomy
[b]Visible in B-allel frequency plot (genotypes)

2. RPMI 1640 medium with L-glutamine 1× (Cellgro; Mediatech Inc., Herndon VA).
3. Starting medium (100 mL): 86-mL RPMI 1640 medium with L-glutamine, 10-mL fetal calf serum, 1-mL penicillin-streptomycin, 1-mL L-glutamine, and 2-mL phytohemagglutinin.
4. Continuing medium (100 mL): 87-mL RPMI 1640 medium with L-glutamine, 10-mL fetal calf serum, 1-mL HEPES buffer, 1-mL penicillin-streptomycin, and 1-mL L-glutamine.
5. Fetal calf serum (Irvine Scientific, Santa Ana CA).

Table 2
Useful methods for confirmation or further characterization of already identified constitutional chromosome abnormalities

	Method								
Abnormality	Karyotyping by GTG banding	SKY/ MFISH	Array CGH	SNP array	Locus specific metaphase FISH	Locus specific interphase FISH	MLPA	Micro-satellite	QF-PCR
Balanced translocation	±	++	–	–	++	–	–	–	–
Unbalanced translocation	±	++	++	++	++	–	++	–	–
Inversion	±	–	–	–	++	–	–	–	–
Insertion	+	++	–	–	++	–	–	–	–
Complex rearrange-ment	±	++	±	±	++	–	–	–	–
Deletion	±	±	++	++	++	+	++	+	+
Duplication	±	±	++	++	–	±	++	+	+
Triplication	±	±	++	++	–	±	++	±	±
Trisomy	++	+[a]	+[a]	+[a]	+	++	++	+	++
Triploidy	++	+[a]	–	±	+	++	–	+	++
Monosomy	++	+[a]	+[a]	+[a]	+	++	++	+	++
Uniparental disomy	–	–	–	+[b]	–	–	–	++	+
Methylation defect	–	–	–	–	–	–	++	–	–
Copy neutral LOH	–	–	–	+	–	–	–	++	+

[a]Costly
[b]Only isodisomy, if parents are included also heterodisomy
[c]Methylation specific MLPA (MS-MLPA)

6. Phytohemagglutinin (45 mg) (Irvine Scientific, Santa Ana CA). Reconstitute in 4.5 mL of sterile distilled H_2O to give a stock solution of 10 mg/mL (store at 4°C).
7. Thymidine solution (15 mg/mL) made up in PBS (aliquot and store at –20°C).
8. 1× Dulbecco's phosphate buffered saline (PBS) (Cellgro; Mediatech Inc., Herndon VA or Gibco BRL).

9. KaryoMAX colcemid (10 μg/mL) (Gibco-BRL; Life Technologies, Grand Island NY).
10. 10 mg/ml Plus One Ethidiumbromide solution (Amersham Biosciences).
11. KCl (0.075 M). Store at room temperature and prewarm at 37°C before use.
12. Freshly made fixative (3:1 methanol:glacial acetic acid).
13. HEPES buffer solution (1 M) (Gibco-BRL; Life Technologies, Grand Island NY).
14. Gurr buffer (Gibco BRL)
15. 2× SSC (BioWhittaker)
16. 2.5% Trypsine 10× (Gibco-BRL)
17. Giemsa (Merck)
18. Leishman (Gurr-BDH)

2.2. Fluorescence In Situ Hybridization (FISH)

1. BioProbe random primed DNA labeling kit art no 42720 (Enzo Life Sciences Inc. NY) or Nicktranslation kit art no N5500 (Amersham Biosciences).
2. 20× SSC stocksolution PH 7.0 (0.3 M sodium citrate, pH 7.0, 3 M NaCI) (Sigma-Aldrich Chemical).
3. 2× SSC dilution made up in H_2O
4. Pepsin Stock solution (1 g pepsin dissolved in 10 ml distilled H_2O, (aliquot and store at −20°C)
5. HCl, 1 N
6. HCL solution 0.001 M
7. 37% Formaldehyde (Sigma)
8. 1% Formaldehyde solution made up in PBS
9. Formamide, deionized (Ambion)
10. Denaturation solution. 70% formamide made up in 2× SSC buffer
11. dH_2O
12. Human Cot-1 (Invitrogen)
13. Cold dehydration solutions (99% ethanol, 80% ethanol and 70% ethanol) kept at −20°C
14. Rubber Cement

2.3. Spectral Karyotyping (SKY)/Multicolor-FISH (M-FISH)

1. SpectralKaryotyping kit (Applied Spectral Imaging Ltd) containing;
   (a) SKYPaint™ probe mixture,
   (b) blocking reagent,
   (c) anti-fade-DAPI reagent,

(d) Cy5 staining reagents
(e) Cy5.5 staining reagent

2. See FISH reagents

2.4. Multiplex Ligation-dependent Probe Amplification (MLPA)

1. SALSA kit: MRC-Holland (see www. http://www.mlpa.com/) or custom designed long oligos (Sigma)
2. Primers, FAM-labeled, HPLC purified (Biolegio or Invitrogen)
3. Size standard: GeneScan, LIZ-500 (Applied Biosystems)
4. Size standard: GeneScan, ROX-500 (Applied Biosystems)
5. Deionized Formamide (Lucron Bioproducts)
6. Applied Biosystems genetic analyzer

2.5. Quantitative Fluorescence Polymerase Chain Reaction (QF-PCR)

1. Aneufast™ QF-PCR Kit (Genomed Ltd, UK)
2. Size standard: GeneScan, ROX-500 (Applied Biosystems)
3. Deionized Formamide (Lucron Bioproducts)
4. Applied Biosystems genetic analyzer

2.6. Array Genomic Hybridization

2.6.1. Array CGH

Agilent materials see manufacture's instruction (http://www.chem.agilent.com) Agilent technologies

2.6.2. SNP Array

Affymetrix materials see the manufacturer's instruction (http://www.affymetrix.com)

Illumina materials see the manufacturer's instruction (http://www.illumina.com).

2.7. Microsatellite Makers for Detection of Parental Origin and UPD

1. Human Linkage Mapping Sets (LMS)/Custom Markers for PCR Genotyping (Applied Biosystems)
2. Ampli Taq®DNA Polymerase Taq Polymerase ((Applied Biosystems).
3. 100 mM dNTP (Amersham Biosciences).

3. Methods

3.1. Chromosome Analysis by G-bands using Trypsine and Giemsa (GTG-banding)

The chromosome banding techniques allow the identification of microscopic numerical and structural aberrations, including translocations, inversions, deletions, and duplications. Conventional chromosome analysis using banding techniques, particularly G-banding (3), are now routine procedures in all cytogenetics laboratories (see Note 1 for a discussion of limitations of this technique).

General procedure:

1. For conventional chromosome analysis, peripheral blood cells are cultured in medium for 48 h, 72 h, or 96 h (0.3–0.5 ml Heparin whole blood in 8 ml medium). Many types of defined medium are used, such as RPMI 1640, MEM, M199, and others. Addition of fetal calf serum is recommended in concentrations between 10 and 25%.
2. Though lymphocytes from blood do not normally divide spontaneously, cells can be induced to proliferate by addition of a mitogen. The most commonly used is phytohaemagglutinin that is supplemented to the medium.
3. To increase the number of cells in the same stage of cell division, the cultures can be synchronized using the chemical blocking agent thymidine (40 μM).
4. Cells are arrested in metaphase by adding the spindle inhibitor colcemid (0.5 μg/ml) for the final culture time. Long colcemid incubation will result in high mitotic index, but short chromosomes, while a short colcemid incubation results in long chromosomes but lower mitotic index. Typically, colcemid at concentration 10 μg/ml is incubated for 15 min for synchronized lymphocytes to 1.5 h for not synchronized lymphocytes. Elongated chromosomes can be achieved by the addition of ethidium-bromide in combination with colcemid.
5. Then cells are treated with a hypotonic buffer containing 75 mM Potassium Chloride and harvested in a 3:1 methyl alcohol:glacial acetic acid (volume:volume) fixative.
6. Metaphase slides are prepared by dropping the cell solution on (wet) slides. The quality of the metaphase spreading is dependent upon a number of factors, including humidity, airflow, room temperature, and cell concentration.
7. Prior to staining the slides need to be air dried and aged. This can be done by overnight incubation at 37°C or a more rapid aging by increasing the temperature to 60°C or 90°C and decreasing the incubation time to 1 h.
8. For staining trypsin treatment is used followed by Giemsa staining that generates a pattern of dark and light bands characteristic for each individual chromosome.
9. Chromosomes are subsequently visualized in a light microscope; images are captured with an automated karyotyping system for chromosome classification.

3.2. Fluorescence In Situ Hybridization (FISH)

FISH is a technique that allows visualization of genetic alterations directly on interphase nuclei and metaphase chromosomes. A fluorescent labeled DNA probe is hybridized onto cells that are fixed and immobilized on a glass slide and detection is performed using a fluorescent microscope. The choice of FISH probes is dependent on the biological question in mind, and the locus of interest must be at least approximately known, for instance microdeletion syndromes. Locus specific probes can be made from BAC (Bacterial Artificial Chromosome), PAC (P1 Artificial Chromosome), cosmid or fosmid clones or from PCR products (see Note 2). The resolution depends on the probe size (>5–10 kb) and the target DNA used. FISH is a powerful technique that allows detection of deletions, duplications, rearrangements, and mapping of translocation breakpoints, but is not optimal for the detection of small tandem duplications. However, it is relatively labor intensive because each locus has to be analyzed separately. Even when combining different fluorescent dyes, the number of different loci that can be visualized simultaneously and reliably distinguished is limited.

General procedure:

1. Freshly prepared metaphase or interphase slides need to be aged before they can be used for FISH, this can be done either by aging a few days-weeks at room temperature or by rapid aging by incubating in 2× SSC at 37°C for 30 min to 2 h.
2. The slides are than pretreated in a freshly prepared pepsin solution (0.1 mg/ml in 0.01MHCl) at 37°C for ±5 min (interphase slides can be treated longer) in order to remove cytoplasm and post fixated in a 1% formaldehyde/PBS solution.
3. Denaturation of labeled probe (see Note 3) and slide can be done separately or by adding the probe directly on the slide, cover the region containing probe with a coverslip and glue the edges to protect it from evaporation.
4. Then codenature both probe and cells together by incubation on a heating block for 2–5 min at 73°C.
5. After denaturation, the probe is hybridized onto the slide by incubation overnight at 37°C in a humidified chamber and protected from light.
6. Posthybridization washes are carried out by taking of the cover slip, dipping the slide in 2× SSC at room temperature to wash of the unbound probe, and incubate the slide in a 2× SSC solution for 2 min at 73°C.
7. The slides are counterstained using DAPI (4′-6-Diamidino-2-phenylindole) and visualized in a fluorescent microscope. The DAPI stain is used to generate a pseudo-G-band image of the metaphase chromosomes.

3.3. Spectral Karyotyping (SKY)/ Multicolor-FISH (M-FISH) (see Fig. 1)

SKY/M-FISH are complementary fluorescent molecular cytogenetic techniques. SKY/M-FISH permits the simultaneous visualization of each human chromosome in a different color, facilitating the identification of chromosomal aberrations. Chromosome-specific probe pools (chromosome painting probes) are generated from flow-sorted chromosomes, and then amplified and fluorescently labeled by degenerate oligonucleotide-primed polymerase chain reaction (see Notes 4–6 for discussion of probe labeling).

SKY or M-FISH permits the detection of interchromosomal aberrations such as translocations, insertions, and complex chromosome rearrangements. In addition, it enables the identification of the chromosomal origin of marker chromosomes. Intra-chromosomal alterations such as inversions, small duplications, or deletions are not detectable. Both SKY and M-FISH have their limitations and Fluorescence flaring appears to be a significant cause for the misclassification of rearrangements involving small chromosomal segments (4).

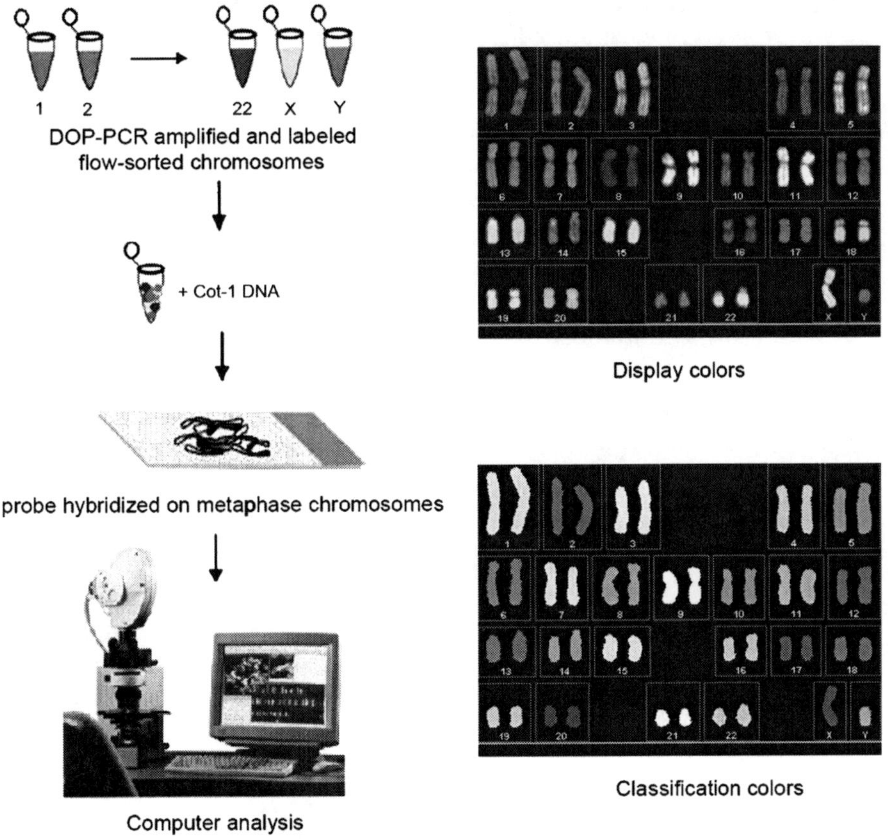

Fig. 1. Spectral karyotyping. Flow-sorted chromosomes are DOP-PCR amplified and labeled using 5 fluorochromes. A probe cocktail containing 24 specific colors for each chromosome (derived from a combination of 1–4 fluorochromes) is hybridized on metaphase chromosomes. Cot-1 DNA is added for blocking of repetitive sequences. For image capturing, a fluorescence microscope equipped with a CCD camera and Spectracube is used. Computer software allows chromosome identification and classification by each chromosome specific spectral color

General procedure:

1. The denaturation of the probe cocktail and metaphase slide, the hybridization, and posthybridization washes can be performed as described above in the FISH procedure.
2. For detection in the SKY procedure, 2 dyes (Cy5 and Cy5.5) are indirectly labeled and labeled antibodies are incubated onto the slides at 37°C for 40 min each before applying the counterstaining.
3. The quality of the metaphases slide (such as good spreaded chromosomes, minimal chromosome overlap, no cytoplasm) is of great importance for a successful SKY/M-FISH analysis.
4. Hybridization procedures and reagents are provided by Applied Spectral Imaging (SKY) and Metasystems (M-FISH).

3.4. Multiplex Ligation-dependent Probe Amplification (MLPA)

MLPA is a method that can detect copy number changes of different DNA targets simultaneously in one reaction. The commercially available MLPA kits are designed in such way that the length of each amplification product has a unique size that is separated by electrophoresis (5). There are several commercial MLPA kits available (www.mlpa.com). For example, there is an MLPA kit that contains one MLPA probe for each subtelomeric region. This kit is specially developed to screen patients with unexplained mental retardation and/or developmental delay, since in 5–7% of the cases, aberrant copy number changes of subtelomeric regions are detected as cause. In addition, there are kits designed to screen for multiple known micro-deletion and duplication syndromes simultaneously.

General procedure (see Notes 7 and 8):

1. The technique of MLPA does not involve amplification of the sample DNA, instead the probe sets that are added to the DNA sample are amplified.
2. Amplification is performed by hybridization of two probes that are designed to bind adjacently to the target specific sequences in such way that they can be joined by the use of ligase (Fig. 2). Only ligated probes will be amplified by a PCR reaction. Each probe consists of one unique synthetic oligonucleotide and one M13-derived oligonucleotide.
3. One of the two probes contains a nonhybridizing stuffer sequence. This sequence varies in size for each probe set which gives the advantage to analyze concurrently up to 45 target sequences size ranging between 130 and 480 bp. Each probe also contains a tagged sequence that is universal for all probe sets, in order to allow simultaneous PCR amplification of all targets in a single reaction by adding a universal primer-pair, including a fluorescent labeled forward primer.

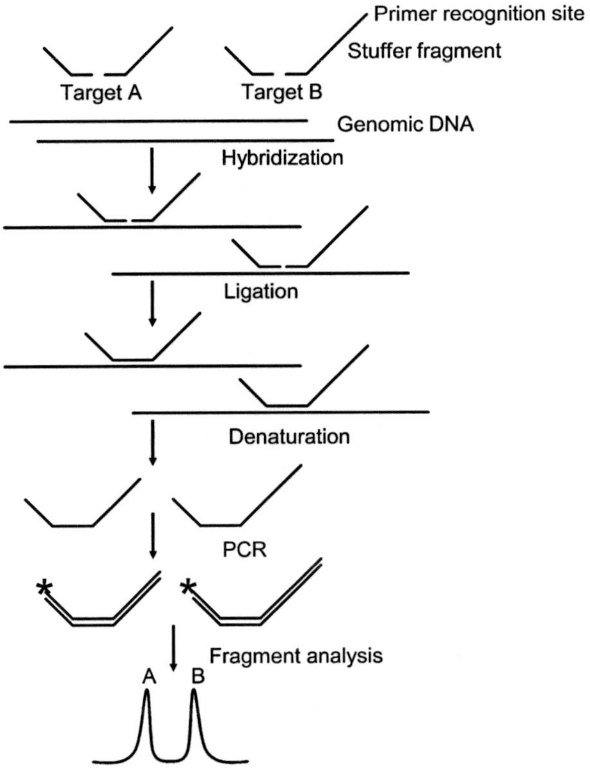

Fig. 2. MLPA. Oligonucleotide half probes hybridize on target DNA A and Target DNA B. Using DNA ligase the two half probes ligate together. In the following PCR reaction, the ligated probes are amplified using a universal primer pair that contain a fluorescent label at one primer and the PCR products are size-separated by electrophoreses

4. The amount of ligated probes is proportional to the copy number of the target sequence in the sample.
5. Comparison of the relative peak heights of each amplification product to a normal control reflects the relative copy number of the target sequence.
6. To avoid false positive and false negative results, several normalizations are applied (e.g. normalized by the mean value of all peaks and normalization by the peak areas of a normal control that are run together with the samples).
7. Methylation defects can be detected using methylation sensitive MLPA (MS-MLPA). Using this method, a methylation sensitive endonuclease digestion of the target DNA is performed prior to the ligation of the probes.

The probes are designed to hybridize on the methylated DNA only, which is not digested and can subsequently be amplified in the PCR-reaction.

3.5. Quantitative Fluorescence Polymerase Chain Reaction (QF-PCR)

QF-PCR is mostly used to detect trisomy 13, 18 and 21 in prenatal samples or directly after birth, because a fast result is obtained (1–2 days) (see Note 9). QF-PCR involves the amplification of chromosome-specific, microsatellite DNA sequences that consist of small arrays of tandem repeats known as short tandem repeats (STRs). STRs are stable and polymorphic, that is, they vary in length between subjects, depending on the number of times the di-, tri-, or tetra-nucleotides are repeated.

1. The sample DNA is amplified by PCR using fluorescent primers, so that products can be visualized and quantified as peak areas of the respective repeat lengths using an automated DNA sequencer.
2. Peak area is proportional to copy number.
3. DNA amplified from normal subjects who are heterozygous (have alleles of different lengths) is expected to show two peaks with the same area.
4. DNA amplified from subjects who are trisomic will exhibit either an extra peak (being triallelic) with the same area, or two peaks (being diallelic), one of them with a twice as large peak area as the other.
5. Subjects who are monosomic will exhibit only one peak (Fig. 3).

3.6. Array Genomic Hybridization

The current method of choice for performing whole-genome scans for detection of submicroscopic copy number variations (deletions and duplications) is array based Comparative Genomic Hybridization (CGH) or array based copy number analysis using SNPs. Several commercial platforms are available, each with strengths and limitations (see Notes 10 and 11 for discussion of these).

3.6.1. Array CGH General Procedure

1. Array-CGH is based on competitive hybridization of test and reference DNA labeled with different fluorochromes on immobilized large genomic clones or oligonucleotides on a glass surface.
2. Copy number detection is carried out with a high resolution (2–10 μm) laser scanner (Fig. 4). Because of the competitive nature of the binding, regions of the test-DNA with an increased copy number are identified by fluorescence as an increase in signal intensity of the test-DNA compared to the reference-DNA. Likewise, regions with genomic loss of the test genome are identified by an increase in signal intensity of the reference-DNA compared to the test-DNA.

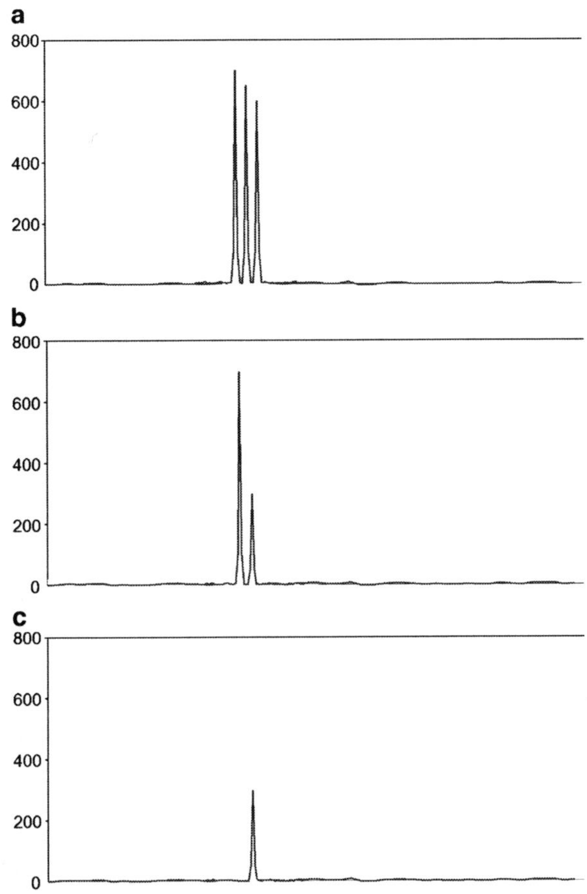

Fig. 3. QF-PCR. Figure (**a**) displays a triallelic trisomy (tree peaks with the same peak area), (**b**) shows a diallelic trisomy (one normal peak and one peak twice as large) while (**c**) shows a monosomy (one single peak)

3. Intensity measurements and ratio calculations are performed using designated software packages.
4. Initially, array CGH was performed using 'in-house' produced bacterial artificial chromosome (BAC) arrays consisting of large insert clones with an initial coverage of approximately one clone per Mb. Currently, arrays are commercially available with different resolutions containing different numbers of probes covering the whole genome (see Note 12 on manufacturers).

3.6.2. SNP Arrays

A special type of oligonucleotide arrays are the SNP arrays that are based on the genome wide detection of SNPs in a high resolution. This method allows the identification of not only amplifications and deletions, but also the SNP-genotype based haplotype

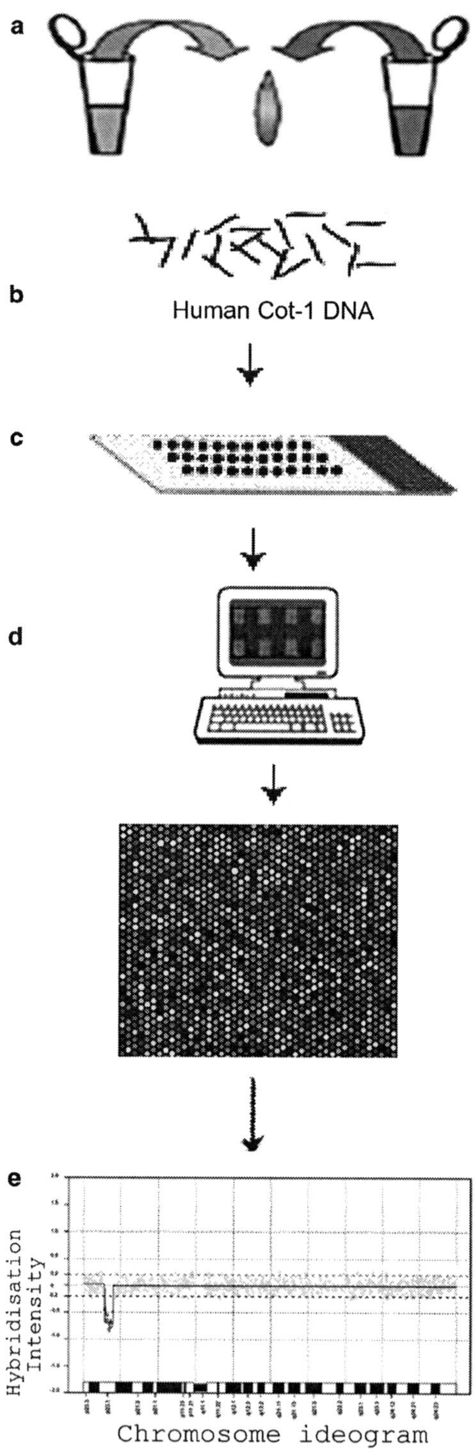

Fig. 4. Array CGH. Differentially labeled DNA is hybridized either on immobilized clones on a slide. Signals are detected using a laser scanner and ratio values between test and reference are quantified for each probe on the array using software packages and plotted according to their genomic location

of the amplified or deleted region. Therefore, it is obvious that the high resolution genome wide SNP array approach will be invaluable for the diagnosis of mental retardation.

The major manufacturers of SNP arrays are Affymetrix and Illumina and both offer arrays that contain more than 1 million SNPs. For both platforms, a different technique of allele discrimination in genotyping is applied. SNP arrays have some limitations for copy number detection; these are discussed in Note 13)

3.6.2.1. Affymetrix Procedure

1. The method of Affymetrix is a single channel (one color) assay based on allele-specific hybridization. 25-mer probes on the array correspond to both of the two possible alleles at each SNP. After hybridizing the target to the array, the resulting signal from the allele-specific probes can be analyzed, and determined whether an SNP is AA, AB, or BB. The signal intensity is quantified and compared to in silico references to determine SNP copy number.
2. About 250 ng of genomic DNA is digested with restriction enzymes and ligated to adaptors recognizing the overhangs (Fig. 5).
3. All fragments resulting from restriction enzyme digestion, regardless of size, are substrates for adaptor ligation.
4. A generic primer, which recognizes the adaptor sequence, is used to amplify ligated DNA fragments, and PCR conditions are optimized to preferentially amplify fragments in the 250–1,000 bp size range.
5. The amplified DNA is labeled and hybridized to GeneChip arrays.
6. The arrays are washed and stained on a GeneChip fluidics station and scanned on a GeneChip Scanner 3000 (http://www.affymetrix.com). Several software packages have been developed to analyze SNP genotypes and to determine copy number.

3.6.2.2. Illumina Procedure

1. The Illumina assay is a single base extension two color assay. Samples are hybridized to 50-mer bead based probes. The probes end one nucleotide before the SNP, so that the different alleles (AA, AB and BB) are scored by a single base extension using differentially labeled terminators.
2. The signal intensity is used to score copy number.
3. 750 ng of genomic DNA is needed for the assay that consists of 4 steps: (1) whole genome amplification, (2) hybridization to a bead array, (3) single base extension SNP scoring assay, and (4) signal amplification (Fig. 5).

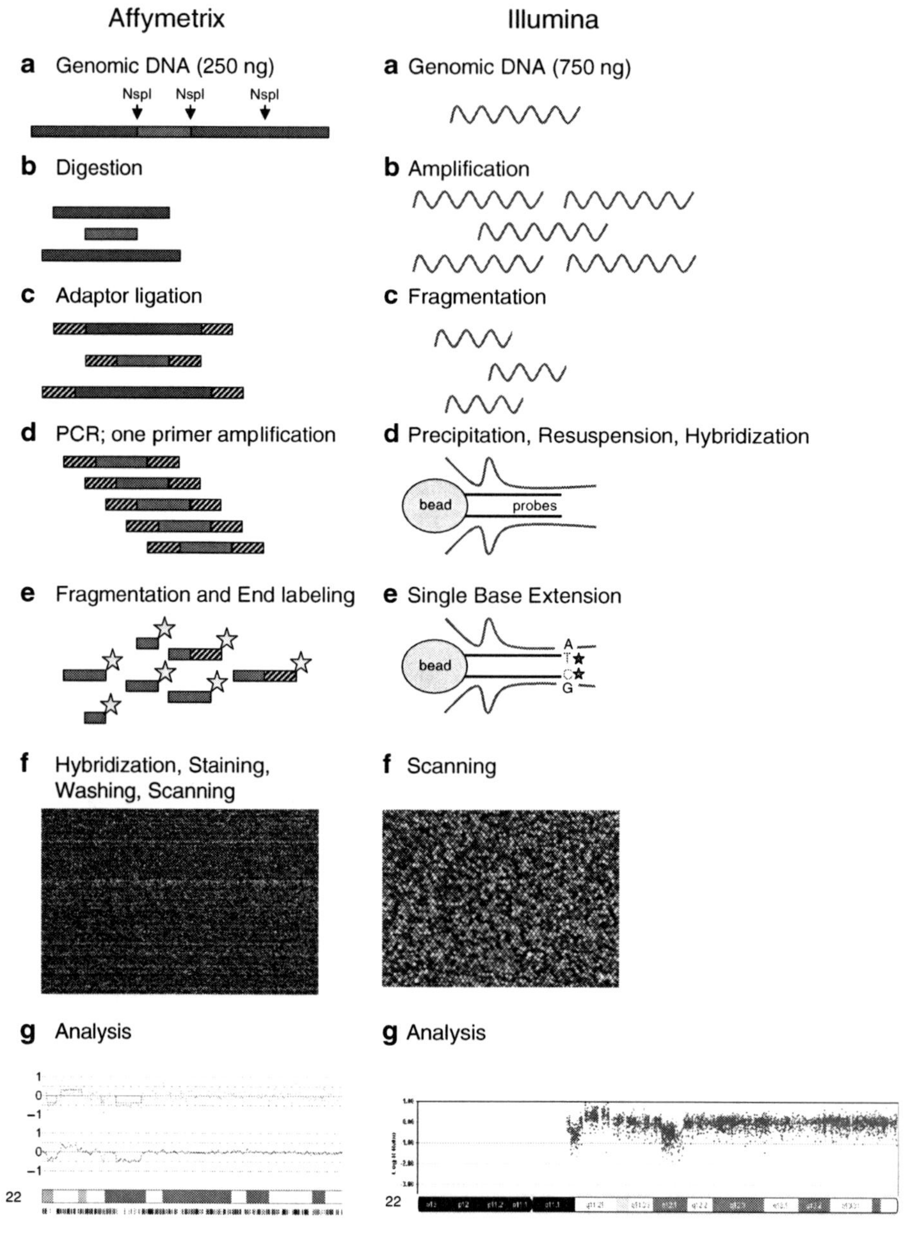

Fig. 5. SNP array procedure. Affymetrix platform (*left*) and Illumina platform (*right*). The two platforms apply different techniques of allele discrimination in genotyping. Affymetrix exploits an allele-specific hybridization, whereas Illumina utilizes a single base extension. For both platforms, probe intensity signals are quantified for each probe and compared to in silico references to analyze DNA copy number. Intensity ratio between test and reference is plotted according to the genomic location of the probe. For both platforms, an analysis profile of chromosome 22 is displayed showing a deletion prior to a duplication and deletion

4. After completion of the assay, the BeadChips are scanned with a two-color confocal Illumina® BeadArrayTM Reader at a 0.84–1.0 μm pixel resolution. Image intensities are extracted and genotypes and copy number are determined using Illumina's BeadStudio software (www.illumina.com).

3.7. Microsatellite Makers for Detection of Parental Origin and UPD

Chromosome sequences may appear normal, but can still be pathogenic if they have the wrong parental origin. Genomic imprinting is an epigenetic mechanism of non-Mendelian inheritance that is unique to mammals. A small number of genes are imprinted and are expressed differently according to their maternal or paternal origin. As chromosomes pass through the male and female germlines, they acquire an imprint to signal a difference between paternal and maternal alleles in the developing organism. Even if the sequence of the gene is not altered, gene-malfunction will occur if two imprinted alleles are inherited from the same parent. An individual with both homologs derived from the same parent (uniparental disomy, UPD) may show symptoms if the chromosome contains imprinted genes. Examples of human genetic diseases linked to UPD are Prader-Willi syndrome (MIM 176270), Angelman syndrome (MIM 105830), and Beckwith–Wiedemann syndrome (MIM 130650). Array-based SNP genotype analysis or microsatellite DNA sequences can be used to distinguish different parental alleles in order to determine whether the patient has both a maternal and a paternal allele. In deleted or duplicated regions that contain imprinted genes, the phenotype might vary depending whether the maternal or paternal copy is missing or gained, such as duplication of 15q11–q13 result in a more severe phenotype when the maternal copy is duplicated (6). Standard methods of Microsatellite analysis and SNP array genotype analysis are used for determination of the parental origin.

4. Notes

1. *GTG banding analysis*: The resolution of this technique is limited. A routinely prepared metaphase contains ~450–550 bands per haploid genome which roughly corresponds to a resolution of 5–10 Mb. High resolution banding techniques (arresting the cell in pro-metaphase) can achieve ~1,000 bands per haploid genome. But this analysis is very labor-intensive and not practical for routine analysis. In addition, chromosome banding analysis has limitations that include the inconsistency with which band resolution can be routinely achieved and the difficulty in visualizing some rearrangements due to staining properties of specific regions of the genome.
2. *Selection of FISH clones*: For FISH-mapping of translocation breakpoints, BAC and PAC clones can be selected based on their location on the physical map on the public accessible genome browsers at http://www.ensembl.org or http://genome.ucsc.edu. The clones can be acquired from BACPAC Resource Center Children's Hospital (Oakland Research Institute, Oakland, CA

http://bacpac.chori.org/), and investigated within a relative short period of time. In addition, there are a large number of commercially available probes that are fluorescently labeled and ready for use.

3. *FISH Probe labeling*: FISH probes can be labeled directly or indirectly by nicktranslation or random priming. The direct labeling employs the integration of a fluorescent labeled nucleotide into the DNA. The indirect method attaches a hapten (biotine, digoxigenin) to the nucleotides of the probe. After hybridization, a labeled binding protein (avidine, streptavidine) is detected by a specific fluorescent antibody. With nicktranslation, a DNA fragment is treated with DNase to produce single-stranded nicks, followed by incorporation of labeled (fluorescent or hapten) nucleotides from the nicked sites by DNA polymerase I. The random prime labeling is based on the method described by Feinberg and Vogelstein (7), in which a mixture of random hexamers containing all possible sequences anneal to denatured DNA template and act as primers for complementary strand synthesis by DNA polymerase (Klenow fragment). Reagent kits for labeling are commercially available both for random priming (e.g. Invitrogen; http://www.invitrogen.com and Enzo life-sciences; http://www.enzo.com) and nicktranslation (e.g. Abbott; http://www.abbott.com, Amersham; http://www.amersham.com).

4. *Preparation of SKY probes*: For SKY, a probe cocktail is prepared from flow-sorted chromosomes, in which each chromosome is labeled with a combination of one to four different fluorochromes. Using five different fluorochromes (Spectrum Orange, Texas Red, FITC, Cy5 and Cy5.5) in 24 different combinations, each chromosome is stained by a specific color (Fig. 1). Labeled probes mix are manufactured by Applied Spectral Imaging (ASI), Both SKY and M-FISH use a combinatorial labeling scheme with spectrally distinguishable fluorochromes, but employ different methods for detecting and discriminating the different combinations of fluorescence after in situ hybridization.

5. In SKY, image acquisition is based on a combination of epifluorescence microscopy, charge-coupled device (CCD) imaging, and Fourier spectroscopy (8). This makes the measurement of the entire emission spectrum possible with a single exposure at all image points.

6. In M-FISH, separate images are captured for each of the five fluorochromes using narrow bandpass microscope filters; these images are then combined by dedicated software. In both techniques, unique pseudo-colors are

assigned to the chromosomes based on their specific fluorochrome signatures.

7. *Comments on MLPA*: MLPA is a fast, sensitive, relatively cheap, and easy to perform technique. It is a powerful technique for detection of copy number changes of sizes too small to be identified by cytogenetic techniques including array, and too large to be detected by PCR and sequencing. However, MLPA reactions are more sensitive to contamination of PCR inhibitors compared to ordinary PCR reactions and it is not suitable for detection of new mutations, since the probe has to be designed at a known locus of interest. In addition, the development of a custom probe mix for each sequence of interest is time consuming. Each probe requires the design and preparation of an M13 clone and the purification and restriction enzyme digestion of that clone. Instead of M13 derived probes, completely synthetic probe sets with variable sizes can be used. The amplified PCR products are of shorter length using synthetic probes (87–132 bp), and this limits the number of targets that can be interrogated simultaneously. Yet, using this approach, the number of targets can be increased by adding a new primer-pair labeled with a different fluorescent color (9).
8. *Designing MLPA probes*: The oligonucleotide sequence of the probes needs to fulfill a number of criteria in order to give a reliable copy number result. Probes need to be unique (not cross hybridize on other genomic regions), contain a GC content of 40–60%, a Tm > 65°C, and a G or C at the junction between the target and the universal primer sequence. The ligation site should not be between CC or GG sequences. To investigate the uniqueness of the probe sequence, it is blast in a genome browser while the other sequences properties can be investigated using the program RAW probe which is free downloadable from MRC Holland at http://www.mrc-holland.com/pages/indexpag.html. To run the in house designed probe set, MRC Holland provides a protocol and reagents kit with all buffers, primers-pairs, and enzymes needed.
9. QF-PCR has the advantage of being much less expensive and allowing the simultaneous processing of much larger numbers of samples compared to FISH.
10. Comments on copy number analysis (detection of deletions and duplications) using array technology: Copy number analysis by array has many potential advantages over the use of chromosomes. It offers rapid genome-wide analysis at high resolution. The resolution is determined by the genomic distance between the probes as well as their sizes, and the

information it provides is directly linked to the physical and genetic map of the human genome. The first generation of genomic arrays contained large inserts clones such as BAC and PAC clones covering the whole genome (tilling path array). The latest developments in genome-wide array CGH technologies using oligonucleotides and single-nucleotide polymorphisms (SNPs) have resulted in a new generation of genome-wide array platforms. These platforms contain a larger number of shorter DNA fragments (oligonucleotides) to further increase the resolution. However, an adequate description of the capability of each platform is difficult to define since the resolution of the array is not only determined by the number and size of probes, but more importantly by the genomic spacing and the hybridization sensitivity of the probes on the array (10–12).

11. It is recommended to evaluate both the specificity and the sensitivity of the arrays in every laboratory offering the diagnostic test (13).

12. The major manufacturers of oligonucleotide arrays are Agilent and Nimblegen and both offer arrays that contain more than 1 million oligonucleotides. For detailed procedure and software packages for analysis, see manufacturers protocol http://www.chem.agilent.com and http://www.nimblegen.com.

13. *Limitations of arrays for diagnostics*: Arrays containing such large number of probes might be too costly and not be the most practical for diagnosing constitutional abnormalities. False-positive calls occur more often when there are more elements on an array and in addition to pathogenic subtle chromosome abnormalities, small benign copy number variants will frequently be detected. Therefore, greater resolution does not necessarily translate into more meaningful data. The interpretation of the array results is complicated by the detection of benign variants that must be distinguished from those that cause disease. The database of normal variants at http://projects.tcag.ca/variation/ is a great tool to determine whether a CNV is benign or pathogenic. Yet, further research is still needed to characterize human CNVs to develop more comprehensive human genetic variation maps. This in turn will facilitate more accurate interpretations of the clinical impact of specific genomic imbalances.

References

1. Shaffer LG, Lupski JR (2000) Molecular mechanisms for constitutional chromosomal rearrangements in humans. *Annual Review of Genetics* **34**:297–329

2. Jacobs PA, Browne C, Gregson N, Joyce C, White H (1992) Estimates of the frequency of chromosome abnormalities detectable in unselected newborns using moderate levels

of banding. *Journal of Medical Genetics* **29**:103–108

3. Seabright M (1971) A rapid banding technique for human chromosomes. *Lancet* **2**:971–972
4. Lee C, Gisselsson D, Jin C, et al. (2001) Limitations of chromosome classification by multicolor karyotyping. *American Journal of Human Genetics* **68**:1043–1047
5. Schouten JP, McElgunn CJ, Waaijer R, Zwijnenburg D, Diepvens F, Pals G (2002) Relative quantification of 40 nucleic acid sequences by multiplex ligation-dependent probe amplification. *Nucleic Acids Research* **30**:e57
6. Cook EH, Jr., Lindgren V, Leventhal BL, et al. (1997) Autism or atypical autism in maternally but not paternally derived proximal 15q duplication. *American Journal of Human Genetics* **60**:928–934
7. Feinberg AP, Vogelstein B (1983) A technique for radiolabeling DNA restriction endonuclease fragments to high specific activity. *Analytical Biochemistry* **132**:6–13
8. Schrock E, du Manoir S, Veldman T, et al. (1996) Multicolor spectral karyotyping of human chromosomes. *Science* **273**: 494–497
9. White SJ, Vink GR, Kriek M, et al. (2004) Two-color multiplex ligation-dependent probe amplification: detecting genomic rearrangements in hereditary multiple exostoses. *Human Mutation* **24**:86–92
10. Coe BP, Ylstra B, Carvalho B, Meijer GA, Macaulay C, Lam WL (2007) Resolving the resolution of array CGH. *Genomics* **89**:647–653
11. Hehir-Kwa JY, Egmont-Petersen M, Janssen IM, Smeets D, van Kessel AG, Veltman JA (2007) Genome-wide copy number profiling on high-density bacterial artificial chromosomes, single-nucleotide polymorphisms, and oligonucleotide microarrays: a platform comparison based on statistical power analysis. *DNA Research* **14**:1–11
12. Zhang ZF, Ruivenkamp C, Staaf J, et al. (2008) Detection of submicroscopic constitutional chromosome aberrations in clinical diagnostics: a validation of the practical performance of different array platforms. *European Journal of Human Genetics* **16**:786–792
13. Vermeesch JR, Fiegler H, de Leeuw N, et al. (2007) Guidelines for molecular karyotyping in constitutional genetic diagnosis. *European Journal of Human Genetics* **15**:1105–1114

Chapter 5

Cancer Genome Analysis Informatics

Ian P. Barrett

Abstract

The analysis of cancer genomes has benefited from the advances in technology that enable data to be generated on an unprecedented scale, describing a tumour genome's sequence and composition at increasingly high resolution and reducing cost. This progress is likely to increase further over the coming years as next-generation sequencing approaches are applied to the study of cancer genomes, in tandem with large-scale efforts such as the Cancer Genome Atlas and recently announced International Cancer Genome Consortium efforts to complement those already established such as the Sanger Institute Cancer Genome Project. This presents challenges for the cancer researcher and the research community in general, in terms of analysing the data generated in one's own projects and also in coordinating and interrogating data that are publicly available. This review aims to provide a brief overview of some of the main informatics resources currently available and their use, and some of the informatics approaches that may be applied in the study of cancer genomes.

Key words: Cancer, Bioinformatics, Genome analysis, Mutation, Comparative genomic hybridisation, Database, Array CGH

1. Introduction

Cancer is a disease that comprises many different types and can arise in many different tissues. It affects approximately 280,000 people each year in the UK alone (1), and according to WHO statistics for 2005, accounted for 7.6 million deaths worldwide (2).

Cancer is fundamentally a genetic disorder, where the prevailing model is that mutations accumulate within a precancerous cell that allow it to overcome the growth restraints and safety mechanisms, which are present in normal cells and prevent uncontrolled cell growth and invasion (3). As such, the study of cancer genomes has long been of interest in the hope that genetic aberrations

Michael R. Barnes and Gerome Breen (eds.), *Genetic Variation: Methods and Protocols*, Methods in Molecular Biology, vol. 628, DOI 10.1007/978-1-60327-367-1_5, © Springer Science+Business Media, LLC 2010

identified may provide clues as to the causes and mechanisms underlying different cancers, and ultimately suggest points for targeting of therapeutics.

There are numerous examples where this approach has borne fruit. Prominent examples include identification of the key tumour suppressor gene P53 (4), and a gene predisposing to familial breast cancer, BRCA1 (5), that has subsequently led to the development of screening approaches to help assess risk for relevant patients. Other examples include the fusion gene BCR/ABL, which results from a chromosomal translocation in certain leukaemias and subsequent targeting with the drug Gleevec has enabled the treatment of patients whose cancers harbour this mutation (6). Similarly, amplification of the ERBB2 gene in breast cancer has led to the development of therapies directed against this receptor (e.g., Herceptin) (7). Finally, more recently, the advent of large-scale sequencing projects has discovered mutations in genes, which have been shown to play a role in the tumourigenic process (e.g., the BRAF gene)(8).

Cancer genomes present the researcher with several additional challenges when attempting to assess genomic alterations and derive hypotheses as to what they may mean. First, the emerging picture appears to be that cancer genome variation presents a noisy background in which causal factors may not be easily distinguished, both at the individual base pair resolution (so called passenger mutations) (9, 10) and the chromosomal level where the karyotypes of epithelial tumours are frequently complex but in which important signals may still reside (11). Second, tumours can be heterogeneous in cellular composition (12) and, when samples are taken for analysis, may be contaminated with normal tissue that can skew results. Third, although when discussing cancer genomes we are often actually meaning primary or secondary tumour characteristics, the germline genetic background and mutations in stromal or histologically normal cells in the environment in which the tumour resides may also be of importance (13). All of this means that, although generation of data regarding cancer genome variation increases, interpretation of these data remains challenging and complex.

Various technological advances have meant that there is now a great deal of information on cancer genomes available to the cancer research community, and generated by groups with access to the appropriate facilities. Reducing costs of DNA sequencing coupled with advances in and availability of technologies such as increased resolution of array CGH platforms means that mutation and DNA copy number data are increasingly provided via databases in various formats to the research community. Newer approaches also offer the potential to provide detailed data on structural genomic aberrations such as balanced translocations that may not be readily detected at high resolution by other means (14).

There are large-scale projects taking a systematic approach to the identification of genetic aberrations in cancer genomes (cell lines and tumours), such as the Sanger Institute Cancer Genome Project (15), TCGA (16) and recently the ICGC (17), which release valuable data into the public domain. These are supplemented by other published studies that may involve large numbers of samples and/or large-scale datasets such as sequencing data or high-resolution DNA copy number data (9, 10, 18, 19).

This brings challenges in which the cancer genome data are of diverse types (e.g., mutation data, gross chromosomal aberrations such as translocations, high-resolution DNA copy number aberrations), from diverse experimental approaches (both traditional and next-generation sequencing, various array CGH platforms), and are collated and made available in many different databases and formats. It should be stressed that in this area informatics resources are constantly evolving both for sharing and analysing data. For example, online databases may become out of date as funding alters or research interests change, and hence some resources are more stable than others. The researcher should bear this in mind when using these resources, particularly if the most recent and comprehensive data are required.

It is beyond the scope of a single review to provide a detailed review of all of these data types, but instead I will strive to provide a brief overview of some of the main cancer genome informatics resources available to the community and some notes on their use. As the research landscape is constantly evolving in shape and scope as are the related informatics approaches and resources, I will attempt to focus on some principles that may be more generic and that may apply in various settings. In this instance, I am not covering epigenetic and other related fields, but over time these data will grow and increasingly add to our understanding of cancer genomes. Naturally, the diversity of this subject area means that I will focus on some examples and omit valuable alternatives, apologies to other researchers for any omissions.

Additionally, various methodologies and software packages (some of which are commercial products and require licenses for use, and which may have different licensing terms for academic and commercial users) are referred to within this chapter – these are examples, which the author is familiar with or aware of and do not constitute a recommendation for these particular products, as the method or software used should always depend on the researchers questions and requirements, level of expertise and understanding of complexities and caveats involved in such analyses.

2. Materials

The use of resources described in this review can be accomplished with most modern desktop PCs and an internet connection, ideally using a broadband connection as many resources involve complex graphics and downloading of large datasets. Similarly, the more RAM memory available the better.

Specific methods for the analysis of array CGH data from raw data, for example, are not covered here; however, in these settings, it is best to consult with the providers of whichever software package(s) are being used as to computing requirements, as these data files can be very large especially with greater sample numbers and so higher performance computers will be required.

3. Methods

3.1. Chromosomal Aberrations

Online resources that collate chromosomal aberration data reported in the scientific literature are available. Often these data will be of lower resolution (i.e., morphological cytogenetic analysis); however, there are also data on fusion proteins that may arise from these aberrations where studies describe this.

A key established database is the Mitelman Database of Chromosome Aberrations in Cancer ((20) and Table 1). This database collates manually curated data on chromosomal aberrations and associated metadata (e.g., genes involved, tumour characteristics) from the scientific literature. Users can search using a variety of criteria such as individual patient cases, reference, clinical features or genes affected. This database has been incorporated into the NCBI cancer chromosomes online resource ((21) and Table 1), which integrates the Mitelman Database with the NCI/NCBI SKY/M-FISH & CGH database ((21) and Table 1) and also the NCI Recurrent Aberrations in Cancer database (Table 1).

The NCI Recurrent Aberrations in Cancer database is a derivative of the Mitelman database, which collates those aberrations occurring in more than one case. The researcher can query for structural or numerical aberrations, and restrict searches by breakpoint or chromosome and other features such as tissue type, tumour morphology or gene involved.

The NCI/NCBI SKY/M-FISH & CGH database collates data describing karyotypic features of cancers (tumours and cell lines) using these technologies. Researchers can submit their data, browse by submitter or tissue, or search the repository contents for data that have been released for query.

Table 1
List of URLs for introduction and methods

Resource	URL[a]
TCGA	http://cancergenome.nih.gov/index.asp
ICGC	http://www.icgc.org/
Sanger Institute Cancer Genome Project	http://www.sanger.ac.uk/research/projects/cancergenome/
Mitelman Database of Chromosome Aberrations in Cancer	http://cgap.nci.nih.gov/Chromosomes/Mitelman
NCBI cancer chromosomes	http://www.ncbi.nlm.nih.gov/sites/entrez?db=cancerchromosomes
NCI Recurrent Aberrations in Cancer database	http://cgap.nci.nih.gov/Chromosomes/RecurrentAberrations
Atlas of Genetics and Cytogenetics in Oncology and Haematology	http://atlasgeneticsoncology.org//
Progenetix	http://www.progenetix.net/progenetix/
HybridDB	http://www.primate.or.kr/hybriddb/
ChimerDB	http://genome.ewha.ac.kr/ChimerDB/
OMIM	http://www.ncbi.nlm.nih.gov/sites/entrez?db=OMIM

[a]See Note 14

Full help on searching these resources is available online, but here are two brief examples – a search for genetic aberrations leading to fusions of the ABL gene, and a search for SKY karyotype data of the NCI60 cancer cell panel.

3.1.1. Querying NCBI Cancer Chromosomes for Recurrent Aberrations Involving the ABL1 Gene in a Cancer Subtype of Interest

1. Navigate to the NCBI cancer chromosomes resource (see Table 1).
2. Click the link at the top of the page to navigate to the recurrent aberrations database. Check the box for Structural Aberrations, leave other options as default except in the Morphology section scroll down the list and select "Acute myeloid leukaemia (all subtypes)", and in the Gene section select "ABL1" (note that this gene list shows those genes with aberrations noted in this resource, and from this list the user can search for relevance in other tissue types for any of these genes). Click the retrieve button.
3. This returns details on the cases retrieved, in this case (at the time of writing) all balanced chromosomal abnormalities.

4. From the results, we can see what bands are involved in the ABL1 recurrent aberrations, the nature of the abnormality, the morphologies in which they are noted, the number of cases for each recurrent aberration, and in the Gene section of the results table we can see that the BCR gene is also represented. Clicking on the case numbers retrieves the relevant references for further study.

3.1.2. Querying SKY-DB for NCI-60 Cell Line Karyotypes

1. Navigate to the NCBI cancer chromosomes resource as above. Follow the link towards the top of the page to the SKY/M-FISH & CGH database.
2. In the quick search box at the top of the page, enter "NCI60" and press GO. This returns 59 cases (at the time of writing), with a table of information in the browser detailing for example the organism, tissue, the author submitting the data, and links to further data.
3. Click on the "Case Details" button for the cell line OVCAR-4 (second row). This shows further information such as clinical features and references, as well as a cytogenetic summary.
4. Click on the "SKYGRAM" button on this page – this returns a karyotype view summarising the SKY data for this cell line, showing for example increased ploidy and rearrangements for various chromosomes, as well as the number of cells counted and aberrations seen.

Other databases of relevance in this area include the Atlas of Genetics and Cytogenetics in Oncology and Haematology ((22) and Table 1), and the Progenetix database ((23) and Table 1).

3.2. Atlas of Genetics and Cytogenetics in Oncology and Haematology

This resource, described on its website as a peer-reviewed online journal, collates summary information on genes (e.g., synonyms, gene and protein features, many links to other resources), and at the time of writing 543 genes linked to cancer were annotated and a further 6551 noted as potentially implicated in cancer. Of note in our context, where annotated the gene pages also contain summarised information on selected noted mutations and other variation. Disease-centric overviews (e.g., breast tumours) are also provided which give information on noted genetic aberrations, as is functionality to browse by chromosome and aberration.

3.3. Progenetix

This is an online resource where the author strives to collate experimental results from published CGH experiments, thus collating potential chromosomal aberrations and associated metadata describing the tumour source, etc. While much of the older data may be from lower-resolution analyses, more recently array CGH data has been noted to be collated.

3.4. HybridDB

This resource has, as its foundation, a computational analysis whereby nucleotide sequences in GenBank have been mapped back to the human genome to identify sequences that are shared between more than one loci (i.e., representing potential intergene transcripts or chromosomal aberrations)((24) and Table 1). The user can search by gene name, chromosome or tissue (including neoplasia ESTs). It should be noted that this contains sequences sourced from non-cancerous tissue also, and so is not a cancer-specific resource. However, information on tissue source of the sequence is displayed where available.

3.5. ChimerDB

This is similar to HybridDB (and is also not cancer specific) and also includes collation of information from the literature, OMIM and other resources, and covers the human, mouse and rat genomes ((25) and Table 1). Gene, protein domain and chromosomal position-based search functionality is provided.

For resources like HybridDB and ChimerDB, some steps have been taken to reduce noise in these data, but chimeric clones may of course still affect such analyses. Due to the nature of the underlying data, coverage is currently limited but we might expect that as data become available from studies using next-generation sequencing approaches, then focused resources like these examples may collate data flagging putative fusion genes if not covered by other existing resources.

See Note 1 for general comment on manually curated databases and resources.

3.6. Array CGH Analysis and Databases

Array CGH technologies provide higher resolution data on DNA copy number, LOH and genotype, depending on the platform used. For example, some platforms may use arrayed genomic clones or oligonucleotides to detect DNA abundance, while others use sequences designed to detect individual SNPs but which also provide information on DNA abundance (so called SNP arrays). Older array designs provided a relatively lower resolution (e.g., one probe per 1 MB); however, more recent developments have given considerable increases in resolution so that it is possible to infer effects more reliably at the gene level.

A detailed description of array CGH data analysis techniques is beyond the scope of this overview (and there is abundant literature on this topic), but a brief description of the general workflow and some related notes are provided below in the first section that may also help with interpretation of data summarised via online resources. The latter part of this section outlines key data repositories where array CGH data may be deposited or retrieved for analysis, and many of which offer a variety of analytical functionality.

3.6.1. Array CGH Data Analysis Workflow (see Fig. 1 for an Overview)

Following scanning of the array, various normalisation procedures may be performed aiming to remove experimental biases (e.g., spatial effects, systematic intensity differences) while retaining biological signal (26), which is critical for subsequent analysis. In addition, there is processing to determine copy number changes

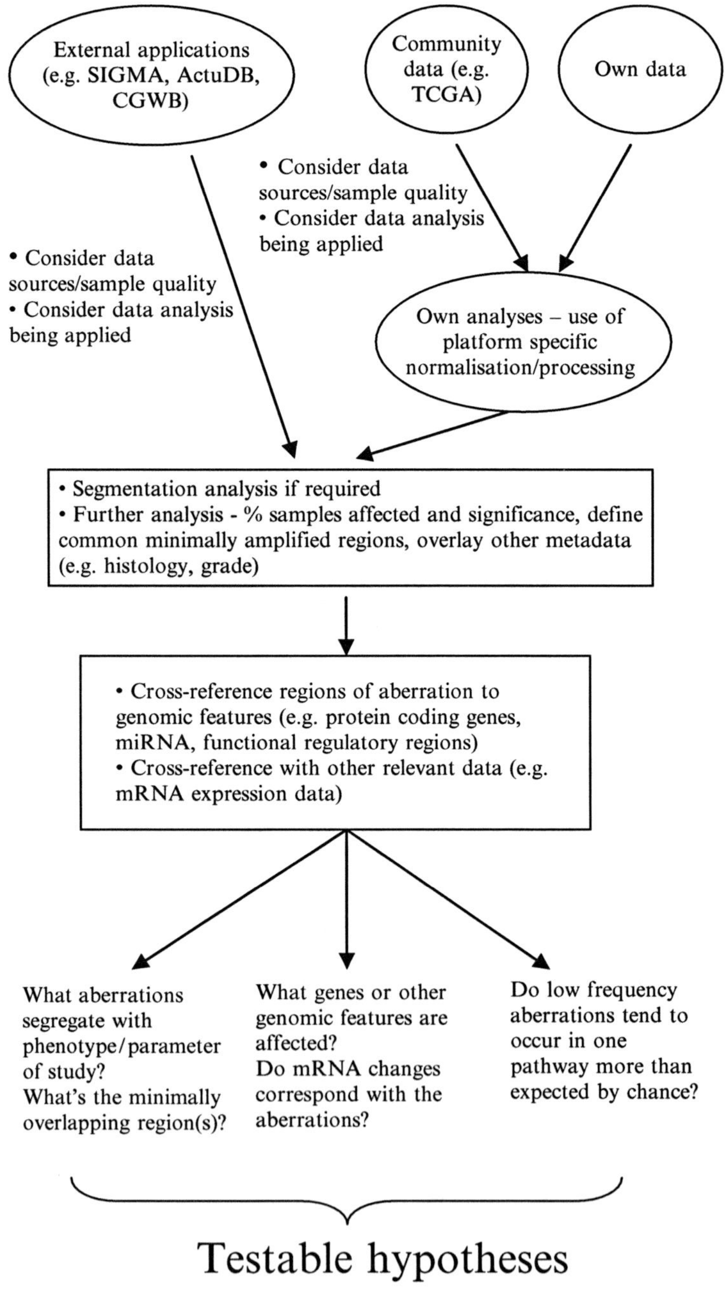

Fig. 1 Overview of array CGH analysis informatics workflow using data from different sources

with respect to either a matched normal DNA sample or unmatched normals (see also Notes 2 and 3). The methods used will depend on the platform involved – the user should be able to receive support for this if using a commercial supplier such as Affymetrix, Agilent or Illumina, alternatively there are other resources available via the R BioConductor suite (27), applications such as GenePattern (28) or various other methods reported in the literature.

Having calculated copy number changes for the individual probes, a segmentation process is often performed to extrapolate from individual probe changes to infer regional copy number changes. This may, for example, assess probe changes within a given window, and if a threshold is surpassed, then a region of copy number change is "called" with the chromosomal boundary coordinates estimated (with chromosomal positioning information depending on mapping files that relate probes (e.g., BAC clones, oligos) to chromosomal position). Again, there are various approaches and software for performing this (e.g., GLAD (29)). There may then be further processing to assess copy number magnitude and frequency within the sample set (e.g., GISTIC (30)) to attempt to highlight aberrations that may be more biologically relevant amidst the widespread genome variation seen in cancer genomes.

Finally, the chromosomal aberrations need to be related to positioning of genes and other genomic features. This can be performed by cross-referencing with files generated from genome browsers such as Ensembl (using the BioMart functionality) (31), UCSC genome browser (32) or NCBI MapViewer (33), or if using a comprehensive software package this may be achieved using links to these resources or using static mapping files (see Note 4).

Other steps that may be taken include cross-checking to see if any of the inferred aberrations overlap with known CNV regions, and integration with other data, such as complementary mRNA expression data generated from the same samples, to help flag genes which may be candidates for driving selection of the aberration (34). Defining the minimally amplified region of recurring aberrations may help focus attention on potential candidates for selection of that aberration. The size of the region and density of genomic features affected will influence feasibility of follow up study, as for large aberrations many genes or other features could be involved, any or a combination of which could be functionally relevant. Other approaches to consider could assess whether low frequency aberrations together hint at a particular pathway being disrupted in the class of interest. Analysis tools such as GeneGO or Ingenuity's PathwayAssist (which are commercial knowledge-exploration resources) and resources, such as the Nature/NCI Pathway Interaction Database, may be useful in this setting (see Table 2 for related links).

Table 2
List of URLs for cancer gene expression and genome analysis

Resource	URL[a]
Affymetrix	http://www.affymetrix.com/index.affx
Agilent (life sciences genomics division)	http://www.chem.agilent.com/Scripts/IDS.asp?lPage=23129
Illumina	http://www.illumina.com/
R BioConductor	http://www.bioconductor.org/
GenePattern	http://www.broad.mit.edu/cancer/software/genepattern/index.html
Ensembl	http://www.ensembl.org/index.html
UCSC genome browser	http://genome.ucsc.edu/
NCBI map viewer	http://www.ncbi.nlm.nih.gov/mapview/
Partek	http://www.partek.com/
Nexus (BioDiscovery product)	http://www.biodiscovery.com/index/nexus
GEO	http://www.ncbi.nlm.nih.gov/geo/
ArrayExpress	http://www.ebi.ac.uk/microarray-as/aer/#ae-main[0]
CaArray	http://caarray.nci.nih.gov/(note – follow link further down the page to caArray 2.0 installation at NCICB to view community datasets; https://array.nci.nih.gov/caarray/)
SMD	http://genome-www5.stanford.edu/
EBI	http://www.ebi.ac.uk/
TCGA data portal	http://cancergenome.nih.gov/data/portal/
SIGMA	http://sigma.bccrc.ca/
ActuDB	http://bioinfo.curie.fr/actudb/
OncoMine	http://www.oncomine.org/
CanGEM	http://www.cangem.org/
GeneService	http://www.geneservice.co.uk/products/
Ingenuity	http://www.ingenuity.com/
GeneGO	http://www.genego.com/
NCI/Nature Pathway Interaction Database	http://pid.nci.nih.gov/
Multiple Myeloma Genomics Portal	http://www.broadinstitute.org/mmgp/home

[a]See Note 14

There are commercial software packages available (e.g., Partek or Nexus (Table 2)) that cater for different parts of this analysis process, with some catering for different platforms and covering different parts of the analysis process to different degrees.

3.7. Key Data Repositories that Include Array CGH Datasets

There are a number of established data repositories that initially were implemented to house the growing mRNA microarray data being published and to promote standards within the field, and have subsequently adapted to also cater for array CGH data. These are summarised below, with links to these resources provided in Table 2. These provide access to publicly available datasets that may be downloaded and analysed as required.

3.7.1. GEO (NCBI)

This resource is hosted by the National Centre for Biotechnology Information (NCBI), and is a repository for gene expression and array CGH datasets with functionality for data deposition and browsing, querying (including via the NCBI Entrez query system) and downloading datasets, and searching for profiles of interest with subsequent visualisation (35). Datasets have a GEO accession number (syntax GSExxxx), which may be referenced in more recent publications if those data are available in GEO.

3.7.2. ArrayExpress (EBI)

This resource, hosted by the European Bioinformatics Institute (EBI), also collates gene expression and array CGH datasets, with similar functionality as for GEO (36).

3.7.3. caArray

This is a data management system that is part of the caBIG initiative. It is implemented at the NCI, and at the time of writing contained 38 experiments with only one of these being publicly available (however, this experiment includes 676 cancer cell line samples analysed on the Affymetrix 250 K SNP array platform).

3.7.4. Stanford Microarray Database (SMD)

This repository is similar to the others above in collating mRNA microarray and array CGH datasets, and with associated analysis and visualisation functionality (37).

3.7.5. TCGA Data Portal

This provides an interface to access the data emerging from the TCGA activities, which include array CGH datasets.

All the repositories listed above are not restricted to these data types and are continually evolving, and offer wider functionality than mentioned briefly here. Please refer to the references for more details. It is also worth noting that these repositories may not contain all available datasets, and that the same data may be provided via more than one online resource. Some data may be made available via specific data portals (e.g., TCGA data and also likely the ICGC data when available), via specific websites catering for medium-scale projects (e.g., Multiple Myeloma Genomics Portal, see Table 2) or data from a particular institute (e.g., Broad Institute datasets) or laboratory/study. Additionally, these main repositories and other data sources (e.g., TCGA portal) may place certain restrictions on some datasets or house pre-published or otherwise restricted data and as such not all data are necessarily publicly available. Clarification can be sought as needed from the repository contacts via details on their websites. Finally, some of these resources also provide analysis functionality, and this may be expected to develop further over time.

3.8. Array CGH Databases

In addition to the main established data repositories, there are other applications emerging that aim to collate array CGH (and other microarray) datasets, and provide analytical functionality. Examples are provided below (links provided in Table 2).

3.8.1. SIGMA

This resource collates array CGH datasets from various platforms and provides an interface in which to search for data using various parameters (e.g., platform, study), and then perform a variety of operations such as single dataset visualisations, group comparisons and platform comparisons (38). Other functionality is provided such as searching for example for a particular gene or chromosomal location, and other visualisation aids.

3.8.2. ActuDB

Similar to SIGMA, this database collates array CGH data and provides an interface for analysis and visualisation (39). In addition, it also collates mRNA and clinical data where available.

3.8.3. OncoMine

OncoMine is a software tool available to the scientific community (licensed for commercial users) that previously focused primarily on collation, analysis, and visualisation of mRNA microarray data; however, they are now also beginning to collate and analyse array CGH data.

3.8.4. CanGem

This is another database resource similar to SIGMA and ActuDB (40).

Finally, it is worth noting that methods for analysing mRNA microarray datasets have also been explored to attempt to use these data as surrogates to flag potential genetic aberrations such as DNA copy number change, epigenetic or chromatin-mediated effects. For example, the TCM method assesses correlations of mRNA expression between genes to see if these correlations are higher for neighboring genes than would be expected by chance (41).

3.9. Detailed Example 1: Assessment of Genomic Features of a Genetic Aberration in Cancer Cell Line Samples

In this hypothetical example, we have identified a set of breast cancer cell lines that have different phenotypic characteristics to others, and want to investigate potential genetic aberrations to attempt to derive a hypothesis for what genes may be functionally involved.

3.9.1. Identification of an Aberration in Multiple Cancer Cell Lines Using SIGMA

[AU1]

1. Launch the SIGMA application. Select two group analysis. When the Experiment Search window launches, in the search drop down menus at the top of the window select the following search criteria; Array_Platform IS LIKE BCCRC Genomic SMRT, AND Tissue IS LIKE Breast, then click the search button.
2. This returns a list of breast cell line experiments collated in SIGMA that have been run on this particular platform. Now select (by clicking on while holding the Ctrl key) the following cell lines; BT474, SKBR3, HCC1954, and UACC893, then in the first box at the right hand side click the Add button. This has selected these four lines as the first group.
3. Then repeat the step above selecting all of the other cell lines, this time adding them to the right hand box. This gives us our two groups for comparison. Then click the Create button.
4. This then launches a New Analysis Wizard. You are asked to input analysis and dataset names, and also select various parameters. For detailed explanation, refer to the SIGMA help guides. In this example, enter analysis and dataset names and in Analysis Parameters set the Assembly to March 2006 (hg18). Then click Finish. The analyses will now appear in the left hand pane.
5. In the left hand pane, expand the Summary folder and double click on Whole Genome. This graphically illustrates a comparison between these two groups chosen, highlighting regions of inferred gain/loss. Note a region on 17q that appears to be gained in our first group. Double clicking on Chromosome 17 [AU2] (Chr:17) in the same section in the left pane now shows a more detailed view, and the user can compare data from each cell line in each group. Using the zoom functionality, this shows a region of gain in our first group at 17q12-q21.1 (note that segmentation is not being performed here).
6. When the cursor is placed over the clone representations on the display, at the bottom left of the window the chromosomal coordinates are displayed. Using this, we can see that the region of gain is at approximately 34.7 MB to 35.8 MB, shown most clearly for SKBR3 and UACC893.
7. The data for Chr:17 for each cell line can be displayed together by double clicking Chr:17 in the Serial folder in the left hand pane.
8. At this resolution, the breakpoints of the aberration (amplification in this instance) are approximately 34.7 MB–35.8 MB. We will take a larger region forward for further analysis at this point, to account for uncertainties in the actual breakpoint positions. We will now explore the genomic context of this region of potential amplification in our group of cell lines. See also Notes 5 and 6, which flag points to bear in mind in this sort of scenario.

3.9.2. Assessment of Genomic Features in this Region Using Ensembl BioMart

9. Navigate to Ensembl (see Table 2). From the "New to Ensembl?" right-hand pane, select the "Mine Ensembl with BioMart" link. This is a powerful information retrieval tool with many options. We will introduce some options here, but I encourage the user to browse the additional options and use the online Help to see the full potential of this feature (note that other genome browsers such as the UCSC and NCBI browsers have data download functionality, and different browsers may feature different data sources so the user should familiarize themselves with these also and select the one most appropriate for their needs).
10. In the drop down menu for "Choose Database", select "Ensembl 56". In the drop down menu for "Choose Dataset", select "Homo sapiens genes (GRCh37)". Note that this version will change over time, as may the query interface. At this point it should be checked that the genome build version used for this procedure corresponds to the genome build used for the array CGH analysis. In this instance it is not consistent due to updates during the course of writing this example and hence we will take a larger region for analysis, but still serves to illustrate the principle. If synchronizing versions proves problematic, sequence alignment of known markers can help to orient the user in the respective genome assemblies (although the user should remain alert to further insertion/deletion changes affecting coordinates).
11. Click on the Filters link that has now appeared under the Dataset area in the left hand panel. In the right hand panel, select the chromosome and input the coordinates of interest in the "REGION" section. In this case select Chromosome 17, base pair gene start 34000000 and gene end 38000000.
12. Scroll down the right hand panel and select other filters as required. For example, under the Protein Domains section the user could restrict to genes annotated with a set of InterPro protein domains, such as kinase domains. In this example, we will export all protein coding genes and mapped miRNAs; open the Gene section and under Gene Type select both "protein coding" and "miRNA" (by holding the Ctrl button when selecting >1 field).
13. In the left hand panel, select the Attributes link. In the right hand panel, select Features and in the Gene section select; Ensembl Gene ID, Associated Gene Name, Gene Start (bp), Gene End (bp). In the Protein Domains section, select Interpro Short Description.
14. This query is now set to retrieve protein coding genes and miRNAs from our region of interest, and return gene names, position, and names of protein domains mapped to those genes. At the top of the left pane, select Count. This informs us that 83/49506 genes are selected. At the top of the left pane now select Results.

15. In the right hand panel, select export all results to file, csv, then click GO. When the download dialog appears, select save as file and name and save to an appropriate location, saving as a text file (.txt).
16. This file can now be opened in Excel (open Excel, select file, open, find the relevant file then when prompted select data type delimited and then comma as the separator), and sorted or filtered (using the autofilter capability for example, to restrict results to those mentioning a certain protein domain). For example, when opened in Excel click on cell A1, and select data, autofilter from the top menu. Click on the filter tab for the Interpro short description column, select custom, then select 'contains' from the drop down menu, type "Tyr_pkinase" into the right hand box, and click OK. This filter shows that there are two genes within this region (ERBB2 and CRKRS) whose proteins are annotated to have kinase domains. Note there is extensive redundancy in this table due to multiple protein domains per gene. This file can be analysed further (e.g., to remove redundancy) by using further Excel functionality such as pivot tables, uploading to a database software, such as MS Access, or the export could just be simplified as required depending on the requirements and expertise of the user.
17. At this stage, these data could be cross-referenced with other data sources, perhaps gene expression data results comparing these cell lines to others, or lists of genes associated with a particular process curated by other means (e.g., list of genes with GO terms relevant to oncogenesis or the phenotype of study). The user should ensure that the different data can be linked by an identifier that can be selected in this export process, and that redundancy is taken into account.

3.9.3. Identification of Clones for FISH Confirmation Experiments

18. Depending on what methodology the researcher follows for developing a FISH probe, different routes could be taken here. In this example, we will find a genomic clone within the region of interest from which we can develop a FISH probe, in this case using the Ensembl genome viewer (although BioMart could be used also). Navigate to Ensembl (see Table 2).
19. In the left hand pane, select Human. In the search box at the top of the page enter "17:34000000-38000000". This displays an overview of chromosome 17, base pairs 34000000 to 38000000, displaying chromosomal features graphically in their context against the genomic sequence. Note that this view can be highly customised to show or hide various features and data. You should then see the clone Tilepath displayed towards the bottom of the display, which shows the genomic clones that constitute the human tilepath set. If not,

this can be switched on using the "Configure this page" functionality (consult Ensembl help pages for full details).

20. In this display, the user can then navigate proximal and distal, and zoom in and out as required using the navigation buttons provided and decide which tilepath clones to use to generate FISH probes for confirmatory experiments. Here we can zoom in to the ERBB2 gene for example, to focus on that gene as a candidate of interest.
21. The clones of interest can (at the time of writing) be ordered from GeneService (see Table 2). Note – the physical size of the clone insert can differ from the sequence submitted to EMBL, and hence displayed in Ensembl. In other words, the physical clone insert may extend proximal/distal to the positions shown in the genome browser. If this is of critical importance, the researcher can attempt to ascertain whether or not the genomic insert ends for that clone were sequenced, and use the Ensembl BLAST facilities to ascertain their positions.
22. Once received the genomic clone can be sequence verified (e.g., using PCR primers designed against the ERBB2 gene) and used to generate FISH probes.

This is a simplistic and subjective example (there is obviously an imbalance between the groups studied for instance), but it just serves to illustrate part of the workflow.

3.10. Mutation Databases and Other Resources

There are two main types of mutation that are relevant to the study of cancer genomes – somatic mutations in sporadic cancers that are acquired by the tumour or related cell(s), some of which have a functional effect (e.g., amplification of ERBB2 gene in some breast cancer patients)(7), and germline mutations that are inherited and may predispose to familial cancer (e.g., BRCA mutation effects in familial breast cancer)(5).

Some of the SNP and emerging copy number variation resources will be covered elsewhere in this volume, so here we will introduce some databases catering for cancer mutations and in particular somatic mutations, in addition to other selected resources as an example of what else may be used in conjunction (see also Notes 7 and 8). Whole genome association studies are recently generating large datasets highlighting regions and genes that may predispose to cancer (e.g., (42, 43)) and while these are not discussed in detail here, a link to a relevant resource (dbGAP) is given in Table 3 for the interested researcher.

3.10.1. COSMIC: Catalogue of Somatic Mutations in Cancer

This database collates somatic mutations in cancer, and unlike others is not centered on a particular gene ((44) Table 3). It serves to collate data emerging from the Sanger Institute Cancer Genome Project (large-scale mutation analysis of cancer cell lines

Table 3
Cancer mutation databases and related resources

Resource	URL[a]
DbGAP	http://www.ncbi.nlm.nih.gov/sites/entrez?db=gap
COSMIC	http://www.sanger.ac.uk/genetics/CGP/cosmic/
HGVS locus specific database listing	http://www.genomic.unimelb.edu.au/mdi/dblist/glsdb.html
EGFR mutations database	http://www.somaticmutations-egfr.org/
OMIM	http://www.ncbi.nlm.nih.gov/sites/entrez?db=OMIM
KinMutBase	http://bioinf.uta.fi/KinMutBase/
Cancer Genome WorkBench	http://cgwb.nci.nih.gov/
Cancer Gene Census	http://www.sanger.ac.uk/genetics/CGP/Census/
EBI UniProtKB	http://www.ebi.ac.uk/uniprot/
UniProt	http://www.uniprot.org
ENCODE	http://www.genome.gov/10005107
EBI sequence analysis tools	http://www.ebi.ac.uk/Tools/sequence.html
EMBOSS align (2 sequence alignments)	http://www.ebi.ac.uk/emboss/align/
ClustalW2 (multiple sequence alignment)	http://www.ebi.ac.uk/Tools/clustalw2/

[a]See Note 14

and tumour samples) and also mutation information curated from the literature for selected genes. As the name suggests this database focuses on somatic mutations, and more recently also includes gene fusion data. At the time of writing, 55,779 mutations, 4,773 genes, and 2,249 fusions were noted. The database can be queried by gene or by tissue.

3.11. Locus-Specific Resources

There are a myriad of databases that collect data on mutations for specific genes both in cancers and in other diseases, and collating somatic or germline mutations. In some cases, there is more than one database for the same gene (e.g., TP53 (45, 46)).

A useful listing of these is provided on a website by the Human Genome Variation Society, although it should be noted that at the time of writing it appeared to be a year since this was updated (see Table 3).

One of the most recent examples of one of these types of database is the EGFR mutations database (see Table 3), which collates published human EGFR somatic mutations in NSCLC and other cancers, reflecting the interest in EGFR mutations due to the ongoing research into the role of EGFR mutation in response to EGFR inhibitor therapies and lung cancer.

Databases such as these can be of value for example if the researcher wants to see if a particular mutation has previously been reported, or to assess the spectrum of mutations in a gene in different cancer types. Some of the non-cancer related resources may also provide clues as to potential functional effects of mutations, as a mutation may have been noted in a disease other than cancer that may flag a functional consequence of that mutation.

3.12. Other Related Resources

The *OMIM* (Online Mendelian Inheritance in Man) database collates information on germline mutations in genes, not restricted to familial cancers but catering for a variety of mendelian inherited disorders (47). This can be of use, however, in ascertaining if a gene of interest has mutations in other disorders and highlighting relevant references that may aid the development of a hypothesis of a functional role of a given mutation (48).

KinMutBase is an example of a database set up to collate mutation data but addressing a specific question (49). In this resource, the authors focus on disease-causing mutations in kinase domains, and again this database is not cancer specific (and the website indicates that it was last updated in 2004). However, resources such as this can help in cross-referencing with other sources as above (see Note 9).

The *Cancer Genome Workbench* (CGWB) is an informatics framework produced by the NCI that provides functionality for analysis of trace data from sequencing projects to identify mutations, and that also integrates data from multiple projects presenting it for analysis via a viewer based on the UCSC genome browser ((50) and Table 3). This includes mutation and copy number data at present, and currently integrates initial data from the TCGA with selected data from COSMIC and other projects. It is an example for how data from multiple sources may be integrated – for example, COSMIC mutations can be viewed alongside TCGA mutations and copy number data where available. A tutorial is provided for this resource via the website. See Notes for caution on interpretation of such integrated data.

The *Cancer Gene Census* provided by the Sanger Institute Cancer Genome Project is a manually curated dynamic list of genes for which mutation has been implicated as causal in cancer ((51) and Table 3). This differentiates it from some other mutation data sources, for which many mutations may lack any form of functional appraisal. The list is in the form of a spreadsheet that can be downloaded, and contains various annotations such as tissue type involved, nature of mutations (e.g., translocation, missense) and whether other germline mutations have been noted.

Although not a data resource, it is worth mentioning that researchers are also exploring text-mining methods for more effective extraction of mutation data from the literature to attempt to aid handling of increasing volumes of information. These processes may aid the task of the database curator; similarly the interested researcher may wish to explore such methods deployed in their own projects. One such example is the program MutationFinder (52) that is designed to identify point mutation references in text. This is still an active area of research with various issues to overcome, but may aid the challenging task of collating what is published and hence the comprehension and accuracy of data in resources such as those noted above.

In summary, a variety of resources are available to the researcher and part of the challenge is in identifying those that are most relevant to the researcher's interests and are suited for answering the questions in mind. The researchers should then ensure that they understand the resources sufficiently to avoid potential pitfalls associated with the complexities of the data (see Notes 10–13).

3.13. Detailed Example 2: Assessment of Somatic Mutations in COSMIC for a Gene of Interest (KIT)

In this hypothetical example, we have a gene of interest and want to know if somatic mutations have been noted in COSMIC and if so, what tissues they are in and what the potential impact of the mutation(s) may be.

3.13.1. Interrogation of Data in COSMIC

1. Navigate to the COSMIC website (see Table 3). You can either use the text search or browse by gene or tissue. In this case in the Detailed Search section select Browse by Gene. Select the tab "Cancer genes (from the Cancer Gene Census)", select "K" and then click on the radio button next to "KIT" and click the Next button. This takes you to a summary page for this gene.
2. The Additional Info section is important for later, as it notes the sequences used as the reference for mapping the curated mutations. This is very important when cross-referencing to other information relating to this gene, as different authors may use different source sequences. We can also view the mutations in relation to other genomic features using the links to Ensembl. To do this, if you have not done this before click the "Click here" link in the DAS section of the Additional Info panel. This launches an Ensembl Detailed View of the KIT gene, showing COSMIC mutations and transcripts in relation to other genomic features. The user can then use the navigational tools to explore the data. In future sessions, the user need only click the "Ensembl Contig View" link in this same section to do this.

3. The References section notes the number of references curated in COSMIC for this gene. In this case, the gene is highly curated with many references. Click the Publications button in this section to list all curated references. At the bottom of the page, click the Export button. Select MS Excel as the file type, and click the Export button. The list of publications can then be saved to the user's PC and the information reviewed (e.g., to see if it includes any key papers of interest, or to see how far the curated references extend back).
4. The Studies section lists Sanger Institute Cancer Genome Project studies in which this gene was included. Follow the links for more information. The Samples section lists the total number of samples curated for this gene, and the number of samples with mutations (see Note 7). Now, we will explore the mutations curated. At the top of the page, click the Histogram button in the Small Intragenic Mutation Summary (note that gene fusions are summarised elsewhere when available – see TMPRSS2/ERG fusions as an example).
5. This shows a graphical representation of different mutation types against the peptide sequence, in relation to noted protein domains. The user can zoom in on particular sections by restricting to amino acid coordinates and clicking the Display button. The "Details for KIT" table below this graph gives further information on mutations noted and sample numbers. Note that this table changes depending on the region being viewed in the graph. In this example, in the Navigation section of the graph, enter AA 550–600 in the relevant entry boxes and click the Display button.
6. This gives a restricted graph and we can see various indels, complex mutations, and substitutions noted. Clicking on the icons in the graph gives further information. In this instance, click the Mutations button just underneath the graph. This lists mutations seen in this view, grouped by type, and numbers of samples these mutations occur in. For instance, at amino acid position 568 three types of substitution are noted, with the Y568D mutation occurring in two samples (i.e., a recurrent mutation and so probably less likely to be a chance occurrence or passenger mutation – but see also note below regarding same samples occurring in different studies). Click on this Y568D mutation link. This now shows the details of this mutation, such as positional context and chromosomal coordinates. In the Associated Samples section, we see that the two samples this mutation occurs in are noted (with different identifiers). The tissue type is noted as soft tissue. Click on the link for sample E13752. This gives information on the sample, such as the related reference and other genes noted to be mutated in this sample if data is present (e.g., for well

studied cell lines this can be extensive). In this case, we can see that there is an additional KIT mutation noted in this sample at AA 572. Of note, in the Tumour Classification section we see that the tissue is noted as gastrointestinal stromal tumour (GIST). This illustrates how awareness of the ontologies being used in resources such as this is important – in this example GIST samples are represented as a sub-category under the broader soft tissue category. Checking the other sample link, we see that this sample is from a different reference, but note that in this example some authors are noted by both references, which raises some caution regarding potential re-analysis of samples.

7. We have now found that there is a Y568D mutation curated from two GIST samples each from a different reference (see caveat above – but just for the purposes of this example we will assume it is in two different samples), with other substitution mutations affecting Y568 also noted. We could now check these references to see what approach was used (whole gene screening or focused approach, sample quality, etc.), and to see if it is possible that the two different references (for Y568D mutation) refer to the same sample as can occur for example in follow-up studies (however, note that the COSMIC curators do strive to guard against this issue when considering samples within the same publication). We can also see from the graph and the mutations table that other deletions and complex mutations affect Y568.

Note that this example gives only a small view of the functionality and complexity of COSMIC, so I urge the reader to explore the functionality further and use the Help features (and/or contact the COSMIC team for more details). See Notes for additional comments on the complexities of ascertaining mutation frequencies.

3.13.2. Assessment of potential mutation effects

8. An important question now is what are the potential consequences of a Y568 mutation in KIT? Here, of course, further experimentation and study is ultimately required, but some further resources can at least be used to cross-reference to see if an obvious hypothesis emerges. First, we can check OMIM (see Table 3) to see if other mutations are noted here or if the gene has been associated with another disorder (other than cancer), which may give a clue as to what biological processes we may want to evaluate impact on. Navigate to OMIM and in the search box at the top of the page type "KIT". Select the appropriate entry from the results returned (in this case entry 164920), and click on the link to view the summary page. In this case, there is a lot of information (as of course it is a well-studied cancer gene).

9. We will now view information in the SwissProt entry for the KIT gene's protein product. Navigate to the UniProt resource (which can also be reached from the EBI website)(see Table 3), ensure that the search drop down menu at the top of the page is set to search UniProtKB, enter the search term "KIT AND Human" in the text box and click the Search button. Many results are found – select the result with Accession P10721 and open the SwissProt summary page by clicking on this Accession number.
10. This shows the SwissProt summary that collates a variety of information. In this case, for example, we can look down the page and see what regions are predicted to contain conserved protein domains (e.g., protein kinase domain at positions 589–937), that there are protein structure data available for this molecule (e.g., PDB X-ray structures), that in the amino acid modifications section Y568 is noted as a potential phosphotyrosine site, and that in the experimental info section several residues have been targeted in mutagenesis experiments.
11. There are several important points to reinforce here. It must be emphasised that when cross-referencing coordinate-based data (e.g., mutation positions) from different sources the researcher should check that the coordinates correspond, or investigate to decide how the data can be cross-referenced if source sequences differ. This can be aided using alignments of sequences in question using one of the various sequence alignment tools available (for examples see Table 3). Second, these sorts of data are not static, and the views and formats may also change over time. Third, while SwissProt summaries are curated, some features may be predicted from similarity to other peptides with the noted feature for example (i.e., not experimentally verified). For full details of the SwissProt annotation process see the UniProt documentation via the online links. Finally, not all genes of interest may have detailed SwissProt entries thus far – however, in these instances one could still for example assess similarity to other paralogues or orthologues for the region in question which may hint at its functional significance. One note of caution – even for conserved residues disruption of them by mutation may not have the same functional effects (e.g., B-raf and C-raf (53)).
12. In summary – we have seen how to find somatic mutations in our gene of interest (KIT) using COSMIC. We have found mutations of interest (at Y568), although clearly KIT is well studied with many other mutations noted. Mutation at this position may disrupt a putative phosphorylation site. Further work could now be performed, for example, to ascertain mutation frequency at this position in GIST or other cancers

as part of a well-focused study, and potential effects on kinase function or phosphorylation at Y568 could be assessed.

13. Finally, we need to acknowledge that ascertainment of functional consequences or the biological relevance of mutations against a backdrop of widespread genomic aberration in cancer is complex, and we can at this point simply attempt to bring to bear as much information as possible to help develop hypotheses and guide further experimentation, while bearing in mind the heterogeneity of available data and how it is presented so as to analyse it effectively. Consideration of the pathways in which genes are mutated as well as how disease subtype-specific some of them may be, is also relevant to appraisal of mutations and their frequencies (54). Pathway analysis tools and resources (as mentioned earlier) can be relevant here, but caution is needed due to passenger mutations which could add considerable noise to such analyses, particularly if affecting well-studied and "well-connected" genes that may bias such approaches.

4. Notes

1. It should be borne in mind that any database based on manual curation may risk not capturing every last fact from the scientific literature given the sheer scale of the task, despite the laudable efforts involved, and indeed it may not even be their intention depending on their approach. For a specific example and if absolute comprehension is key, it may be worth considering a supplementary search of the literature in addition to use of such online resources (which can include the use of other mutation collation databases).
2. If not using matched normal samples, it should be remembered that the collection of normal samples used may have its own pattern of structural variation, due to copy number polymorphism within a "normal" population. The extent and nature of copy number polymorphism is still an area of active research (55).
3. It should be noted that the assessment of cancer genomes using array CGH will generate lists of aberrations and coordinates. However, some information has been lost by this stage, such as spatial resolution, which may describe tumour heterogeneity or admixture with surrounding normal tissue or stroma. The extent of this admixture will have some impact on magnitude of copy number change seen, and complicates the picture when trying to compare copy number changes between arrays. Similarly, one cannot of course distinguish

from these data between high-level amplification in a small number of tumour cells versus lower-level amplification in a larger number of cells.

4. When relating array CGH probes to chromosomal coordinates and subsequently to genomic features such as genes, versioning is critical as genome assemblies can still change and genomic features are continually updated as data and knowledge will continue to grow (e.g., recent inclusion of microRNAs and regulatory feature data from the initial ENCODE projects (56)). If using static mapping files rather than direct links, this is clearly an issue that needs to be borne in mind. Similarly, if moving between software and analyses, it should be checked whether the genome assemblies being used are synchronised, and that if obtaining genomic data from different resources that the numbering used is consistent. This can also be checked manually using a known genomic feature. Some resources such as Ensembl also make available archive versions of their data which can be useful if needing to cross-reference with older datasets. If there is any doubt, contact the respective provider of information (e.g., Ensembl helpdesk).
5. Some older datasets, in particular from BAC arrays, are likely to be from lower resolution platforms (e.g., 1 MB resolution). As such, computational inference of breakpoints for copy number gain/loss is naturally not going to give an exact picture at the genomic level, as for our example shown earlier. In this scenario, if interrogating regions for genes contained within regions of gain/loss, it is worth considering genes proximal/distal to the coordinates provided from segmentation analysis to an extent dictated by the resolution of the platform, the interest of the researcher, and resources for the follow-up experimentation.
6. When using data analyses from online resources or downloading datasets, you may not necessarily have full details for samples used and data processing (e.g., detailed histology and tumour content of the samples, details on normalisation parameters etc.). This should be borne in mind in terms of the conclusions based on such analyses, and further information sought as required.
7. Each database will have its own intended usage, and users should assess resources according to their needs and what types of queries are being asked (e.g., COSMIC does not aim to cover all genes, but does also include data from the Sanger Cancer Genome Project). For example, if wanting to calculate mutation frequencies within a given cancer type the user needs to beware that in some databases (such as COSMIC), data from different screening procedures are collated, which can potentially provide a skewed picture if not accounted for

(e.g., whole-gene screening versus screening of particular codons). In addition, there can be multiple samples per patient (e.g., primary tumour and metastatic tumour samples), or the same samples could potentially occur in more than one study. All of this can complicate the seemingly simple question of estimating mutation frequency within a target tumour/population. This is just as relevant when integrating data from various different sources (e.g., CGWB). Of course, this just reflects the complexity of the issues facing the database administrators/curators tackling these difficult tasks, and one resource can rarely cover all potential querying aspects (57). If the user is in any doubt as to suitability for their purposes or needs help with appropriate querying, it is recommended that they contact the database administrators/authors to get the most up to date information, and who may be able to provide further expert help.

8. Databases can come and go depending on funding and research interests, as such it is worth bearing in mind that this is an evolving landscape. Even for the more established wider databases, some large datasets may not always be captured in these databases if they are not within their remit and so could potentially remain "hidden" in supplementary data tables or other locations.
9. Similarly, some databases may still be available online but may not have been updated for some time (e.g., KinMutBase) – in these cases the researcher needs to consider how important it is for them to have the most up to date and most comprehensive information. In these instances, these resources may still provide a useful start point to supplement further manual review.
10. Different databases will have different processes for collating and curating data, which will naturally influence the content of these repositories. Some databases may, for example, focus more on comprehension of data at the expense of coverage (in terms of genes), seen at its logical extreme in locus specific databases. This should be taken into account when querying. As before, clarification with the database administrators/authors is always a worthwhile step.
11. The emerging picture from larger-scale studies is that there are many somatic mutations in cancer genomes and while some will be causal or otherwise functional, others are likely to be "passenger" mutations, which may not have a functional relevance but are instead occurring in those tumour cells, which are subsequently selected for during the process of tumourigenesis. Methods are being explored to attempt to address this issue as more data are generated from high-throughput approaches (e.g., TCGA), but this complexity remains an issue

in interpreting these data and should be borne in mind as data become richer and more widely available (58).

12. For all databases that curate information from the literature, it may be necessary to relate terminology used by the authors describing key features (e.g., tissue and histology) to the ontologies used within the database structure. This may not always be straightforward and again if critical to the analyses, clarification may be sought via reference to the source data or database authors.
13. In order to try and keep abreast of resources, such as KinMutDB, that may be quite niche and may be less stable or updated less frequently, the Nucleic Acids Research database issue published each year is a useful start point.
14. URLs cited in this review may become out of date as resources may change website or the resources themselves become out of date. A search of the internet using the resource name should help in this instance.

Acknowledgements

The author would like to thank Dr. Pall Jonsson and Dr. Tim French for their helpful comments and suggestions during the preparation of this manuscript.

References

1. CRUK published statistics - http://publications.cancerresearchuk.org/WebRoot/crukstoredb/CRUK_PDFs/CSINC08.pdf
2. World Health Organisation (WHO) cancer factsheet - http://www.who.int/mediacentre/factsheets/fs297/en/index.html
3. Hanahan D, Weinberg RA. (2000) The hallmarks of cancer. Cell, **100**, 57–70.
4. Steele RJ, Thompson AM, Hall PA, Lane DP. (1998) The p53 tumour suppressor gene. British Journal of Surgery, **85**, 1460–1467.
5. Miki Y, Swensen J, ShattuckEidens D, Futreal PA, Harshman K, Tavtigian S et al. (1994) A strong candidate for the breast and ovarian cancer susceptibility gene BRCA1. Science, **266**, 66–71.
6. Druker BJ, Lydon NB. (2000) Lessons learned from the development of an abl tyrosine kinase inhibitor for chronic myelogenous leukemia. Journal of Clinical Investigation, **105**, 3–7.
7. Albanell J, Baselga J. (1999) The ErbB receptors as targets for breast cancer therapy. Journal of Mammary Gland Biology & Neoplasia, **4**, 337–351.
8. Davies H, Bignell GR, Cox C, Stephens P, Edkins S, Clegg S et al. (2002) Mutations of the BRAF gene in human cancer. Nature, **417**, 949–954.
9. Sjoblom T, Jones S, Wood LD, Parsons DW, Lin J, Barber TD et al. (2006) The consensus coding sequences of human breast and colorectal cancers. Science, **314**, 268–274.
10. Greenman C, Stephens P, Smith R, Dalgliesh GL, Hunter C, Bignell G et al. (2007) Patterns of somatic mutation in human cancer genomes. Nature, **446**, 153–158.
11. Tomlins SA, Rhodes DR, Perner S, Dhanasekaran SM, Mehra R, Sun XW et al. (2005) Recurrent fusion of TMPRSS2 and ETS transcription factor genes in prostate cancer. Science, **310**, 644–648.
12. Khalique L, Ayhan A, Weale ME, Jacobs IJ, Ramus SJ, Gayther SA. (2007) Genetic intratumour heterogeneity in epithelial ovarian cancer and its implications for molecular diagnosis of tumours. Journal of Pathology, **211**, 286–295.

13. Tang X, Shigematsu H, Bekele BN, Roth JA, Minna JD, Hong WK et al. (2005) EGFR tyrosine kinase domain mutations are detected in histologically normal respiratory epithelium in lung cancer patients. Cancer Research, **65**, 7568–7572.
14. Campbell PJ, et al. (2008) Identification of somatically acquired rearrangements in cancer using genome-wide massively parallel paired-end sequencing. Nature Genetics, **40**, 722–729.
15. http://www.sanger.ac.uk/genetics/CGP/
16. http://cancergenome.nih.gov/index.asp
17. http://www.icgc.org/
18. Bardelli A, Parsons DW, Silliman N, Ptak J, Szabo S, Saha S et al. (2003) Mutational analysis of the tyrosine kinome in colorectal cancers. Science, **300**, 949.
19. Weir BA, Woo MS, Getz G, Perner S, Ding L, Beroukhim R et al. (2007) Characterizing the cancer genome in lung adenocarcinoma. Nature, **450**, 893–898.
20. Mitelman Database of Chromosome Aberrations in Cancer (2008). Mitelman F, Johansson B, and Mertens F (Eds.), http://cgap.nci.nih.gov/Chromosomes/Mitelman
21. Knutsen T, Gobu V, Knaus R, PadillaNash H, Augustus M, Strausberg RL et al. (2005) The interactive online SKY/M-FISH & CGH database and the Entrez cancer chromosomes search database: linkage of chromosomal aberrations with the genome sequence. Genes, Chromosomes & Cancer, **44**, 52–64.
22. Huret JL, Senon S, Bernheim A, Dessen P. (2004) An Atlas on genes and chromosomes in oncology and haematology. Cellular & Molecular Biology, **50**, 805–807.
23. Baudis M, Cleary ML. (2001) Progenetix. net: an online repository for molecular cytogenetic aberration data. Bioinformatics, **17**, 1228–1229.
24. Kim DS, Huh JW, Kim HS. (2007) HYBRIDdb: a database of hybrid genes in the human genome. BMC Genomics, **8**, 128.
25. Kim N, Kim P, Nam S, Shin S, Lee S. (2006) ChimerDB–a knowledgebase for fusion sequences. Nucleic Acids Research, **34**, 21–4.
26. Baross A, Delaney AD, Li HI, Nayar T, Flibotte S, Qian H et al. (2007) Assessment of algorithms for high throughput detection of genomic copy number variation in oligonucleotide microarray data. BMC Bioinformatics, **8**, 368.
27. Gentleman RC, Carey VJ, Bates DM, Bolstad B, Dettling M, Dudoit S et al. (2004) Bioconductor: open software development for computational biology and bioinformatics. Genome Biology, 5, R80.
28. Reich M, Liefeld T, Gould J, Lerner J, Tamayo P, Mesirov JP. (2006) GenePattern 2.0. Nature Genetics, **38**, 500–501.
29. Hupe P, Stransky N, Thiery JP, Radvanyi F, Barillot E. (2004) Analysis of array CGH data: from signal ratio to gain and loss of DNA regions. Bioinformatics, **20**, 3413–3422.
30. Beroukhim R, Getz G, Nghiemphu L, Barretina J, Hsueh T, Linhart D et al. (2007) Assessing the significance of chromosomal aberrations in cancer: methodology and application to glioma. Proceedings of the National Academy of Sciences of the United States of America, **104**, 20007–20012.
31. Flicek P, Aken BL, Beal K, Ballester B, Caccamo M, Chen Y et al. (2008) Ensembl 2008. Nucleic Acids Research, **36**, 707–14.
32. Karolchik D, Kuhn RM, Baertsch R, Barber GP, Clawson H, Diekhans M et al. (2008) The UCSC Genome Browser Database: 2008 update. Nucleic Acids Research, **36**, 773–9.
33. Wheeler DL, Barrett T, Benson DA, Bryant SH, Canese K, Chetvernin V et al. (2008) Database resources of the National Center for Biotechnology Information. Nucleic Acids Research, **36**, 13–21.
34. Garraway LA, Widlund HR, Rubin MA, Getz G, Berger AJ, Ramaswamy S et al. (2005) Integrative genomic analyses identify MITF as a lineage survival oncogene amplified in malignant melanoma. Nature, **436**, 117–122.
35. Barrett T, Troup DB, Wilhite SE, Ledoux P, Rudnev D, Evangelista C et al. (2007) NCBI GEO: mining tens of millions of expression profiles–database and tools update. Nucleic Acids Research, **35**, 760–5.
36. Parkinson H, Kapushesky M, Shojatalab M, Abeygunawardena N, Coulson R, Farne A et al. (2007) ArrayExpress–a public database of microarray experiments and gene expression profiles. Nucleic Acids Research, **35**, 745–50.
37. Demeter J, Beauheim C, Gollub J, HernandezBoussard T, Jin H, Maier D et al. (2007) The Stanford Microarray Database: implementation of new analysis tools and open source release of software. Nucleic Acids Research, **35**, 766–70.
38. Chari R, Lockwood WW, Coe BP, Chu A, Macey D, Thomson A et al. (2006) SIGMA: a system for integrative genomic microarray analysis of cancer genomes. BMC Genomics, **7**, 324.
39. Hupe P, La Rosa P, Liva S, Lair S, Servant N, Barillot E. (2007) ACTuDB, a new database for the integrated analysis of array-CGH and clinical data for tumors. Oncogene, **26**, 6641–6652.
40. Scheinin I, Myllykangas S, Borze I, Bohling T, Knuutila S, Saharinen J. (2008) CanGEM: mining gene copy number changes in cancer. Nucleic Acids Research, **36**, 830–835.

41. Reyal F, Stransky N, BernardPierrot I, VincentSalomon A, de Rycke Y, Elvin P et al. (2005) Visualizing chromosomes as transcriptome correlation maps: evidence of chromosomal domains containing co-expressed genes--a study of 130 invasive ductal breast carcinomas. Cancer Research, **65**, 1376–1383.
42. Gudmundsson J, Sulem P, Rafnar T, Bergthorsson JT, Manolescu A, Gudbjartsson D et al. (2008) Common sequence variants on 2p15 and Xp11.22 confer susceptibility to prostate cancer. Nature Genetics, **40**, 281–283.
43. Easton DF, Pooley KA, Dunning AM, Pharoah PD, Thompson D, Ballinger DG et al. (2007) Genome-wide association study identifies novel breast cancer susceptibility loci. Nature, **447**, 1087–1093.
44. Forbes S, Clements J, Dawson E, Bamford S, Webb T, Dogan A et al. (2005) Cosmic 2005. British Journal of Cancer, **94**, 318–322.
45. Hamroun D, Kato S, Ishioka C, Claustres M, Beroud C, Soussi T. (2006) The UMD TP53 database and website: update and revisions. Human Mutation, **27**, 14–20.
46. Petitjean A, Mathe E, Kato S, Ishioka C, Tavtigian SV, Hainaut P et al. (2007) Impact of mutant p53 functional properties on TP53 mutation patterns and tumor phenotype: lessons from recent developments in the IARC TP53 database. Human Mutation, **28**, 622–629.
47. Online Mendelian Inheritance in Man, OMIM (TM). McKusick-Nathans Institute of Genetic Medicine, Johns Hopkins University (Baltimore, MD) and National Center for Biotechnology Information, National Library of Medicine (Bethesda, MD).
48. Pollock PM, Gartside MG, Dejeza LC, Powell MA, Mallon MA, Davies H et al. (2007) Frequent activating FGFR2 mutations in endometrial carcinomas parallel germline mutations associated with craniosynostosis and skeletal dysplasia syndromes. Oncogene, **26**, 7158–7162.
49. Ortutay C, Valiaho J, Stenberg K, Vihinen M. (2005) KinMutBase: a registry of disease-causing mutations in protein kinase domains. Human Mutation, **25**, 435–442.
50. Zhang J, Finney RP, Rowe W, Edmonson M, Yang SH, Dracheva T et al. (2007) Systematic analysis of genetic alterations in tumors using Cancer Genome WorkBench (CGWB). Genome Research, **17**, 1111–1117.
51. Futreal PA, Coin L, Marshall M, Down T, Hubbard T, Wooster R et al. (2004) A census of human cancer genes. Nature Reviews Cancer, **4**, 177–183.
52. Caporaso JG, Baumgartner WA, Jr., Randolph DA, Cohen KB, Hunter L. (2007) MutationFinder: a high-performance system for extracting point mutation mentions from text. Bioinformatics, **23**, 1862–1865.
53. Emuss V, Garnett M, Mason C, Marais R. (2005) Mutations of C-RAF are rare in human cancer because C-RAF has a low basal kinase activity compared with B-RAF. Cancer Research, **65**, 9719–9726.
54. Annunziata CM, Davis RE, Demchenko Y, Bellamy W, Gabrea A, Zhan F et al. (2007) Frequent engagement of the classical and alternative NF-kappaB pathways by diverse genetic abnormalities in multiple myeloma. Cancer Cell, **12**, 115–130.
55. Perry GH, BenDor A, Tsalenko A, Sampas N, RodriguezRevenga L, Tran CW et al. (2008) The fine-scale and complex architecture of human copy-number variation. American Journal of Human Genetics, **82**, 685–695.
56. ENCODE Project Consortium, Birney E, Stamatoyannopoulos JA, Dutta A, Guigo R, Gingeras TR et al. (2007) Identification and analysis of functional elements in 1% of the human genome by the ENCODE pilot project. Nature, **447**, 799–816.
57. Soussi T, Ishioka C, Claustres M, Beroud C. (2006) Locus-specific mutation databases: pitfalls and good practice based on the p53 experience. Naturev Reviews Cancer, **6**, 83–90.
58. Torkamani A, Schork NJ. (2008) Prediction of cancer driver mutations in protein kinases. Cancer Research, **68**, 1675–1682.

Chapter 6

Copy Number Variations in the Human Genome and Strategies for Analysis

Emily A. Vucic, Kelsie L. Thu, Ariane C. Williams, Wan L. Lam, and Bradley P. Coe

Abstract

The structure and sequence of the genome is immensely variable in the human population. Segmental copy number variants (CNVs) contribute to the extensive phenotypic diversity among humans and have been shown to associate with disease susceptibility. In this article, we provide a detailed review of human genetic variations and the experimental approaches used to discover, catalog, and genotype CNVs.

Key words: Copy number variation, Copy number polymorphism, Single nucleotide polymorphism, Array CGH, Tiling array, Next generation sequencing

1. Introduction

There is a considerable degree of sequence and structural variation (SV) in the human genome. Advancement of array based comparative genomic hybridization (aCGH) and sequencing technologies has resulted in a great appreciation for the normal variation that exists in the human genome. Individuals can vary at several levels, from single nucleotides to long stretches of repeated DNA sequences and, as most recently discovered, by segmental copy number changes termed copy number variations (CNVs) (1) (see Table 1). Some variations translate to differences in protein sequence while others do not but may nonetheless affect gene regulation in a number of ways. Our knowledge as to how this variation affects an individual's susceptibility to disease (HIV and cancer); contributes to genetic disorders (autism, mental retardation); affects an individual's response to therapy (metabolism); and contributes to phenotypic variation in humans continues to

Michael R. Barnes and Gerome Breen (eds.), *Genetic Variation: Methods and Protocols*, Methods in Molecular Biology, vol. 628, DOI 10.1007/978-1-60327-367-1_6,

Table 1
Structural and sequence variations in the human genome

Features	Description
Segmental duplication (SD)	Duplicons that have >90% sequence homology to another region in the genome, and are prone to nonallelic homologous recombination
Copy number variants (CNVs)	DNA segments >1 kb in length, whose copy number varies with respect to a reference genome. Copy number polymorphisms (CNPs) have a >1% frequency in the population
Single nucleotide polymorphism (SNP)	SNPs are sequence variations occurring at the single base pair level and at a population frequency >1%. There are approximately 12 million SNPs cataloged so far
Tandem repeats	Tandem sequences repeated in either orientation. They include minisatellites and microsatellites and represent ~10% of the genome. Commonly found in heterochromatic DNA, e.g., short tandem repeats of TTAGGG units at telomeres
Interspersed repeated elements	Long interspersed nuclear elements (LINEs) and short interspersed nuclear elements (SINEs) constitute a significant fraction of the human genome

advance with new technologies. The magnitude and distribution of genotypic variation in the human genome is broad, and the extent of this variation and consequences are not yet fully understood (2). Difficulties in describing these regions, even with next generation sequencing platforms, are due to the inherent challenges in analyzing vast stretches of repetitive DNA (3). A more complete understanding of how this variation manifests phenotypically is crucial to our understanding of disease etiology and treatment. A summary of current knowledge with respect to genetic variation in human populations and the impact this variation has on disease etiology and treatment will be discussed in this chapter. Technologies used to describe genetic variation, along with future directions for CNV analysis, will also be discussed.

1.1. Variation in the Human Genome

The genomes of individuals vary at the sequence and structural level. Variations at the single base pair level occurring at a population frequency >1% are termed single nucleotide polymorphisms (SNPs), and they represent the most common type of genomic variation. Over 10 million of these SNPs have been cataloged in the haplotype map (HapMap) project and other databases (1, 2). Estimates of genetic variation in humans based on the analysis of protein sequences from several individuals in the 1960s led to a gross underestimation of the extent of genetic variation at the base pair level, due to the exclusion of 5′ regulatory sequences of genes as well as DNA not coding for protein. Additionally, SNPs that result in amino acid changes are much less common than those that occur in gene introns and noncoding regions of DNA, likely due to selection against functional changes to proteins. SNPs in regulatory regions of genes, such as promoters and enhancers, have become a major focus in the study of disease etiology and evolution. SNPs can be used in linkage disequilibrium studies to elucidate association with certain phenotypes, disease susceptibility, and evolutionary adaptation (4–7).

In addition to SNPs, there are many other classes of variation in the human genome. These include high to low copy repeats (which form the majority of the DNA sequence), segmental duplications (SDs), CNVs, and other variant sequences. These are detailed in Table 1.

The prevalence of CNVs in normal populations has been realized with the advent of array based genomic hybridization platforms and sequencing technologies (8–12). CNVs are defined as a segment of DNA >1 kb in length whose copy number varies with respect to the reference genome and are termed copy number polymorphisms (CNPs) when they occur in the population at a frequency >1% (2, 8, 11). CNVs, while less prevalent than SNPs, affect a larger portion of the genome due to their size (thousands to millions of base pairs per individual CNV). Approximately, 12% of the genome is thought to be copy number variable, with 10–60% of these variations encompassing genes (8, 13).

1.2. CNV Origin and Distribution

CNVs are strongly associated with segmental duplications (SDs) and repetitive regions of the genome, including *Alu* elements and LINES. This is likely due to high recombination rates between stretches of nonallelic homologous flanking repeats. This is thought to result in the structural rearrangements characteristic of CNVs: deletions, duplications, and inversions (8, 11). The probability of misalignment between nonallelic homologous regions in the genome appears to be strongly influenced by several factors, for example: sequence homology, length, and orientation (8). In fact, much of the aberrant recombination associated with human genomic disorders is characterized by 10–400 kb stretches of duplicated sequence. The larger and more homologous the duplicon, the more prevalent the genomic disorder (8, 12, 14). As described by Sharp et al., a search for novel regions of copy number variance can be performed by focusing on regions of recombination hot spots or SD (11). SDs, also referred to as duplicons or low copy repeats, are defined as structurally variable portions of genomic DNA that have high levels of homology (>90%) with another region >1 kb in the genome. CNVs can be distinguished by their size and degree of divergence, as SDs are theoretically fixed CNVs (7). Some SDs in the reference genome are not fixed and are actually CNVs, calling into question the definition of SD based on a single reference sequence (8).

The distribution of CNVs in the genome has been further described by Redon and others, who showed duplications to be favored over deletions in gene-rich areas (8). Interestingly, Redon et al. observed a significantly higher proportion of duplications than deletions affecting disease related genes. CNVs are under-represented in highly conserved and functionally important regions. Genes found in CNV enriched regions have been shown to be involved in sensory perception (notably olfactory genes), rhesus type, metabolism, cell adhesion, neurophysiological processes, and disease (7–9, 15).

2. The Role of CNVs in Disease

The balance of specific gene products in cells is critical to the dynamics of several cellular pathways. Gain or loss of even a single copy of a gene can result in the deregulation and dysfunction of cellular function. The functional effects of abnormal copy number at various loci throughout the human genome have significant impacts on disease (16). The roles CNVs play in normal variation and functional consequences of pathogenic CNVs are outlined below.

2.1. Benign and Pathogenic CNVs

There is a significant contribution of CNVs to human variation at the level of normal and disease phenotype. A large fraction of the variation in human gene expression has been linked to CNVs, demonstrating their importance (17). Evolutionary studies continue to accumulate evidence for the role of positive selection in the establishment of differential frequencies of certain CNVs in the human population, indicating the importance of genotypic variation in generating phenotypes with differential fitness (15).

Generally, CNVs can be classified as benign or pathogenic (2, 18). Benign CNVs often do not associate with observable phenotypic effects as they typically associate with nonfunctional regions of the genome. However, some occur in genes as mentioned above, such as visual pigment genes resulting in impaired red–green color vision (7, 9, 19, 20). Although such CNVs are associated with phenotypic effects, they can be considered benign in the context of disease as they do not negatively impact the health of the individual. Benign CNVs are typically heritable and can be identified in normal healthy individuals (2, 14). CNVs have also been shown to modify complex phenotypes such as inflammation, immune response, drug response, and cell signaling (10, 20).

In contrast, CNVs affecting functional DNA can be pathogenic, increase disease susceptibility, or drive development of genomic diseases and disorders. The burden of pathogenic CNVs on health can be immense. Studies including those by Wong et al. have reported an overlap of CNVs with Online Mendelian Inheritance in Man (OMIM) genes, which are known to have significant impact on disease (9). Well-known examples of pathogenic CNVs include the discovery that copy number gain of *CCL3L1* reduces HIV/AIDS susceptibility, and copy number loss of *SMN1* causes spinal muscular atrophy type III (21, 22). Table 2 summarizes a number of CNVs that have been associated with susceptibility and disease. It is believed that most pathogenic CNVs arise de novo; however, some are inherited (14). Distinguishing between pathogenic and benign CNVs can be complicated due to variable manifestation of pathogenic phenotypes between individuals with similar CNVs. For example, de Vries et al. identified a CNV that is benign in some individuals but appears to cause mental retardation in others (23). Reports detailing newly discovered CNVs are emerging at a rate faster than they can be accurately categorized. Thus, in addition to benign and pathogenic CNV categories, there is a growing class consisting of CNVs with undetermined clinical significance (2). Further investigation into these CNVs is required to elucidate their potential phenotypic consequences, if any.

2.2. Physical Properties of CNVs Influencing Pathogenicity

Several properties influence the phenotypic effects of CNVs on individuals. Inherited CNVs are typically benign while those arising de novo tend to be causative in disease (2, 14). Since it is now apparent that CNVs are responsible for a large amount of genetic variation in the human genome, there is increasing interest in

Table 2
Examples of CNVs associated with disease

CNV locus	Location	Disease	Reference
PMP22	17p11.2	Charcot-Marie-Tooth neuropathy type I	(49)
CCL3L1	17q11.2	HIV susceptibility	(22)
NRXN1	2p16.3	Autism spectrum disorder	(50)
SMN1	5q13	Spinal muscular atrophy	(21)
MBD5, RAI1	2q23.1, 17p11.2	Mental retardation	(37, 51)
C4A, C4B	6p21.3	Systemic lupus erythematosus	(27)
PRODH2, DGCR6	22q11	Schizophrenia susceptibility	(52)
DEFB4	8p23.1	Crohn disease susceptibility	(53)

studying the impact that CNVs can have on disease (18). CNVs spanning regulatory sequences of DNA, genes important to development, or dose-sensitive genes are more likely to contribute to disease development or susceptibility. Thus, the location of a CNV within the genome has a significant impact on phenotype (2). The size of the CNV is another important factor in pathogenicity as large CNVs potentially affect multiple genes, whereas smaller CNVs generally affect fewer genes.

The nature of CNVs is also important as genomic gains in copy number are postulated to have smaller pathogenic impact on disease than genomic losses (2). For example, the *SMN1* locus has been implicated in causing the autosomal recessive disease spinal muscular atrophy (SMA), and deletion of this locus on chromosome 5q13 is found in over 96% of SMA patients (24). However, individuals carrying two or more copies of the *SMN1* gene are generally healthy (25, 26). Interestingly, for some phenotypes, the absolute number of gene copies conferred by CNVs appears to be an important factor. This is apparent in a study by Yang et al., which demonstrated that increased susceptibility to systemic lupus erythematosus (SLE) is conferred by genotypes with less than two copies of the *C4* complement gene, whereas susceptibility for SLE is lower when more than 5 copies of *C4* genes are present (27). Similarly, susceptibility to HIV/AIDs is mediated by *CCL3L1*, where low copy numbers increase and high copy numbers decrease disease susceptibility (22).

One disease in which CNVs are emerging as important pathogenic factors is cancer. Recently, reports have found that some CNVs correspond to well-known tumor suppressors and oncogenes, highlighting a potential role for CNVs in cancer susceptibility and potential targets for cancer therapy (9). Additionally, CNVs affecting drug metabolism genes have also been shown to affect patient response to chemotherapy (28, 29).

2.3. Challenges Associated with CNV-Disease Association Studies

The advent of array based and sequencing technologies has resulted in an abundance of databases and genetic maps cataloging CNVs. The availability of this information has provided the opportunity to perform large-scale CNV-disease association studies (14). However, it has proven difficult to deduce the roles of CNVs in disease partially due to the challenge of elucidating the complex relationships between CNV genotype and apparent phenotype. There are several confounding factors that can influence the phenotypic expression of CNVs, such as variable penetrance and environmental factors. For instance, some disease phenotypes associated with certain CNVs seem to be context dependant, where benign CNVs in some individuals are pathogenic in others and vice versa (14, 23, 30). Penetrance of CNV associated phenotypes can be influenced by genomic imprinting, haploinsufficiency, and the presence of other genetic alterations; thus, delineating pathogenicity of CNVs can be a very cumbersome endeavor (14).

Additionally, logistical impediments further confound CNV disease association studies, including a lack of validated CNV databases (many are discordant), and confounding factors such as variable penetrance and environmental contribution to phenotype (2, 14). Therefore, a comprehensive and accurate human genomic database of normal CNVs is mandatory to facilitate such studies and improve our ability to differentiate between potential disease-causing and benign CNVs (30).

3. Current Tools for the Analysis of Copy Number Variation

Prior to the advent of high throughput tools for the analysis of genomic structures, the number of CNVs in the human genome was largely underestimated. Basic techniques such as restriction mapping allowed identification of a few loci, such as the α-globulin locus, which is triplicated in some individuals, but progress was slow due to the low throughput of such methods (31). Once array CGH (aCGH) gained popularity as a tool, it was finally feasible to scan the genome for these large copy number variations. Targeted profiling efforts began to identify hundreds to thousands of loci, which demonstrate variable copy number between individuals, with a significant percentage clustering near segmental duplications (10, 32).

Currently, there are many tools available to the CNV researcher, and choosing between them is not always straightforward. Low resolution array based techniques are affordable and allow rapid generation of reference CNV data which is particularly useful for studies of disease using the same platform. Similarly, high resolution arrays offer more insight at a slightly increased cost. However, all array techniques suffer from an inability to define the sequence of a breakpoint or identify structural alterations, such as inversions, which do not affect copy number (20). Sequencing techniques have been recently adopted to bypass these disadvantages; however, these projects range from moderately to extremely expensive.

3.1. Array Based Approaches

Array based strategies are one of the more cost effective methods for cataloging CNVs in normal and diseased individuals. These techniques all rely on hybridization of DNA from an individual to a matrix of DNA fragments. There are three main categories of DNA fragments used in array CGH: bacterial artificial chromosomes (BACs), short oligonucleotides, and long oligonucleotides. The unique attributes of these platforms will be discussed in the appropriate sections. In all cases, copy number is analyzed by examining the intensity with which labeled patient DNA binds each DNA fragment on the array and comparing this measurement with that observed for a normal diploid individual(s). Most of these technologies cohybridize a sample and reference at the same time using different fluorescent dyes, while some oligonucleotide platforms use single channel methods where normal samples are examined on separate arrays.

3.1.1. Marker Based Platforms

Early studies made use of low resolution array platforms, such as 1 Mb resolution BAC platforms, to identify CNVs in the normal population. Despite their low resolution, use of these platforms detected hundreds of altered clones in panels of normal individuals (32). This observation was far beyond the level of variation anyone had expected to observe. Although these tools were instrumental in early CNV studies and are still useful in many research contexts, the primary utility of these arrays in the analysis of CNVs today is as reference hybridizations for studies utilizing these platforms. Discovery of new CNVs is best left to high resolution platforms.

3.1.2. Medium to High Resolution Platforms

Medium to high resolution array platforms include whole genome tiling path (WGTP) BAC arrays and several commercial and in lab oligonucleotide arrays. These tools vary in cost but represent an affordable tool for screening large sample sets, allowing detection of alterations >10 kb depending on the platform (33).

WGTP BAC arrays offer the ability to detect alterations >50 kb and define breakpoints to within 150 kb (the size of a BAC clone). These platforms have a few distinct advantages. First, sensitivity is higher than most oligonucleotide platforms and a single element is often sufficient to detect a gain. This increased sensitivity leads to a higher true positive rate than some oligonucleotide platforms. Another benefit of increased sensitivity to gains is that deletion variants are strongly underrepresented in the subset of high frequency CNVs. Additionally, the large genomic segments allow coverage of genomic regions with complex structure and repeat/SD content (8, 9, 32–35). The primary disadvantage of WGTP platforms lies in their limited ability to define the breakpoints of an alteration, and the maximal detection sensitivity of 50 kb.

Alternative technologies are based on either short (25 mers) or long (>60 mers) oligonucleotides. A prime example of short oligonucleotide arrays are the Affymetrix genome-wide SNP platforms. Although primarily designed for genotyping, these arrays are capable of copy number profiling by utilizing hybridization intensities. Noise tends to be high in these platforms due to the short oligonucleotides and restriction digest meditated PCR required in sample preparation. As a result, multiple oligonucleotides must be altered to allow reliable detection of CNVs (estimates range from 3 to 20 SNP probes depending on array version and DNA quality) (8, 33, 36, 37). Utilization of the 500 K platform (detection of 75 kb alterations, breakpoint mapping to 75 kb) in combination with WGTP arrays in a study by Redon et al. proved an effective strategy to merge the high sensitivity of BAC platforms with the high breakpoint mapping precision of high density oligonucleotide platforms (8). Affymetrix has recently released the SNP6.0 platform, which improves upon the density of the 500 K platform; however, no detailed CNV analysis has yet been published using this tool.

Three main long oligonucleotide platforms are commonly applied to the study of CNVs. These include the ROMA platform and commercial solutions form Nimblegen and Agilent. The ROMA platform uses a restriction digest based protocol, similar in concept to that of Affymetrix, utilizing long oligonucleotides to improve noise. This platform has been the basis of several CNV studies (10, 38).

Commercial solutions from Agilent offer the highest sensitivity of any oligonucleotide platform (1 oligonucleotide can detect an alteration). As a result, despite densities less than those offered by other platforms (max 244 K elements), the array demonstrates very high performance, detecting 36 kb alterations and mapping breakpoints to within 56 kb (33, 39). In a recent study by de Smith et al., a 185 K Agilent platform was utilized in conjunction with a targeted (see Subheading 3.1.3) 244K array to detect thousands of CNV loci (39).

Other common commercial long oligonucleotide arrays applied to CNV research are those offered by Nimblegen. They offer platforms ranging from 385 K to >2 million element arrays. Although the Nimblegen array demonstrates higher noise than some competing platforms, the high density and uniform oligonucleotide distribution allows very fine mapping of breakpoints (24 kb for the 385 K platform) (33). Additionally, the photolithography method allows rapid production of custom arrays (this is a significant benefit in the development of targeted arrays). One other significant benefit of Nimblegen is their production of tiling oligonucleotide platforms, which offer the highest resolution possible with array technology (see Subheading 3.1.4) (35).

One of the disadvantages of oligonucleotide platforms is that off the shelf platforms tend to underrepresent segmental duplications that are associated with a significant (~50%) proportion of CNVs (8, 10, 35). The advantages of oligonucleotide arrays include the consistency and improvement in commercial offerings and the incredible flexibility in array design. The resolution of oligonucleotide arrays is only limited by the number of elements that can be placed on a slide, and the ability to design higher density arrays with very precise breakpoint mapping is a significant advantage in the field of CNV research.

3.1.3. Targeted Arrays

In many cases arrays, are utilized to fine map CNV loci by covering only specific regions of the genome with high density probes. This has advantages in array multiplexing (fitting multiple arrays on one slide is cost effective) and defining very precise alteration boundaries. Although discovery studies do occasionally begin with targeted arrays, as demonstrated by Locke et al. who utilized a BAC array targeted to segmental duplications followed by fine mapping with custom Nimblegen arrays to refine breakpoints, they are mainly used to validate and fine map regions discovered with more broad platforms (35). The Nimblegen platform is the most commonly used array in this manner due to its high probe content. However, Agilent targeted arrays have also been used with great success, such as the study by de Smith et al. who used 185 K Agilent arrays to discover new CNVs and then built a targeted 244 K platform to cover known and novel CNV loci (39). Another benefit is in the production of affordable region(s)-specific chips for diagnosis of disease and screening large panels of samples.

3.1.4. Tiling Oligo Arrays

An alternative to targeted arrays is to represent the entire genome with tiled (adjacent) oligonucleotides. This is an expensive methodology and requires multiple arrays to examine a single individual. However, this allows the highest resolution of any array technology at the cost of a significant DNA requirement. Despite the high resolution, this technique does not allow the determination of structure and can perform poorly around repeat rich regions due to the short probes. As such, these arrays are mainly applied in fields outside CNV discovery such as chromatin immunoprecipitation and transcript mapping (40, 41).

3.2. SNP Strategies

After the release of the first array studies of CNVs, the HapMap consortium released its study of SNPs in a panel of normal individuals. While analyzing the data, it became apparent that many loci were deleted in the normal samples. This allowed the discovery of many novel CNVs and promises to be a useful data set in future variation analysis (42).

However, SNP genotyping has a few disadvantages in CNV detection. Deletions are less common than gains, and gains are not readily detectable in SNP data. Also, CNPs tend to have low SNP density, making analysis of linkage challenging, and many commercial array based genotyping platforms undersample regions with complex structure and variation, which are known to coincide with many CNVs (3, 8, 35). Additionally, sequence based genotyping can be an expensive protocol, especially with conventional sequencing tools.

3.3. Sequencing

The advent of new sequencing tools in the past few years has afforded the ability to use sequence based strategies to identify CNVs. Although these are expensive techniques best suited to small sample sizes, they offer several distinct advantages over array based tools. First, they are the best approach for detection of small CNVs (<10 kb); this is especially significant as Korbel et al. identified an increase in SV prevalence at smaller alteration sizes (43). Second, sequence information allows the detection of not only the extra or missing copies of DNA but also information as to their genomic arrangement and general genome structures such as inversions and SNPs (20). Thus, sequencing studies can identify more alterations than array studies.

3.3.1. Fosmid End Mapping

One sequencing strategy of note is the utilization of fosmid (10–20 kb insert clones) end sequences compared to a reference genome. By comparing mapping of fosmid ends to a reference genome vs. fosmid, size alteration >8 kb can be readily detected. Additionally, differential mapping (inverse, different locations) of fosmid ends is indicative of structural abnormalities (20).

3.3.2. Paired End Mapping

Expanding upon the concept of end sequencing are two other strategies termed paired end mapping. Unlike fosmid based strategies, these methods do not require time consuming library construction steps. Korbel et al. describe a strategy based on 3 kb DNA fragments to allow detection of small CNVs and structural variants (43). This strategy detected a significantly larger number of variations compared to previous studies, with over 1,300 differences identified between two individuals. They also observed that SV frequency increases at smaller sizes. Data was validated using a targeted tiling CGH array approach.

Another technique using smaller (400 bp) fragments was recently implemented in the analysis of cancer genomes (44). Although not yet applied to CNVs, this methodology allows detection of very small alterations and more accurate assessment of breakpoint position. However, the short fragments require significantly more sequence data to cover the same proportion of the genome as larger fragments. One of the primary benefits of short fragments is the ease of PCR amplification of the small fragments for conventional sequencing based profiling of breakpoints.

3.3.3. Complete Sequencing

The next logical progression of end sequencing is complete genome sequencing. Using new sequencing technologies, it is now possible to generate a complete sequence for an individual at a significantly reduced cost compared to conventional technologies. However, this technology currently costs around one million dollars per individual (45). Additionally, despite improvement in sequencing throughput, shotgun sequencing has inherent difficulties in assembling through repeat rich regions. This is apparent from comparisons of early Celera and hg genome builds where the clone ordered strategies identified significantly more SDs and repeat rich regions than shotgun sequence data (46). Another issue of concern that also applies to other sequencing technologies is the reliability of the reference sequence. Many SDs are defined as such due to their presence in the reference genome, and they are actually CNVs according to modern studies (8, 47). This adds a further complication to the adoption of complete resequencing at this time. Thus, it is likely that future application of sequencing in the detection of CNVs and SVs will rely on a combination of techniques, with clones used to span complex structures and shotgun sequencing to fill in the rest of the genome (46).

3.3.4. Challenges in Sequencing

Current high throughput sequencing technologies face a few challenges that will need to be addressed in order to generate a complete picture of human variation. Repeat content represents a significant portion of the genome and thus many reads will not represent informative data. This repeat content also leads to challenges inherent to shotgun sequencing where assembly can be difficult across regions of highly similar repeats (20, 43, 44, 46). This is particularly challenging when dealing with large, highly similar duplications that do not contain sufficiently divergent sequence to allow unique mapping. Although read depth can be used to identify duplications, it will be challenging to identify the precise structure of such regions.

4. Future Studies

The accelerated progress in developing new technologies for the study of CNVs has led to a rapid increase in the amount of available data. Publically accessible databases such as the Database of Genomic Variants (http://projects.tcag/variation) and the International HapMap Project (www.hapmap.org) have attempted to catalog common genetic variants that occur in human beings (1, 2). However, the field of structural variation is still immature and it is clear that current databases are by no means complete. Most arrays that have been used to identify CNVs have lacked sufficient resolution to reliably detect CNVs less than 50 kb, making small CNVs underrepresented in these databases. The use of high resolution oligonucleotide arrays and sequencing technologies have shown that most variants are actually smaller in size than is recorded, resulting in an exaggeration of the amount of the genome that is defined as structurally variant (43, 48). The increasing use of new sequencing technologies enables the precise characterization of CNV breakpoints, as well as structural alterations such as inversions, which do not affect copy number (20). Currently, these technologies range from moderately to extremely expensive, but in the future will become amenable to population based studies. A systematic effort to validate individual CNVs at an increased resolution will thus be necessary in the future to eliminate falsely annotated sites and better define the boundaries of annotated CNVs. The completion of a reliable next-generation CNV map will add a new dimension to current genomic platforms and provide a much needed baseline of benign and pathogenic CNVs. This will allow benign CNVs to be more accurately and efficiently distinguished from those that are pathogenic, facilitating a better understanding of the clinical impact of specific genomic imbalances. The diagnostic range of genomic assays will thus be improved, and it is likely to help implement their widespread use for diagnosing genetic disorders and pathogenic genomic aberrations.

Acknowledgments

This work was supported by funds from the Canadian Institutes for Health Research, Canadian Breast Cancer Research Alliance, and National Institute of Dental and Craniofacial Research (NIDCR) grant R01 DE15965.

References

1. Beckmann, J.S., Estivill, X., and Antonarakis, S.E. (2007) Copy number variants and genetic traits: closer to the resolution of phenotypic to genotypic variability. *Nat Rev Genet* **8**, 639–46.
2. Lee, C., Iafrate, A.J., and Brothman, A.R. (2007) Copy number variations and clinical cytogenetic diagnosis of constitutional disorders. *Nat Genet* **39**, S48–54.
3. Kidd, J.M., Cooper, G.M., Donahue, W.F., Hayden, H.S., Sampas, N., Graves, T., *et al.* (2008) Mapping and sequencing of structural variation from eight human genomes. *Nature* **453**, 56–64.
4. Abraham, J., Earl, H.M., Pharoah, P.D., and Caldas, C. (2006) Pharmacogenetics of cancer chemotherapy. *Biochim Biophys Acta* **1766**, 168–83.
5. Craig, D. W., and Stephan, D. A. (2005) Applications of whole-genome high-density SNP genotyping. *Expert Rev Mol Diagn* **5**, 159–70.
6. Shastry, B.S. (2007) SNPs in disease gene mapping, medicinal drug development and evolution. *J Hum Genet* **52**, 871–80.
7. Kim, P.M., Lam, H.Y., Urban, A.E., Korbel, J.O., Affourtit, J., Grubert, F., *et al.* (2008) Analysis of copy number variants and segmental duplications in the human genome: Evidence for a change in the process of formation in recent evolutionary history. *Genome Res* **18**, 1865–74.
8. Redon, R., Ishikawa, S., Fitch, K.R., Feuk, L., Perry, G.H., Andrews, T.D., *et al.* (2006) Global variation in copy number in the human genome. *Nature* **444**, 444–54.
9. Wong, K.K., deLeeuw, R.J., Dosanjh, N.S., Kimm, L.R., Cheng, Z., Horsman, D.E., *et al.* (2007) A comprehensive analysis of common copy-number variations in the human genome. *Am J Hum Genet* **80**, 91–104.
10. Sebat, J., Lakshmi, B., Troge, J., Alexander, J., Young, J., Lundin, P., *et al.* (2004) Large-scale copy number polymorphism in the human genome. *Science* **305**, 525–8.
11. Sharp, A.J., Locke, D.P., McGrath, S.D., Cheng, Z., Bailey, J.A., Vallente, R.U., *et al.* (2005) Segmental duplications and copy-number variation in the human genome. *Am J Hum Genet* 77, 78–88.

12. Ji, Y., Eichler, E.E., Schwartz, S., and Nicholls, R.D. (2000) Structure of chromosomal duplicons and their role in mediating human genomic disorders. *Genome Res* **10**, 597–610.
13. Shianna, K.V., and Willard, H.F. (2006) Human genomics: in search of normality. *Nature* **444**, 428–9.
14. Sharp, A.J. (2009) Emerging themes and new challenges in defining the role of structural variation in human disease. *Hum Mutat* **30**, 135–44.
15. Nguyen, D.Q., Webber, C., and Ponting, C.P. (2006) Bias of selection on human copy-number variants. *PLoS Genet* **2**, e20.
16. Feuk, L., Carson, A.R., and Scherer, S.W. (2006) Structural variation in the human genome. *Nat Rev Genet* **7**, 85–97.
17. Stranger, B. E., Forrest, M. S., Dunning, M., Ingle, C. E., Beazley, C., Thorne, N., *et al.* (2007) Relative impact of nucleotide and copy number variation on gene expression phenotypes. *Science* **315**, 848–53.
18. Ionita-Laza, I., Rogers, A.J., Lange, C., Raby, B.A., and Lee, C. (2008) Genetic association analysis of copy-number variation (CNV) in human disease pathogenesis. *Genomics* **93**, 22–6.
19. Neitz, M., and Neitz, J. (1995) Numbers and ratios of visual pigment genes for normal red-green color vision. *Science* **267**, 1013–6.
20. Tuzun, E., Sharp, A.J., Bailey, J.A., Kaul, R., Morrison, V.A., Pertz, L.M., *et al.* (2005) Fine-scale structural variation of the human genome. *Nat Genet* **37**, 727–32.
21. Rodrigues, N.R., Owen, N., Talbot, K., Ignatius, J., Dubowitz, V., and Davies, K.E. (1995) Deletions in the survival motor neuron gene on 5q13 in autosomal recessive spinal muscular atrophy. *Hum Mol Genet* **4**, 631–4.
22. Gonzalez, E., Kulkarni, H., Bolivar, H., Mangano, A., Sanchez, R., Catano, G., *et al.* (2005) The influence of CCL3L1 gene-containing segmental duplications on HIV-1/AIDS susceptibility. *Science* **307**, 1434–40.
23. de Vries, B.B., Pfundt, R., Leisink, M., Koolen, D.A., Vissers, L.E., Janssen, I.M., *et al.* (2005) Diagnostic genome profiling in mental retardation. *Am J Hum Genet* **77**, 606–16.
24. Hahnen, E., Forkert, R., Marke, C., Rudnik-Schoneborn, S., Schonling, J., Zerres, K., and Wirth, B. (1995) Molecular analysis of candidate genes on chromosome 5q13 in autosomal recessive spinal muscular atrophy: evidence of homozygous deletions of the SMN gene in unaffected individuals. *Hum Mol Genet* **4**, 1927–33.
25. Chen, W.J., Wu, Z.Y., Wang, N., Lin, M.T., and Mu-rong, S.X. (2005) Quantitative studies on SMN1 gene and carrier testing of spinal muscular atrophy. *Zhonghua Yi Xue Yi Chuan Xue Za Zhi* **22**, 559–602.
26. Feldkotter, M., Schwarzer, V., Wirth, R., Wienker, T.F., and Wirth, B. (2002) Quantitative analyses of SMN1 and SMN2 based on real-time lightCycler PCR: fast and highly reliable carrier testing and prediction of severity of spinal muscular atrophy. *Am J Hum Genet* **70**, 358–68.
27. Yang, Y., Chung, E.K., Wu, Y.L., Savelli, S.L., Nagaraja, H.N., Zhou, B., *et al.* (2007) Gene copy-number variation and associated polymorphisms of complement component C4 in human systemic lupus erythematosus (SLE): low copy number is a risk factor for and high copy number is a protective factor against SLE susceptibility in European Americans. *Am J Hum Genet* **80**, 1037–54.
28. Ingelman-Sundberg, M., Sim, S.C., Gomez, A., and Rodriguez-Antona, C. (2007) Influence of cytochrome P450 polymorphisms on drug therapies: pharmacogenetic, pharmacoepigenetic and clinical aspects. *Pharmacol Ther* **116**, 496–526.
29. Meijerman, I., Sanderson, L.M., Smits, P.H., Beijnen, J.H., and Schellens, J.H. (2007) Pharmacogenetic screening of the gene deletion and duplications of CYP2D6. *Drug Metab Rev* **39**, 45–60.
30. Hegele, R.A. (2007) Copy-number variations and human disease. *Am J Hum Genet* **81**, 414–5.
31. Goossens, M., Dozy, A.M., Embury, S.H., Zachariades, Z., Hadjiminas, M.G., Stamatoyannopoulos, G., and Kan, Y.W. (1980) Triplicated alpha-globin loci in humans. *Proc Natl Acad Sci U S A* **77**, 518–21.
32. Iafrate, A.J., Feuk, L., Rivera, M.N., Listewnik, M.L., Donahoe, P.K., Qi, Y., *et al.* (2004) Detection of large-scale variation in the human genome. *Nat Genet* **36**, 949–51.
33. Coe, B.P., Ylstra, B., Carvalho, B., Meijer, G.A., Macaulay, C., and Lam, W.L. (2007) Resolving the resolution of array CGH. *Genomics* **89**, 647–53.
34. Kehrer-Sawatzki, H. (2007) What a difference copy number variation makes. *Bioessays* **29**, 311–3.
35. Locke, D.P., Sharp, A.J., McCarroll, S.A., McGrath, S.D., Newman, T.L., Cheng, Z., *et al.* (2006) Linkage disequilibrium and

heritability of copy-number polymorphisms within duplicated regions of the human genome. *Am J Hum Genet* **79**, 275–90.

36. Hoyer, J., Dreweke, A., Becker, C., Gohring, I., Thiel, C.T., Peippo, M.M., *et al.* (2007) Molecular karyotyping in patients with mental retardation using 100K single-nucleotide polymorphism arrays. *J Med Genet* **44**, 629–36.
37. Wagenstaller, J., Spranger, S., Lorenz-Depiereux, B., Kazmierczak, B., Nathrath, M., Wahl, D., *et al.* (2007) Copy-number variations measured by single-nucleotide-polymorphism oligonucleotide arrays in patients with mental retardation. *Am J Hum Genet* **81**, 768–79.
38. Sebat, J., Lakshmi, B., Malhotra, D., Troge, J., Lese-Martin, C., Walsh, T., *et al.* (2007) Strong association of de novo copy number mutations with autism. *Science* **316**, 445–9.
39. de Smith, A.J., Tsalenko, A., Sampas, N., Scheffer, A., Yamada, N. A., Tsang, P., Ben-Dor, A., Yakhini, Z., Ellis, R.J., Bruhn, L., Laderman, S., Froguel, P., and Blakemore, A.I. (2007) Array CGH analysis of copy number variation identifies 1284 new genes variant in healthy white males: implications for association studies of complex diseases. *Hum Mol Genet* **16**, 2783–94.
40. Bertone, P., Stolc, V., Royce, T.E., Rozowsky, J.S., Urban, A.E., Zhu, X., *et al.* (2004) Global identification of human transcribed sequences with genome tiling arrays. *Science* **306**, 2242–6.
41. Euskirchen, G.M., Rozowsky, J.S., Wei, C.L., Lee, W.H., Zhang, Z.D., Hartman, S., *et al.*. (2007) Mapping of transcription factor binding regions in mammalian cells by ChIP: comparison of array- and sequencing-based technologies. *Genome Res* **17**, 898–909.
42. McCarroll, S.A., Hadnott, T.N., Perry, G.H., Sabeti, P.C., Zody, M.C., Barrett, J.C., *et al.* (2006) Common deletion polymorphisms in the human genome. *Nat Genet* **38**, 86–92.
43. Korbel, J.O., Urban, A.E., Affourtit, J.P., Godwin, B., Grubert, F., Simons, J.F., *et al.* (2007) Paired-end mapping reveals extensive structural variation in the human genome. *Science* **318**, 420–6.
44. Campbell, P.J., Stephens, P.J., Pleasance, E.D., O'Meara, S., Li, H., Santarius, T., *et al.* (2008) Identification of somatically acquired rearrangements in cancer using genome-wide massively parallel paired-end sequencing. *Nat Genet* **40**, 722–9.
45. Wheeler, D.A., Srinivasan, M., Egholm, M., Shen, Y., Chen, L., McGuire, A., *et al.* (2008) The complete genome of an individual by massively parallel DNA sequencing. *Nature* **452**, 872–6.
46. She, X., Jiang, Z., Clark, R.A., Liu, G., Cheng, Z., Tuzun, E., *et al.* (2004) Shotgun sequence assembly and recent segmental duplications within the human genome. *Nature* **431**, 927–30.
47. Cheung, J., Estivill, X., Khaja, R., MacDonald, J.R., Lau, K., Tsui, L.C., and Scherer, S.W. (2003) Genome-wide detection of segmental duplications and potential assembly errors in the human genome sequence. *Genome Biol* **4**, R25.
48. Perry, G.H., Ben-Dor, A., Tsalenko, A., Sampas, N., Rodriguez-Revenga, L., Tran, C.W., *et al.* (2008) The fine-scale and complex architecture of human copy-number variation. *Am J Hum Genet* **82**, 685–95.
49. Patel, P.I., Roa, B.B., Welcher, A.A., Schoener-Scott, R., Trask, B. J., Pentao, L., *et al.* (1992) The gene for the peripheral myelin protein PMP-22 is a candidate for Charcot-Marie-Tooth disease type 1A. *Nat Genet* **1**, 159–65.
50. Szatmari, P., Paterson, A.D., Zwaigenbaum, L., Roberts, W., Brian, J., Liu, X.Q., *et al.* (2007) Mapping autism risk loci using genetic linkage and chromosomal rearrangements. *Nat Genet* **39**, 319–28.
51. Potocki, L., Bi, W., Treadwell-Deering, D., Carvalho, C.M., Eifert, A., Friedman, E.M., *et al.* (2007) Characterization of Potocki-Lupski syndrome (dup(17)(p11.2p11.2)) and delineation of a dosage-sensitive critical interval that can convey an autism phenotype. *Am J Hum Genet* **80**, 633–49.
52. Liu, H., Heath, S.C., Sobin, C., Roos, J.L., Galke, B.L., Blundell, M.L., *et al.* (2002) Genetic variation at the 22q11 PRODH2/DGCR6 locus presents an unusual pattern and increases susceptibility to schizophrenia. *Proc Natl Acad Sci U S A* **99**, 3717–22.
53. Fellermann, K., Stange, D.E., Schaeffeler, E., Schmalzl, H., Wehkamp, J., Bevins, C.L., *et al.* (2006) A chromosome 8 gene-cluster polymorphism with low human beta-defensin 2 gene copy number predisposes to Crohn disease of the colon. *Am J Hum Genet* **79**, 439–48.

Chapter 7

A Short Primer on the Functional Analysis of Copy Number Variation for Biomedical Scientists

Michael R. Barnes and Gerome Breen

Abstract

Recent studies have highlighted the potential prevalence of copy number variation (CNV) in mammalian genomes, including the human genome. These studies suggest that CNVs may play a potentially important role in human phenotypic diversity and disease susceptibility. Here, we consider some of the in silico challenges of characterizing genomic structural variants. While the phenotypic impact of the vast majority of CNVs is likely to be neutral, some CNVs will clearly impact phenotype. Here, we review some of the key databases hosting CNV data and discuss some of the caveats in the analysis of CNV data. The task is now to translate some of the initial associations between CNVs and disease into causal variants.

Key words: Genome, CNV, Deletion, Duplication, Copy number, Bioinformatics, Variation, CNV

1. Introduction

Genomic copy number variation (CNV) has the potential to impact gene expression and translation with significant phenotypic and functional impact. This has been demonstrated in diverse species, including humans (1–3). The new generations of high throughput array-based techniques that are currently being applied to genome wide association studies (GWAS) (4) are also being used to screen for structural variants. These studies have begun to identify candidate loci harbouring copy variants that are associated not only with known genomic disorders (for example, Velo–Cardiofacial Syndrome (5)), but also with complex diseases (6). Arguably, many CNV studies in complex diseases are somewhat inconclusive, as most have identified rare copy number variants in disease subjects, but the causality of these variants cannot be surely determined, either because they are de novo events or

Michael R. Barnes and Gerome Breen (eds.), *Genetic Variation: Methods and Protocols*, Methods in Molecular Biology, vol. 628,
DOI 10.1007/978-1-60327-367-1_7,

there are no family subjects to study the segregation of CNV and phenotype. In addition to this, while the common disease-common variant model of complex disease has been the prevailing hypothesis for the recent past, there is mounting support for the idea that at least for some loci, rare variants collectively may contribute to common disease phenotypes (7, 8). CNVs appear to be lending a great deal of support to this hypothesis. Moreover, each disease gene may contain a spectrum of common alleles with low effect sizes and rarer alleles with larger effect sizes, perhaps even large enough for diagnostic utility. This situation presents a complex range of challenges for geneticists and informaticians – to translate initial associations between variants and phenotypes into proven causal variants.

DNA copy number variants (CNVs) have been known to geneticists for a long time in various forms. At the most basic level, a CNV is defined as a stretch of DNA larger than 1 kb that displays copy number differences within normal populations (9). It is now known that submicroscopic structural variation of the human genome is also common (10). This results in the deletion, duplication, and rearrangement of parts of genes, whole genes or even multi-gene regions. Where CNV events lead to the duplication of entire genes, this may lead to a difference in "gene dosage" between individuals with consequent effects on protein expression and overall function, which might provide a salient explanation of some complex phenotypes (11). There is also evidence that genes in CNV regions are expressed at lower and more variable levels than genes mapping elsewhere, which actually manifests as an extended global influence on the transcriptome (2). Analysis of CNVs has recently become feasible at the whole genome level, and we now know of several thousand CNVs that probably represent the larger end of structural variation (10) (http://projects.tcag.ca/variation). Smaller variants, e.g. ~2–20 kb, also appear to be common (12). The recent explosion of publications describing common CNVs across the genome was largely unanticipated by the initial analysis of the human genome, first by sequencing, and then later by SNP genotyping. Subsequent analyses using different SNP densities have shown that higher density arrays can lead to a 30-fold enrichment of short-CNV regions (13) and smaller elements are likely to have been artificially collapsed (and hidden) during bioinformatic assembly of the genome. This improved the understanding of the nature of CNVs in the genome that has lead to further improvements in the design of the latest SNP genotyping arrays so that most CNVs >20 Kb are in the range of detection (14, 15).

2. Interpretation of Lab CNV Data

2.1. Early Phase: Data Mining Methods

Now that most genotyping high density genotyping arrays are suitable for the detection of CNVs, it is possible to mine genotyping data for CNV information; this involves detection of relative loss or gain of signal intensity, representing a possible underlying loss or gain of genetic material. Initially, a normalisation step is required to remove any systematic differences in the observed signal intensities. Systematic differences may be caused by variation in DNA quality between samples, making this a particularly important step when samples come from a wide variety of sources. Calculation of relative gain or loss of material between samples after normalisation requires measurement of signal intensity for each probe/allele for a given genotype within a given window, which can then be used to call CNV status in a given region. Hidden Markov models may be used to optimise CNV calling rates (16). Many other CNV calling algorithms exist (17, 18), each slightly different, which can lead to different rates of calling. In addition, the analysis can be confounded by the quality of the initial sample and the limitations of the microarray such as sensitivity (9). One solution is to use multiple programs for consensus CNV calling (see Note 1).

2.2. Mid Phase: Putative CNV Identification and Statistical Analysis

As outlined above, the identification of CNVs salient to the explanation of a complex disease can be difficult. Lack of standardisation of "normal" samples analysed for structural genomic variation has proven to be problematic; however, there have been some encouraging findings in Autism and Schizophrenia that have used a combination of methods for more rigorous detection (19, 20). Figure 1 illustrates a typical result from a CNV calling algorithm, with red and green areas denoting loss and gain of genetic material, respectively. Yau and Holmes (21) offer a good review of statistical approaches for calling CNVs from SNP genotype data.

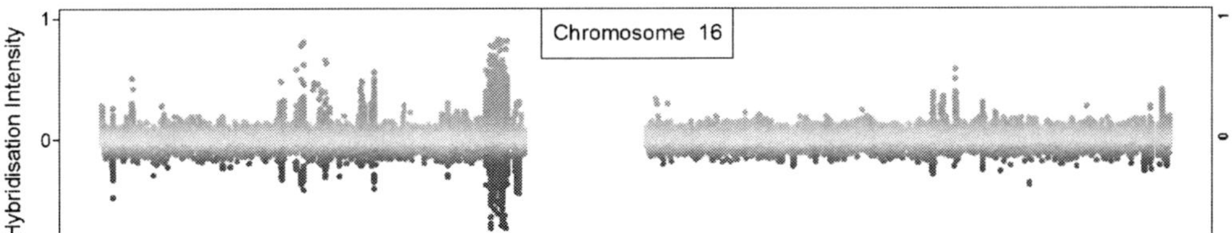

Fig. 1. A typical result of CNV analysis where negative scored (red) and positive scored (green) areas would denote loss and gain of genetic material, respectively

2.3. Late Phase: Validation of Putative CNVs

As CNV calling from SNP genotypes may be prone to error, if possible, it is important to seek an independent method for validation of the CNV. Quantitative PCR approaches looking directly for allele loss or gain may represent the best balance between efficiency and accuracy. Specifically, quantitative multiplex PCR of short fragments (QMPSF) (22) or SYBR-Green I-based real time quantitative PCR (qPCR) with controls at appropriate loci (23). Other avenues may include MAPH (Multiplex Amplifiable Probe Hybridisation) and MLPA (Multiplex Ligation-dependent Probe Amplification) (24).

2.4. De Novo Versus Segregating Deletions

Identification of a CNV in a disease subject is not sufficient to prove causality. Many CNVs may arise as the result of de novo events, in the absence of family data it is difficult to show evidence of a founder effect; however, if the boundaries of a CNV are the same between apparently unrelated individuals in a population, a founder effect is a possibility. Sometimes, a CNV will be seen to clearly segregate between affected and unaffected family members. In the case of large, rare, CNVs with delirious phenotypic impact, strong negative selection pressure (possibly leading to reduced fecundity) would probably lead to rapid removal of the CNV from the population. It is also interesting to note that in some diseases the nature of CNVs may be sex-specific. For example, CNVs associated with Autism show a tendency to familial segregation in Males, while in Female Autistic subjects, observed CNVs are most commonly de novo in origin (25). The reasons for this are unknown.

2.5. Considerations of Sample Quality in CNV Analysis

As described earlier, sample quality can play a critical role in CNV analysis. Whole genome amplification of samples may pose problems for CNV calling, in particular. In a recent study, Pugh et al. (26) extracted DNA from fresh frozen tissue samples, and carried out whole genome amplification and analysis on the Affymetrix 500 k SNP array platform and then performed CNV analysis. They showed that the amplification procedure introduces hundreds of potentially confounding CNV artifacts. They were able to correct for these artifacts, by pair-wise comparison of amplified products, this considerably reduced the number of apparent artifacts and partially rescued the ability to detect real CNVs. Their results that suggest the whole genome amplified material is appropriate for copy number analysis within sample designs. In addition, differing DNA source material, DNA extraction methods and other batch effects have all been shown to cause difficulties in calling CNVs. Sample degradation can also lead to spurious results, for example, formalin-fixed paraffin-embedded (FFPE) DNA samples can yield spurious CNV calls in real-time qPCR assays (27). Cukier et al. (28) also reported that sample degradation under standard laboratory storage conditions generates a significant increase in false-positive CNV results. They suggested that biased degradation might occur among certain genomic regions, further emphasizing the need to assess sample integrity before analysis.

2.6. Miscellaneous Observations That Might Suggest the Need for CNV Analysis

Data generated during the course of non-CNV focused genomic studies can sometimes yield results, which indicated the need for CNV analysis. Before carrying out this analysis, it may be worth consulting pre-existing resources describing CNV identified in population-based studies (http://projects.tcag.ca/variation, Table 1).

2.6.1. Extended Homozygosity

At the simple level of genotypes, a heterozygous CNV deletion would manifest as a stretch of homozygous genotype calls. Observation of extended homozygosity across kilobases or even

Table 1
Tools for genomic characterisation of CNV data

Tool	URL
Genome visualisation	
UCSC Genome Browser	genome.ucsc.edu
ENSEMBL	www.ensembl.org
NCBI MapViewer	www.ncbi.nlm.nih.gov/mapview/map_search.cgi/
LD and haplotype data	
HapMap website	www.hapmap.org
HapMap Genome Browser	www.hapmap.org/cgi-perl/gbrowse/gbrowse/
SNAP	www.broad.mit.edu/mpg/snap/
Structural genome analysis	
Db of Genomic Variants	projects.tcag.ca/variation/
Structural Variation db	Humanparalogy.gs.washington.edu/
Evaluating disrupted gene function	
GNF SymAtlas	symatlas.gnf.org/SymAtlas/
HUGE Navigator	www.hugenavigator.net/
Stanford SOURCE	source.stanford.edu
DAVID Bioinformatics	david.abcc.ncifcrf.gov/home.jsp
PANTHER	www.pantherdb.org/
Expasy Proteomics tools	www.expasy.ch/tools/
STITCH (Pathways)	stitch.embl.de/
ESE finder	rulai.cshl.edu/tools/ESE/
UniProt	www.uniprot.org

megabases of sequence might suggest the presence of a CNV; however, extended homozygosity is not a particularly uncommon phenomenon in the genome. In a study of 1411 Caucasian subjects, Curtis et al. (29) showed that regions of extended homozygosity over 1 Mb are common, with an average of 35.9 occurring per subject, and containing an average of 73 homozygous markers.

2.6.2. Shotgun Sequence Assembly

One of the earliest hints to the genome-wide scale of copy number variation was observed when Celera assembled the human genome using a shotgun sequencing method. When sequences were assembled, they observed regions of the genome with much deeper coverage by shotgun sequence reads. Although this might be explained by a methodological bias of the shotgun sequencing process, it quickly became apparent that these regions might represent copy number variations (30). This information was later used to create a "Segmental Duplications" track at the UCSC genome browser (Table 1).

3. Interpretation of CNVs in a Genomic Context

3.1. Differentiation of Neutral Polymorphic CNVs from Pathological CNVs

When a region of DNA is gained or lost in a CNV event, there is potential for both direct and indirect impact on many genes, with a subsequent impact on cellular function. Deletion or duplication of a region of DNA can obviously impact a gene, which is fully contained within a polymorphic region; however, in many cases deletions will be restricted to introns of genes, or they will delete one or more exons of a gene. When a CNV directly disrupts an exon, the loss of one of the splice sites usually leads to premature termination at the first available in-frame stop codon. A transcript may be transcribed from the truncated gene locus, possibly leading to the translation of a truncated protein product; however, there is a good chance that the prematurely terminated transcript will be destroyed in vivo by the process of nonsense mediated decay (31). A deletion or duplication of one or more entire exons could lead to the production of a functional transcript, depending on the coding phase of the remaining exons.

3.1.1. The Influence of Exon Phase and Alternative Splicing on CNV Impact

Alternative splicing in eukaryotes is a process that enables multiple (multifunctional) gene products to be encoded by a single gene locus. Alternative splicing is also an important gene regulatory mechanism, which is estimated to be employed in more than 90% of multi-exon human genes (32). Evaluation of the potential splice variants of a gene is potentially useful for the evaluation of CNV impact. For a given gene, it is possible to identify all theoretically possible exon–exon junctions by considering all exons, which could be spliced together while keeping the frame of translation.

The frame of translation of an exon is known as the exon phase. Figure 2 illustrates the concept of exon phase. If an exon is spliced after the third base of the codon triplet, it is termed phase 0, after the first base it is termed phase 1, and after the second base it is termed phase 2. If a CNV deletes or duplicates on or more entire exons, the remaining exons could still be spliced into a functional transcript if the splice sites flanking the CNV are in the same phase. Figure 3, illustrates some of the potential permutations that might be observed with different CNVs.

The impact of CNVs on gene function can vary widely. At one extreme, deletion of an intronic region would generally be expected to benign, although the possibility of deleting regulatory elements in intronic regions should not be discounted, also duplication of an intronic region might alter the efficiency of splicing.

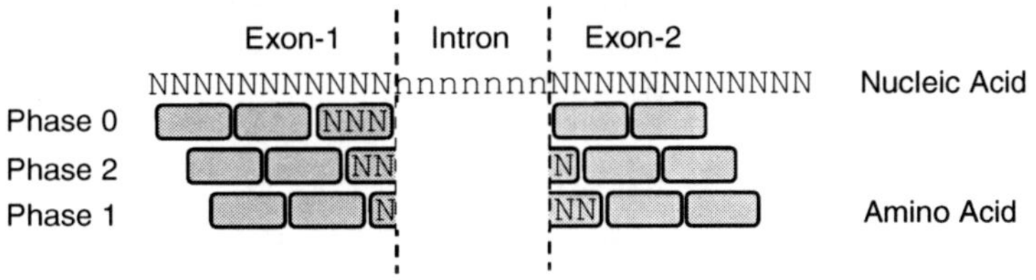

Fig. 2. Illustration demonstrating the concept of exon phase. If an exon is spliced after the third base of the codon triplet, it is termed phase 0, after the first base it is termed phase 1, and after the second base it is termed phase 2. If a CNV deletes or duplicates on or more entire exons, the remaining exons could still be spliced into a functional transcript if the splice sites flanking the CNV are in the same phase

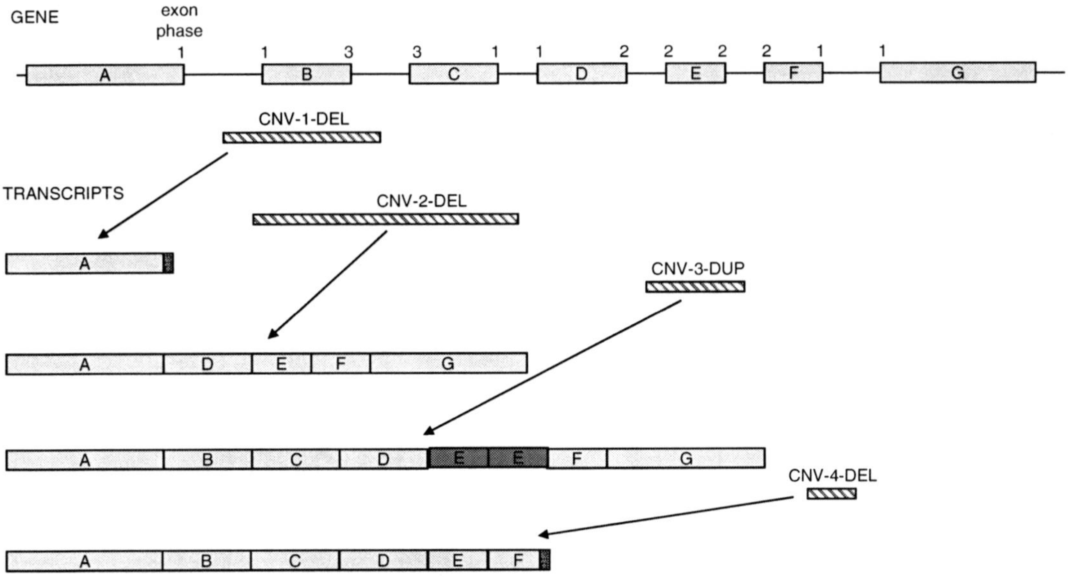

Fig. 3. Illustration of some of the potential permutations of CNV impact on gene transcription that might be observed with CNVs impacting different locations

Deletions or duplications might be modestly deleterious, e.g. if an alternatively spliced exon is deleted, the remainder of the gene and all the other splice variants would probably remain functional. However, if the deleted exon constitutes a critical splice variant to a particular process, the impact might be more deleterious. At other extremes, the deletion of whole or partial exons early in the gene can lead to a completely non-functional transcript bearing premature termination codons. Deletion or duplication of one or more exons would generally be expected to be deleterious, although consideration should be made to the likely composition of a transcript transcribed from the remaining exons. Figure 4 considers some of the questions that need to be considered to evaluate the impact of a CNV on gene function.

3.2. Case Study: Analysis of CNVs in the Neurexin-1 Gene

To illustrate the detailed process for the analysis of CNV impact on gene function, we use a previous study from our laboratory (33). In the study, the Neurexin-1 gene, NRXN1, was screened for CNVs in 2,977 schizophrenia patients and 33,746 controls from seven European populations using high density SNP genotype data. The study identified 66 deletions and 5 duplications, including several de novo deletions. 12 deletions and 2 duplications occurred in cases (0.47%) compared to 49 and 3 (0.15%) in controls. The NRXN1 gene encodes two major isoforms, Neurexin-1 alpha (NRXN1a) and Neurexin-1 beta (NRXN1b), from alternative

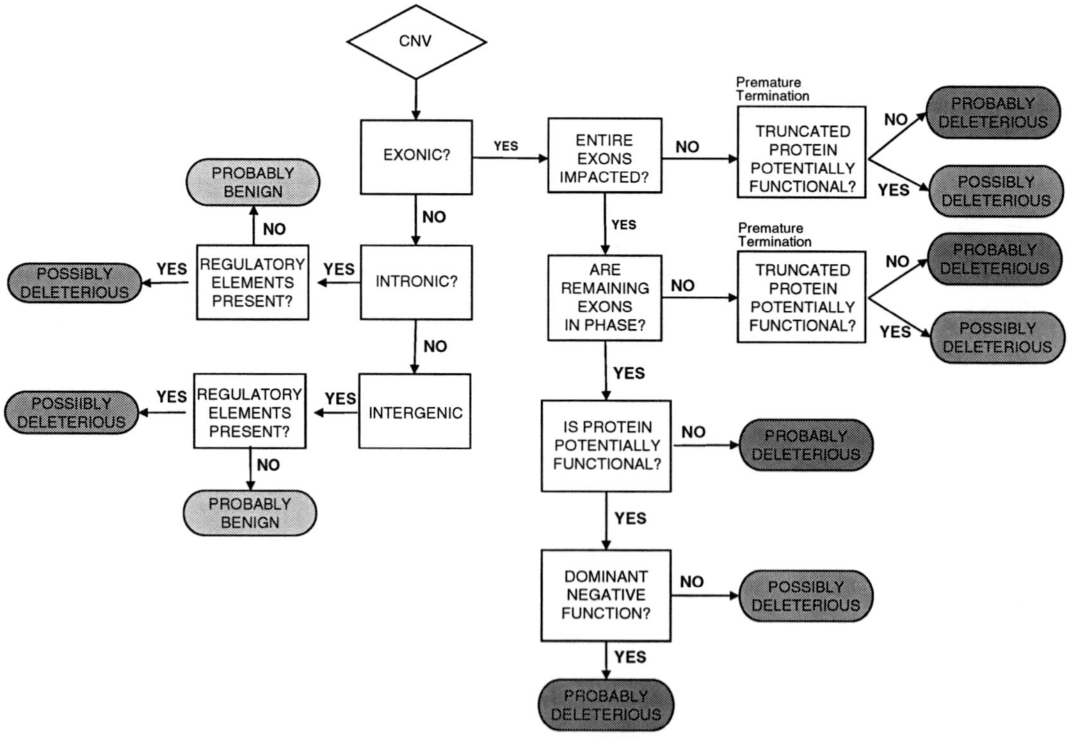

Fig. 4. A decision tree for the analysis of the functional impact of a CNV

first exons with distinct promoters. In order to review the location of the CNVs, we presented the information a genomic context by loading the novel CNVs as a UCSC custom track (see Note 2) (Fig. 5). Analysis of the location of the CNVs revealed no common breakpoints and the CNVs varied from 18 kb to 420 kb. No direct association was seen between CNVs and the phenotype ($P=0.13$; OR = 1.73; 95% CI 0.81–3.50). However, as the functional impact and penetrance of CNVs could vary according to the location of the CNV, an analysis was carried out restricted to CNVs that disrupted exons (0.17% of cases and 0.020% of controls). In this case, several CNVs seen in both cases and controls disrupted the first two or three exons of the Neurexin-1 alpha isoform. When analysed separately, exon disrupting CNVs were significantly associated with schizophrenia with a high odds ratio ($P=0.0027$; OR = 8.97, 95% CI 1.8–51.9). From this, we concluded that among the heterogeneous CNVs in NRXN1, exonic deletions conferred a significant risk of schizophrenia (33).

This study is a very good example of how CNV impact on gene function can vary according to gene location and splicing complexity. In addition to the two major isoforms, Neurexin-1 has previously been shown to regulate by a complex range of alternative splicing. Six splice sites are used alternatively, with the potential to generate over 2,000 Neurexin splice variants, encoding variant extracellular domains, with varying function (34). In this context, the nature of impact of a deletion of the first few exons of NRXN1a is not clear. What is clear is that only one CNV (Verona-1) disrupted the NRXN1b isoform, so we expect that the expression of this isoform with its distinct promoter is likely to be unaffected by most of the CNVs. After reviewing all available genomic data across the NRXN1 locus, we also identified the evidence of alternative NRXN1 splice variants and identified EST

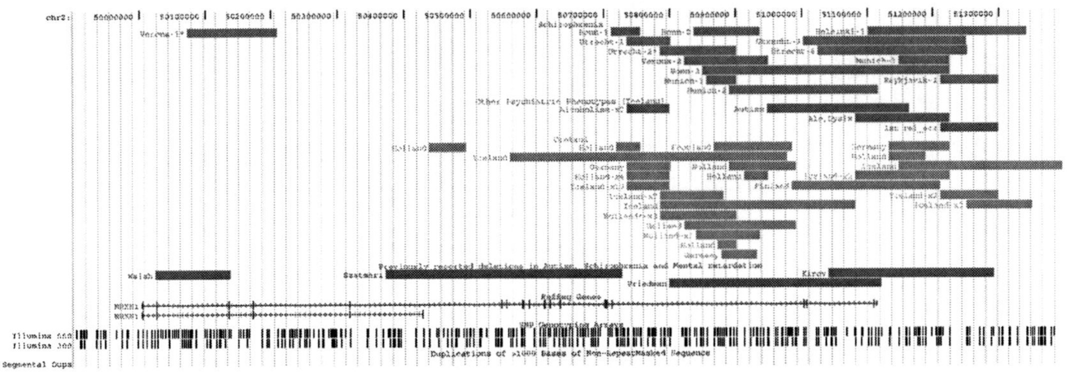

Fig. 5. UCSC browser output showing the positions of the 2p16.3 CNVs from previous studies relative to the NRXN1 gene, and the CNVs discovered in the study by Rujescu et al. (33). The majority of the discovered CNVs are deletions and asterisks indicate duplications. Markers from the Illumina 300 K and 550 K arrays, segmental duplications of >1,000 bp as well as LD structure of the Hapmap CEU sample (r^2) is also shown

and cDNA transcript evidence to support the possible presence of two additional isoforms, which share many exons with NRXN1a, which we termed NRXN1a-2 and NRXN1a-3. These were loaded as a UCSC custom track and are displayed in Fig. 6.

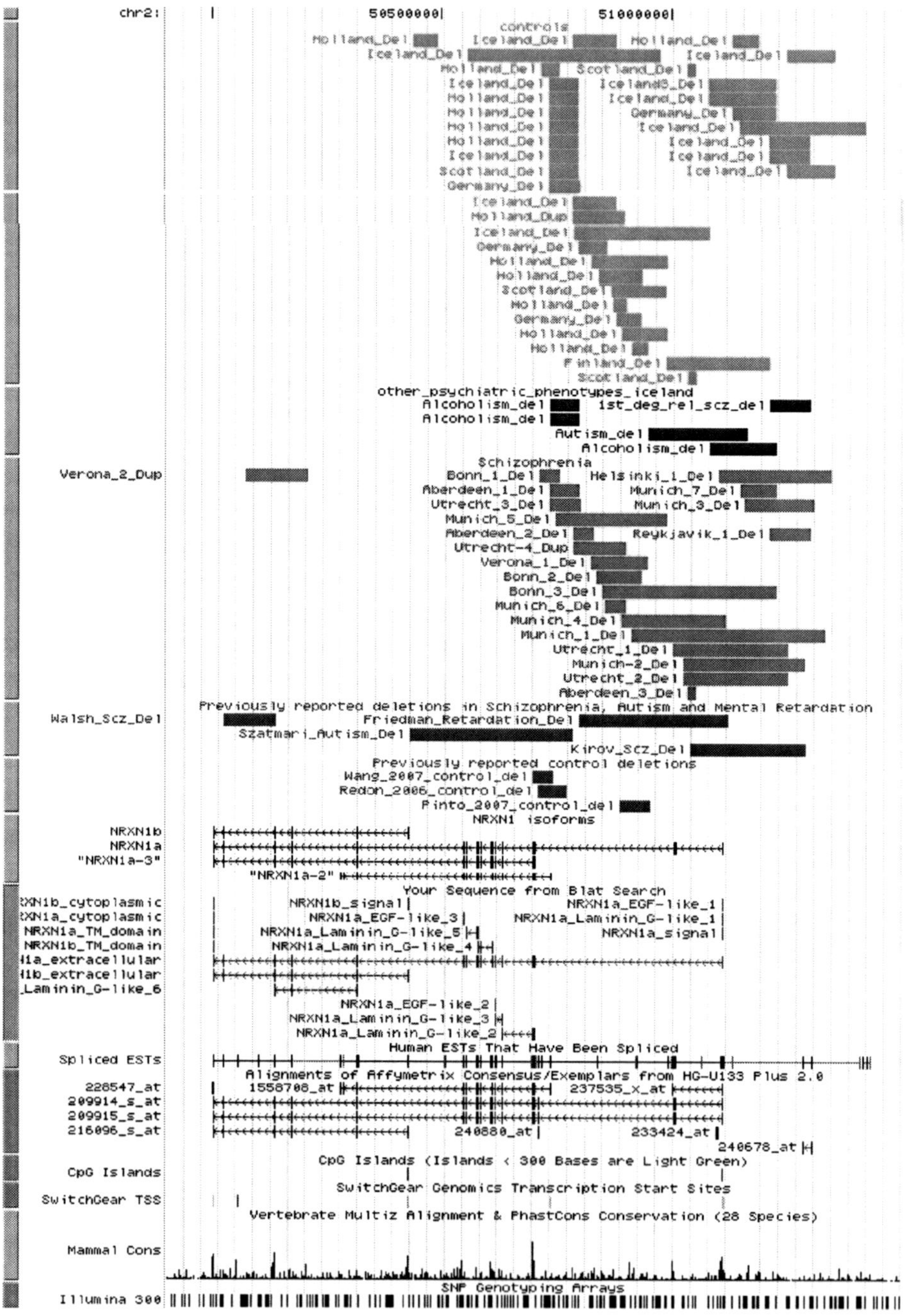

Fig. 6. UCSC browser output showing the positions of exon-disrupting CNVs discovered in the study by Rujescu et al. (33) relative to previous studies and known. The four putative Neurexin isoforms are shown below the deletions, along with protein domains aligned to genomic sequence

It is possible that the expression of these alternative NRXN1 isoforms may compensate for the loss of the major NRXN1a transcript. The NRXN1a-2 isoform shares with exons 6–14 of NRXN1a, but employs alternative first and last exons. These alternative exons allow for distinct expression analysis of the NRXN1a-2 isoform, using probe 1558708_at on the Affymetrix HG-U133_Plus_2.0 GeneChip (see Note 3). This shows a very low expression of NRXN1a-2, predominantly in the brain, compared to much higher brain specific expression of NRXN1a and NRXN1b (33). An understanding of the protein domain structure of these isoforms would help to infer their function; therefore, using the sequence data from the UCSC genome browser, we reconstructed the sequences of each of the putative isoforms (see Note 4). We then used the SMART domain annotation tool to annotate known functional domains on the translated protein sequence (see Note 5). In Fig. 7 the SMART predicted domain structures of each of the Neurexin isoforms are shown. Analysis of the predicted protein sequences from these isoforms reveals that the NRXN1a-2 isoform has no signal peptide and no transmembrane domain; however, it does contain Laminin G domains 2–4 and EGF domains 2 and 3. As most known Neurexin interactions are extracellular, it is unclear if NRXN1a-2 has any function, although if it is secreted, it may exert a dominant negative effect by competitive binding with NRXN1a interactors, alternatively improper cellular localisation of NRXN1a-2 may also be deleterious. It is interesting that the first exon of NRXN1a-2 seems to be deleted in the small founder deletion observed in 24 subjects most of which were controls. Considering the very low levels of expression of NRXN1a-2 and its frequent deletion in controls, it seems likely that NRXN1a-2 may have a very limited biological role in Neurexin signalling. However, activity of the NRXN1a-2 isoform may become significant against a background of deletion of the other Neurexin isoforms. In such circumstances, even low-level dominant negative activity may become biologically relevant.

It is difficult to evaluate the expression of the NRXN1a-3 isoform, as it shares all exons with exons 4–18 of NRXN1a, making differentiation of probes impossible. There is a high level of mammalian conservation before the first exon of NRXN1a-3, and there is also EST evidence to support the existence of this isoform, although these ESTs cannot be clearly differentiated from NRXN1a transcripts. The putative initiation methionine of the NRXN1a-3 isoform correlates with NRXN1a Met321 roughly a third of the way into Laminin G domain 2 of the protein (Fig. 7). This truncated domain may not be functional as it has lost a critical Ca2+ binding residue Asp306, which also mediates interactions with Neurexophilins (34). Loss of Neurexophilin interactions may be particularly relevant to schizophrenia pathology, as Neurexophilin-3 knockout mice show no anatomical defects, but

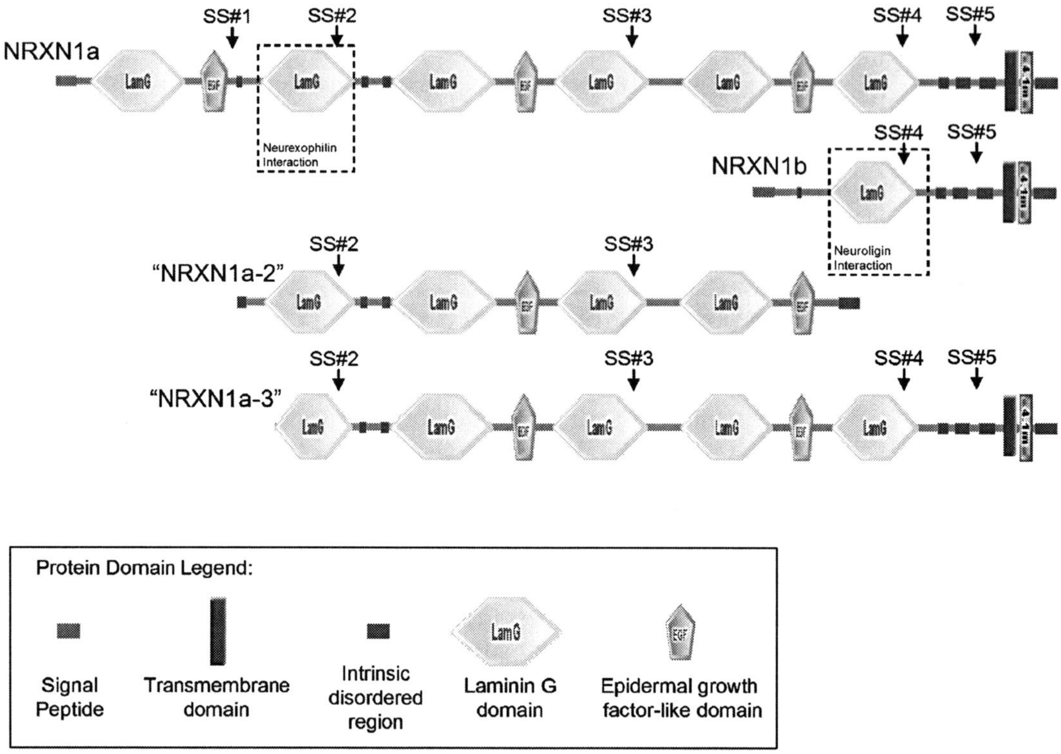

Fig. 7. Protein domain organisation of Neurexin 1 isoforms. NRXN1a contains an N-terminal signal peptide, six laminin G (LamG) domains, three epidermal growth factor-like (EGF) domains, a transmembrane domain, and a short cytoplasmic domain. NRXN1b contains a different signal peptide, but shares the last LamG domain, transmembrane domain and cytoplasmic domain with NRXN1a. NRXN1a-2 shares the second, third, fourth and fifth LamG domains and the second and third EGF domains with NRXN1a, but does not contain a signal peptide or transmembrane domain. NRXN1a-3 has no signal peptide and has a truncated version of the second LamG domain but shares all remaining domains with NRXN1a. The five regions in the NRXN1 gene where alternative splicing occurs, leading to insertion or deletion of amino acids, are indicated by arrows (SS#1–5). Protein domain annotation was generated using SMART (http://smart.embl-heidelberg.de/), using swissprot accessions Q9ULB1 (NRXN1a) and P58400 (NRXN1b) and translations from the genbank transcript AK093260 (NRXN1a-2), and Ensembl transcript ENST00000331040 (NRXN1a-3)

remarkable functional abnormalities in sensory information processing and motor coordination evident by increased startle response, reduced prepulse inhibition, and poor rotarod performance (35). One could speculate that the disruption of NRXN1/ Neurexophilin interactions, which might be seen in the exonic deletions reported in this study could explain some of the pathology of schizophrenia.

In addition to all the potential that CNVs have for the disruption of splicing and domain structure, the association between NRXN1 deletions and schizophrenia could also result from haploinsufficiency of the gene. Rujescu et al. (33) saw the strongest association when analysis was conditioned on exon-disrupting CNVs. The most common and smallest deletion of intron-only sequence showed a founder effect and was thus unlikely to be under strong negative selection pressure. It may be that all the deletions identified at the NRXN1 locus have the same biological effect – disrupting the function of NRXN1. Intron-only deletions could possibly disrupt splicing or regulatory elements, such as exonic splicing enhancer sequences, which despite the name can occur in exons or introns (see Note 6). Ultimately, in silico analysis can only hint at the possible functional impact of a CNV, and the best evidence will always require further laboratory analysis, e.g. to analyse mRNA and protein expression to evaluate the relative allelic expression of non-deleted exons in NRXN1 mRNA.

3.3. Using Pathway Analysis to Distinguish Putative Disease Causing CNVs

As discussed previously, one of the major difficulties in CNV analysis is the determination of causality in the absence of data to support the segregation of the CNV with phenotype. A number of publications have now reported genome-wide surveys of CNVs in different disease populations, including Schizophrenia (6, 8) and Autism (25). Identification of a large gene disrupting CNV in a subject with a disease phenotype does not in any way prove causality of the CNV. In principle, given a large selection of CNVs identified in subjects with a specific phenotype, one might expect to see some enrichment of pathways and cellular processes, which are biologically relevant to the disease compared to CNVs identified in control subjects. This assumption appears to hold up in some studies. For example, in their study of CNVs in Schizophrenic subjects, Walsh et al. (6) showed that genes disrupted by structural variants in cases were significantly overrepresented in pathways important for brain development, including neuregulin signalling, extracellular signal-regulated kinase/mitogen-activated protein kinase (ERK/MAPK) signalling, synaptic long-term potentiation, axonal guidance signalling, integrin signalling, and glutamate receptor signalling. Genes disrupted in controls were not overrepresented in any pathway. They were able to compare genes disrupted in cases versus those disrupted in controls using PANTHER (Table 1). Another good public domain tool, which can be used for this purpose is the DAVID Bioinformatics resource (see Table 1). All these programs are well validated in the field of gene expression (see Note 7) and enable the user to determine in an undirected manner if an experimentally derived set of genes are overrepresented in functionally defined pathways and molecular processes, as compared with all known genes (36). Similar methods are currently being applied to the interpretation of SNP-based whole genome association studies of complex disease with some success (37), so we expect these methods to be equally fruitful in the field of CNV analysis.

3.4. Conclusions

This review has highlighted some of the issues that may present to researchers studying copy number variation both at the bench and in silico. Our case study evaluated the potential impact of copy number variation on Neurexin function in schizophrenia. Analysis of a CNV in the full context of the data annotated by tools like the UCSC genome browser can immediately cast new light on the molecular nature of a CNV. Review of the data across the region helped to predict the full impact of CNVs on Neurexin function identifying a number of routes of further investigation in the lab. On a genome-wide scale there is emerging evidence of wide spread involvement of CNVs in the pathology of many common diseases; however, a full understanding of the true impact of CNVs on human disease is probably some way off.

4. Notes

1. As knowledge of CNVs evolves, so do the CNV calling algorithms, so we recommend that the most appropriate platforms for the analysis of CNV data should be re-reviewed at the time of the analysis.
2. For help with creating and viewing user defined data with the UCSC genome browser, see the excellent UCSC Custom track help documentation: http://genome.ucsc.edu/goldenPath/help/customTrack.html
3. Expression of different splice variants and gene isoforms can sometimes be evaluated using public domain expression data. The UCSC genome browser maps probes from the Affymetrix HG-U133 Plus 2.0 gene expression analysis chip. This track can be used to identify probes, which hybridise to specific splice variants or isoforms. The expression of individual probes can be evaluated with expression analysis tools like GNF Symatlas (Table 1).
4. The UCSC genome browser has a very useful DNA export function, which can be used to export and annotate genomic sequence using EST and mRNA data. To use the DNA export tool, first go to the locus of interest, then press the "DNA" link at the top of the page. This returns an export page. To just export the sequence across the locus, press the "get DNA" button. To annotate the sequence further, press the "extended DNA options" button. This should now display all the currently visible tracks, allowing the user to annotate any data to the sequence. To return annotated exons and CNV sequences, select the desired tracks using bold, underline or italic, and press the submit button. This returns the fully annotated sequence. A text editor can be used to join exon sequences and generate a transcript sequence.

5. The ExPASy (Expert Protein Analysis System) proteomics server at the Swiss Institute of Bioinformatics (SIB) is dedicated to the analysis of protein sequences and structures. The institute maintains an excellent, up-to-date list of tools for all types of protein and nucleic acid analysis (www.expasy.ch/tools/). In our case, we used the SMART tool to identify functional domains in different NRXN1 isoforms (smart.embl-heidelberg.de/). Many other tools are also listed on the server, which could further enhance this type of analysis.
6. Intronic and Exonic sequences can be evaluated for regulatory potential with a number of tools. The UCSC genome browser contains many tracks under the "regulation" section of the track selection menu. Evolutionary conservation is also a key indicator of regulatory potential in intronic regions. DNA sequence from conserved regions can be exported and analysed for splice regulatory elements using the ESE finder tool (see Table 1).
7. Although GSEA methods may prove valuable tools for the analysis of genes identified by CNV analysis, the user should be aware of some caveats, which are more applicable to genes identified by genetic methods than those identified by changes in expression. The most likely cause of "enrichment" of genes identified by genetic methods is the gene size. This is highly relevant in the case of SNP-based studies where large genes are more likely to show association by chance, but should also apply in the case of CNV studies where large genes are also more likely to be disrupted by CNVs. In the case of association analysis, this can be partially corrected by the use of permutation testing; a similar permutation framework could also be applied to CNV data.

References

1. Lee, A.S., Gutierrez-Arcelus, M., Perry, G.H., Vallender, E.J., Johnson, W.E., Miller, G.M., Korbel, J.O. and Lee, C. (2008) Analysis of copy number variation in the rhesus macaque genome identifies candidate loci for evolutionary and human disease studies. *Hum. Mol. Genet.*, **17**, 1127–1136.
2. Henrichsen, C.N., Chaignat, E. and Reymond, A. (2009) Copy number variants, diseases and gene expression. *Hum. Mol. Genet.*, **18**, R1–R8.
3. Freeman, J.L., Perry, G.H., Feuk, L., Redon, R., McCarroll, S.A., Altshuler, D.M., *et al.* (2006) Copy number variation: new insights in genome diversity. *Genome Res.*, **16**, 949–961.
4. Wellcome Trust Case Control Consortium (2007) Genome-wide association study of 14,000 cases of seven common diseases and 3,000 shared controls. *Nature*, **447**, 661–678.
5. McDermid, H.E. and Morrow, B.E. (2002) Genomic disorders on 22q11. *Am. J. Hum. Genet.*, **70**, 1077–1088.
6. Walsh, T., McClellan, J.M., McCarthy, S.E., Addington, A.M., Pierce, S.B., Cooper, G.M., *et al.* (2008) Rare structural variants disrupt multiple genes in neurodevelopmental pathways in schizophrenia. *Science*, **320**, 539–543.
7. Pritchard, J.K. (2001) Are rare variants responsible for susceptibility to complex diseases? *Am. J. Hum. Genet.*, **69**, 124–137.

8. Stefansson, H., Rujescu, D., Cichon, S., Pietilainen, O.P., Ingason, A., Steinberg, S., *et al.* (2008) Large recurrent microdeletions associated with schizophrenia. *Nature*, **455**, 232–236.
9. Scherer, S.W., Lee, C., Birney, E., Altshuler, D.M., Eichler, E.E., Carter, N.P., Hurles, M.E. and Feuk, L. (2007) Challenges and standards in integrating surveys of structural variation. *Nat. Genet.*, **39**, S7–S15.
10. Redon, R., Ishikawa, S., Fitch, K.R., Feuk, L., Perry, G.H., Andrews, T.D., *et al.* (2006) Global variation in copy number in the human genome. *Nature*, **444**, 444–454.
11. Emanuel, B.S. and Saitta, S.C. (2007) From microscopes to microarrays: dissecting recurrent chromosomal rearrangements. *Nat. Rev. Genet.*, **8**, 869–883.
12. Khaja, R., Zhang, J., MacDonald, J.R., He, Y., Joseph-George, A.M., Wei, J., *et al.* (2006) Genome assembly comparison identifies structural variants in the human genome. *Nat. Genet.*, **38**, 1413–1418.
13. Fredman, D., White, S.J., Potter, S., Eichler, E.E., Den Dunnen, J.T. and Brookes, A.J. (2004) Complex SNP-related sequence variation in segmental genome duplications. *Nat. Genet.*, **36**, 861–866.
14. Shen, F., Huang, J., Fitch, K.R., Truong, V.B., Kirby, A., Chen, W., *et al.* (2008) Improved detection of global copy number variation using high density, non-polymorphic oligonucleotide probes. *BMC Genet.*, **9**, 27.
15. Peiffer, D.A. and Gunderson, K.L. (2009) Design of tag SNP whole genome genotyping arrays. *Methods Mol. Biol.*, **529**, 51–61.
16. Colella, S., Yau, C., Taylor, J.M., Mirza, G., Butler, H., Clouston, P., Bassett, A.S., Seller, A., Holmes, C.C. and Ragoussis, J. (2007) QuantiSNP: an Objective Bayes Hidden-Markov Model to detect and accurately map copy number variation using SNP genotyping data. *Nucleic Acids Res.*, **35**, 2013–2025.
17. Wang, K., Li, M., Hadley, D., Liu, R., Glessner, J., Grant, S.F., Hakonarson, H. and Bucan, M. (2007) PennCNV: an integrated hidden Markov model designed for high-resolution copy number variation detection in whole-genome SNP genotyping data. *Genome Res.*, **17**, 1665–1674.
18. Lin, C.H., Huang, M.C., Li, L.H., Wu, J.Y., Chen, Y.T. and Fann, C.S. (2008) Genome-wide copy number analysis using copy number inferring tool (CNIT) and DNA pooling. *Hum. Mutat.*, **29**, 1055–1062.
19. Xu, B., Roos, J.L., Levy, S., van Rensburg, E.J., Gogos, J.A. and Karayiorgou, M. (2008) Strong association of de novo copy number mutations with sporadic schizophrenia. *Nat. Genet.*, **40**, 880–885.
20. Marshall, C.R., Noor, A., Vincent, J.B., Lionel, A.C., Feuk, L., Skaug, J., *et al.* (2008) Structural variation of chromosomes in autism spectrum disorder. *Am. J. Hum. Genet.*, **82**, 477–488.
21. Yau, C. and Holmes, C.C. (2008) CNV discovery using SNP genotyping arrays. *Cytogenet. Genome Res.*, **123**, 307–312.
22. Casilli, F., Di Rocco, Z.C., Gad, S., Tournier, I., Stoppa-Lyonnet, D., Frebourg, T. and Tosi, M. (2002) Rapid detection of novel BRCA1 rearrangements in high-risk breast-ovarian cancer families using multiplex PCR of short fluorescent fragments. *Hum. Mutat.*, **20**, 218–226.
23. Wu, Y.L., Savelli, S.L., Yang, Y., Zhou, B., Rovin, B.H., Birmingham, D.J., Nagaraja, H.N., Hebert, L.A. and Yu, C.Y. (2007) Sensitive and specific real-time polymerase chain reaction assays to accurately determine copy number variations (CNVs) of human complement C4A, C4B, C4-long, C4-short, and RCCX modules: elucidation of C4 CNVs in 50 consanguineous subjects with defined HLA genotypes. *J. Immunol.*, **179**, 3012–3025.
24. Sellner, L.N. and Taylor, G.R. (2004) MLPA and MAPH: new techniques for detection of gene deletions. *Hum. Mutat.*, **23**, 413–419.
25. Sebat, J., Lakshmi, B., Malhotra, D., Troge, J., Lese-Martin, C., Walsh, T., *et al.* (2007) Strong association of de novo copy number mutations with autism. *Science*, **316**, 445–449.
26. Pugh, T.J., Delaney, A.D., Farnoud, N., Flibotte, S., Griffith, M., Li, H.I., Qian, H., Farinha, P., Gascoyne, R.D. and Marra, M.A. (2008) Impact of whole genome amplification on analysis of copy number variants. *Nucleic Acids Res.*, **36**, e80.
27. Bediaga, N.G., Alfonso-Sanchez, M.A., de Renobales, M., Rocandio, A.M., Arroyo, M. and de Pancorbo, M.M. (2008) GSTT1 and GSTM1 gene copy number analysis in paraffin-embedded tissue using quantitative real-time PCR. *Anal. Biochem.*, **378**, 221–223.
28. Cukier, H.N., Pericak-Vance, M.A., Gilbert, J.R. and Hedges, D.J. (2009) Sample degradation leads to false-positive copy number variation calls in multiplex real-time polymerase chain reaction assays. *Anal. Biochem.*, **386**, 288–290.

29. Curtis, D., Vine, A.E. and Knight, J. (2008) Study of regions of extended homozygosity provides a powerful method to explore haplotype structure of human populations. *Ann. Hum. Genet.*, **72**, 261–278.

30. Bailey, J.A., Yavor, A.M., Massa, H.F., Trask, B.J. and Eichler, E.E. (2001) Segmental duplications: organization and impact within the current human genome project assembly. *Genome Res.*, **11**, 1005–1017.

31. Silva, A.L. and Romao, L. (2009) The mammalian nonsense-mediated mRNA decay pathway: to decay or not to decay! Which players make the decision? *FEBS Lett.*, **583**, 499–505.

32. Pan, Q., Shai, O., Lee, L.J., Frey, B.J. and Blencowe, B.J. (2008) Deep surveying of alternative splicing complexity in the human transcriptome by high-throughput sequencing. *Nat. Genet.*, **40**, 1413–1415.

33. Rujescu, D., Ingason, A., Cichon, S., Pietilainen, O.P., Barnes, M.R., Toulopoulou, T., *et al.* (2009) Disruption of the neurexin 1 gene is associated with schizophrenia. *Hum. Mol. Genet.*, **18**, 988–996.

34. Rowen, L., Young, J., Birditt, B., Kaur, A., Madan, A., Philipps, D.L., *et al.* (2002) Analysis of the human neurexin genes: alternative splicing and the generation of protein diversity. *Genomics*, **79**, 587–597.

35. Beglopoulos, V., Montag-Sallaz, M., Rohlmann, A., Piechotta, K., Ahmad, M., Montag, D. and Missler, M. (2005) Neurexophilin 3 is highly localized in cortical and cerebellar regions and is functionally important for sensorimotor gating and motor coordination. *Mol. Cell Biol.*, **25**, 7278–7288.

36. Liu, Q., Dinu, I., Adewale, A.J., Potter, J.D. and Yasui, Y. (2007) Comparative evaluation of gene-set analysis methods. *BMC Bioinformatics*, **8**, 431.

37. Baranzini, S.E., Galwey, N.W., Wang, J., Khankhanian, P., Lindberg, R., Pelletier, D., *et al.* (2009) Pathway and network-based analysis of genome-wide association studies in multiple sclerosis. *Hum. Mol. Genet.* **18**, 2078–2090.

Chapter 8

Computational Methods for the Analysis of Primate Mobile Elements

Richard Cordaux, Shurjo K. Sen, Miriam K. Konkel, and Mark A. Batzer

Abstract

Transposable elements (TE), defined as discrete pieces of DNA that can move from one site to another site in genomes, represent significant components of eukaryotic genomes, including primates. Comparative genome-wide analyses have revealed the considerable structural and functional impact of TE families on primate genomes. Insights into these questions have come in part from the development of computational methods that allow detailed and reliable identification, annotation, and evolutionary analyses of the many TE families that populate primate genomes. Here, we present an overview of these computational methods and describe efficient data mining strategies for providing a comprehensive picture of TE biology in newly available genome sequences.

Key words: Computational methods, Transposable element, Insertion, Identification, Classification, Consensus sequence, Subfamily, Phylogenetic reconstruction, Transpositional activity, Primate, Genome evolution

1. Introduction

Transposable elements (TE), defined as discrete pieces of DNA that can move from one site to another site in genomes, have long been considered as nonsignificant components of genomes. This view started to change, however, when whole genome sequences became available. Hence, nearly half of the human genome is now recognized as being of TE origin (1). It is likely that this is an underestimation because some ancient TEs in the genome may have degraded beyond recognition by current methods. Primates constitute an excellent taxonomic group to analyze TE diversity and evolution because, in addition to humans, complete genome sequences of the chimpanzee and

Michael R. Barnes and Gerome Breen (eds.), *Genetic Variation: Methods and Protocols*, Methods in Molecular Biology, vol. 628, DOI 10.1007/978-1-60327-367-1_8,

rhesus macaque are now available (2, 3) with more genome sequences on the way. Comparative genome-wide analyses have revealed the considerable, structural, and functional impact of TE families on primate genomes.

The primary mode of TE-mediated instability is de novo integration of new elements, which can have a variety of functional consequences (4). However, additional changes in local sequence architecture arising as a by-product of TE activity include, but are not limited to, insertion-mediated deletions (5, 6), recombination-mediated deletions (7, 8), segmental duplications (9, 10), inversions (11, 12) and inter- or intra-chromosomal transduction of host genomic sequence (13, 14). Paradoxically, TE activity is not associated with genomic instability alone; retrotransposon mRNAs can also occasionally serve as molecular bandages for repairing potentially lethal DNA double-strand breaks (15, 16). Another interesting aspect of TE biology in primate genomes has been the discovery that functions encoded by TEs originally for their own purposes can be efficiently adapted by host genomes into unrelated beneficial roles (17, 18). This process of so-called molecular domestication illustrates that TEs may on occasion share a mutualistic relationship with their host genomes, and that the "parasite" tag historically attached to TEs may be somewhat unfair in some cases.

In a broader sense, these observations raise the question of the nature of the host-TE relationship throughout evolution. A popular opinion is that within the evolutionary timescale of the primate radiation, most TE families have been slightly deleterious or at best neutral within the genome and have achieved their high numbers through a finely tuned strategy of parasitism (19–21). However, contrary to this viewpoint, various analyses have proposed different functional roles for some TE families, such as origins of replication, gene expression regulators, agents of DNA repair and X-chromosome inactivation, or scaffolds for meiotic replication (22–24). These views need not be reciprocally exclusive, and it may be overly simplistic to treat the interactions between TE families and primate genomes as being a zero-sum game. Indeed, a systems biology approach wherein interactions between host genomes and TEs are seen in the context of an ecosystem may be a suitable way of representing this complex relationship (25, 26). In any event, addressing these questions requires exhaustive and reliable identification, annotation, and evolutionary analyses of the many TE families that populate primate genomes. A number of computational methods have been developed to this end, which are reviewed in the following protocol.

2. Materials

Computational TE analyses can be performed on a local desktop machine with internet access. However, large-scale studies require a local software installation, typically in a UNIX environment (see Note 1) with considerable memory (preferably 4 GB, 16 GB, or more RAM, depending on the study size). Common (bio-) computational skills should be sufficient for successful use and implementation of the required software.

3. Methods

3.1. TE Identification

In this section, we describe methods to identify: (1) TEs for which prior sequence knowledge exists, (2) TEs with no prior information available (i.e., de novo identification), and (3) TEs which are differentially inserted among genomes (i.e., polymorphic for presence or absence).

3.1.1. Identification of Known TEs

1. TE library: to identify known TEs in a target sequence, we rely on an existing TE library containing the consensus sequences (see Subheading 3.2.2) of multiple TE families. The most comprehensive database of eukaryotic TEs is Repbase (http://girinst.org/) (27, 28). Repbase can be searched for consensus sequences directly, or a desired library can be downloaded.
2. Selection of genome sequences: human genomic sequences can be retrieved from UCSC (http://genome.ucsc.edu; select genomes and species of interest) (see Note 2).
3. TE annotation: using the selected TE library as reference, TEs in the query sequence are identified by similarity searches and annotated using RepeatMasker (http://repeatmasker.org) (see Note 3). Analysis of a relatively small dataset can be performed online at http://www.repeatmasker.org/cgi-bin/WEBRepeatMasker. For larger analyses (e.g., whole genomes), we suggest a local installation of RepeatMasker (http://www.repeatmasker.org/RMDownload.html) (see Note 4).
4. Submission of query sequences to RepeatMasker: RepeatMasker requires files to be in the FASTA format (see Note 5). Submission of several sequences at once is possible. There is no explicit maximum size constraint for

query sequence(s). However, lengthy sequences often are slow to process, accompanied by the risk of an error message caused by connection time-out. The query sequence can be uploaded or pasted into the sequence window on the RepeatMasker web site. Select "Cross_match" as the search engine, and "slow" as the speed/sensitivity to ensure a search with the highest level of TE annotation (highest sensitivity; see Note 6). A DNA source that determines the choice of TE library is then selected. We suggest selecting "Repetitive sequences in lower case" from the masking option bar to show the annotated repetitive sequence in the output file in lower case (see Notes 7 and 8).

5. Results: the output presents the annotation of repeats in the query sequence. The general output indicates what search options were selected; which (if any) and how many TEs are identified; what percentage of the query sequence contains TEs; and several result files that can be saved or reviewed in the web interface. The HTML version of the results gives detailed information about the identified TEs, including length, orientation, TE-subfamily, and matching region. Another important analysis output is the ID number(s) of the identified TEs. This indicates whether multiple TEs or a single element with interruptions have been identified (see Note 9). In addition, an alignment of the identified TE to the TE subfamily consensus sequence for which the sequence was identified as the best match is available.

3.1.2. De Novo TE Identification by Genome Self-Alignment

De novo identification of repeats has proven challenging, especially for large and TE-rich genomes. So, a single dominant method for this task is not yet established. Commonly used software packages include PILER (29), ReAS (30), RECON (31), and RepeatScout (32). Below we describe the use of RepeatScout (http://repeatscout.bioprojects.org/) (see Note 10):

1. Prerequisites: preferably, a computer with LINUX or UNIX and at least 4 GB (ideally more) of RAM and a C compiler (typically freely available on UNIX machines) are needed.
2. Downloading and installing RepeatScout: RepeatScout_1.0.0 is available from http://repeatscout.bioprojects.org/. The software should be extracted and compiled with a command such as:

   ```
   tar -zxf RepeatScout-1.0.2.tar.gz; cd Repeat-Scout-1; make
   ```

   This yields two executable files: `build_lmer_table` and `RepeatScout-v1`.
3. Genome download: assembled genomes can be obtained from NCBI (ftp://ftp.ncbi.nih.gov/genomes) or UCSC (http://hgdownload.cse.ucsc.edu/downloads.html). For a

full-genome analysis, download the chromFa.tar.gz file (see Note 11).

4. Repeat identification: first, an "l-mer" table is constructed; "l" (which defaults to 3) represents the length of the l-mer seeds and should be adjusted to meet the specific needs of the analysis. The following setting for l is suggested (see Note 12): `ceil(log_4(L)+1)` with

   `ceil(x)=smallest integer greater than x; log_4(x)=log base 4 of x; L: length of input sequence`

   A typical execution sequence to build an l-mer table begins with a command like:

   ```
   build_lmer_table -sequence source.fa -freq source.freq
   ```

   This calculates the frequency of l-mers in the specified source.fa DNA sequence. Next, an output file containing the de novo identified repeats is created. RepeatScout-v1 is executed with the built l-mer table (`source.freq`) and the sequence (`source.fa`) in the following manner:

   ```
   RepeatScout-v1 -sequence source.fa -freq source.freq -output repeats.fa
   ```

5. Filtering out non-TE sequences: repetitive elements include TEs as well as low-complexity elements, segmental duplications, or exons. Non-TE sequences may be filtered out with further processing. Low-complexity repeats may be removed with the perl script "filter-stage-1.prl." Next, RepeatMasker (see Subheading 3.1.1) is run with the filtered RepeatScout-v1 library. The "filter-stage-2.prl" excludes all repeats with very low copy numbers (default < 10). Lastly, segmental duplications and exons are identified and may be erased from the library by using the locations identified by RepeatMasker and matching them with gff files containing segmental duplications and exons.

3.1.3. De Novo Identification of Polymorphic TEs by Genome Alignment to Another Genome

1. Preconditions: two genome sequences are required (see Note 13). This approach has been successfully implemented for human *Alu* (33) and LINE-1 (34) elements. A computer with the UNIX or LINUX operating system (or compatible variants) is needed. The user should be comfortable working at the command line. The ability to write programs in Perl, Python, and/or shell scripts is also valuable.
2. Local installation of BLAST (Basic Local Alignment Search Tool) (35): BLAST, downloadable from ftp://ftp.ncbi.nih.gov/blast/, exists as a pre-compiled program suitable for many operating systems.
3. Selection and download of genomes: while we will provide a detailed description of this method for two human genomes obtained from NCBI at ftp://ftp.ncbi.nih.gov/genomes/

H_sapiens/ (see Note 14), in principle any two genomes can be used for this analysis. In our case, the first genome (hereafter genome A) is the human reference genome (ref_genome in NCBI). The second human genome (hereafter genome B) is the publicly available version of the Celera genome (alt_ genome in NCBI).

4. Download of TE consensus sequence: a TE consensus sequence of interest (here *Alu*) is downloaded from Repbase as a query sequence (see Note 15).
5. Identification of TEs and extraction of all matching TEs from genome A: genome A is queried with the *Alu* consensus sequence using the local installation of BLAST, and all candidate elements, including 300 bp of flanking sequence on either side are extracted from genome A sequence.
6. Querying genome B with extracted loci from genome A: each extracted locus from genome A is used as a query sequence against genome B. If the query sequence matches in length and identity to a level of ≥98%, the locus is disqualified as a polymorphic candidate and discharged. In contrast, if either the *Alu* element alone or the flanking sequence is identified as a best match, the locus is a potential polymorphic candidate, and is used for a second, more detailed analysis. For the second analysis, we take the *Alu* element out of the sequence and attach the flanking sequences to each other. This can be done with several BioPython commands such as:

```
flankSize = 300                    #choose a flanking sequence of 300bp
seqSize = len(mySeq)               #find the length of DNA sequence mySeq
flankHead = mySeq[0:flankSize]     #extract the head flanking portion
flankTail = mySeq[seqSize-flankSize:seqSize] #extract the tail
joinedSeq = flankHead + flankTail #assemble the two fragments together
```

The flanking sequence of each locus is again queried against genome B to identify close-to-perfect matches of the flanking sequence. Close-to-perfect matches correspond to loci considered to contain polymorphic *Alu* elements present in genome A and absent in genome B. Other loci are discarded.

7. Identification of TEs from genome B absent in genome A: genome A is swapped with genome B, and steps 5 and 6 are repeated.

8. Comparison of confirmed polymorphic TEs to dbRIP: polymorphic human retrotransposons can be checked for novelty using the dbRIP database, a database of polymorphic human retrotransposons, by submitting the candidate loci to http://falcon.roswellpark.org:9090/searchRIP.html (36).
9. Confirmation of computational results: apart from a detailed manual confirmation of the dataset, we recommend performing wet-bench PCR analyses on a panel of individual genomic DNA samples to confirm that the identified TEs are indeed polymorphic for insertion presence or absence (see Chapter 9, in this issue).

3.2. TE Classification

In this section, we describe methods: (1) to classify TEs into groups of closely related copies (termed subfamilies), and (2) to construct consensus sequences of TE subfamilies.

3.2.1. TE Subfamily Classification

A transpositionally active TE in a genome can produce novel copies of itself, each of which is initially identical in nucleotide sequence to the copy that generated it. Therefore, any sequence feature present in the ancestral TE copy will be shared with its "progeny". TE subfamilies are thus defined as collections of TE copies exclusively sharing diagnostic sequence features. Such features typically include nucleotide substitutions located at homologous sites in all copies within a subfamily, termed "shared sequence variants" (SSV). SSVs are distinguishable from post-insertional random substitutions, which would show no site preference. Efficient SSV identification forms the basis for computational classification of TE copies into discrete subfamilies. A schematic algorithm for this purpose is described below:

1. Generation of a multiple alignment of TE copies of interest: this can be achieved by running the ClustalW alignment program (see Note 16), using a FASTA file of the TE sequences as input. Visually, inspect the alignment and make further refinements using a suitable sequence alignment editor, such as BioEdit (http://www.mbio.ncsu.edu/BioEdit/bioedit.html) or Megalign (http://www.dnastar.com/products/lasergene.php). The alignment forms the input for the algorithms mentioned in the next step.
2. Automated TE subfamily classification: to the best of our knowledge, only two specialized algorithms exist for this purpose: (a) MASC (Multiple Aligned Sequence Classification) (37) hierarchically and recursively splits the multiple alignment into smaller groups of two, continuing till the absence of multiple SSVs invalidates further subdivision. Although MASC is not currently available as a binary distribution, the original algorithm has been described in detail elsewhere (38) and reasonable competence with bioinformatics programming

should enable users to adapt it for their specific analyses. (b) A second approach would be to use a modification of the MULTIPROFILER algorithm (39) to scan the multiple alignment for groups of TEs characterized by overrepresented *n*-tuples of SSVs (where *n* has an integral value > 1), followed by a final step where subfamilies differing from their closest relatives by a single SSV are identified using a probability-based approach. Although this approach has till now been used for the construction of consensus sequences only for the *Alu* family (40), a set of Perl and C programs is available at http://www-cse.ucsd.edu/groups/bioinformatics/software.html#alu-subfam that should in principle be modifiable for other TE families.

3.2.2. Construction of TE Subfamily Consensus Sequences

Over time, TE copies of a "source" TE for any particular subfamily each accumulate random substitutions, and for even moderately old subfamilies, individual members may be quite different from the original source TE. However, the same random nature of these substitutions means that, for any particular subfamily, most elements will retain the original nucleotide of the ancestral TE copy at individual positions along the length of the TE. Thus, by using a majority-rule algorithm that also accounts for increased mutation frequencies at CpG dinucleotides (i.e., wherever a C is followed by a G in 5′ to 3′ orientation), it is possible to accurately reconstruct the ancestral sequence that gave rise to the members of a TE subfamily. We describe a schematic algorithm below:

1. Construct a multiple alignment of TE copies grouped together as a subfamily (see Subheading 3.2.1): quality of the multiple alignment of TE copies will directly influence the accuracy of the reconstructed consensus sequences, and manual curation of the initial computational alignment will almost always result in a better finished product. Higher numbers of copies in the alignment will result in a consensus sequence with greater statistical support.
2. For each position, determine the majority nucleotide. Most multiple alignment software suites allow this to be done in a few clicks (e.g., in BioEdit, click `alignment>positional frequency summary`, or in MegAlign, click `view>alignment report`).
3. CpG dinucleotides have sixfold higher mutation rates compared to other dinucleotides, mostly through transitions at one of the two positions leading to either CpA or TpG (41). However, post-insertional substitutions mimicking CpA or TpG dinucleotides present in the ancestral consensus sequence can be sorted out on the basis of the proportion of subfamily members that carry a particular dinucleotide. If the ancestral state was either CpA or TpG, most copies will retain this state and the consensus sequence will tend to be unequivocal.

If, however, a CpG in the original consensus sequence mutates to CpA or TpG, the ancestral and derived states will be present in almost equal proportions, and the resulting ambiguity at the dinucleotide position can be used to correct the consensus sequence to CpG.

4. The accuracy of the consensus sequence reconstructed using the above two steps can be tested using the following formula: $S = S_1 S_2 + (1 - S_1)(1 - S_2)/3$, where S_1 and S_2 represent sequence similarities between TE elements 1 and 2 of a particular family and the reconstructed source element, and S represents the mutual sequence similarity between the two copies (42). Close correspondence between the observed and expected values of S indicates that the consensus sequence is an accurate reconstruction (43) (see Note 17).

3.3. Analyses of TE Evolution

To decipher the evolutionary history of TE subfamilies and address questions about, e.g., their timing of transpositional activity, several approaches can be used. For example, very recently active TEs are expected to exhibit differential distribution among individuals, i.e., individual copies will be polymorphic for presence or absence at orthologous genomic sites among the compared samples. The method described in Subheading 3.1.3 allows the identification of such differentially inserted TE loci. TE insertions that are responsible for genetic disorders are examples of active subfamilies for which copies have inserted in the genome within the recent past. At a deeper timescale, TE subfamilies that have been active at different evolutionary periods are also expected to be differentially inserted among species. The timing of subfamily activity can thus be deduced from the timing of divergence of the host genomes that carry or lack copies of the TE subfamily of interest (44). In this section, we describe further computational approaches: (1) to estimate the age (i.e., the timing of transpositional activity) of TE subfamilies independently of the genomic location of the copies, and (2) to infer TE amplification dynamics by reconstructing phylogenetic relationships among the members of TE subfamilies.

3.3.1. Inference of TE Subfamily Ages

Because a subfamily consensus sequence (as obtained in Subheading 3.2.2) represents the putative sequence of the active TE copy that gave rise to other copies in the subfamily, and because individual copies gradually diverge from the "source" copy across time, the quantity of sequence divergence accumulated by individual copies relative to their reconstructed consensus sequence can be used to infer the approximate age of the TE subfamily, provided that the substitution rate is known for the lineage being investigated.

Average sequence divergence of individual copies to their consensus sequence can be obtained by creating a multiple alignment containing TE copies from a subfamily together with the subfamily consensus sequence. Pairwise genetic distances between the consensus sequence and each individual copy are calculated, and then averaged. Such calculations can be performed with various software packages for evolutionary and phylogenetic analyses, such as MEGA (45) (see Note 18):

1. Open a FASTA alignment with the text editor implemented in MEGA and convert the alignment to the MEGA format (containing a .meg extension). The converted file can then be opened with the data analyses module of MEGA.
2. Create a group containing the consensus sequence and another group containing all individual subfamily copies: click `Data>Setup/Select taxa & groups`
3. Calculate average divergence between the two groups: click `Distances>Compute Between Groups Means`
4. Subfamily age is calculated as the average divergence from the consensus sequence divided by the substitution rate (see Note 19).

3.3.2. Phylogenetic Analyses

Phylogenetic analyses can be performed to infer the relationships between individual copies within a subfamily and explore subfamily amplification dynamics. Several major methods of tree reconstruction that differ in their underlying philosophy, including distance-, parsimony- and probability-based methodologies are available. Each method has its own advantages and drawbacks, and no single method is the best for all analyses. A number of software suites are available to conduct phylogenetic analyses, including MEGA. A comprehensive list of phylogenetic packages available for download or usable via a web interface can be found at http://evolution.genetics.washington.edu/phylip/software.html. Phylogenetic reconstruction starts with a multiple alignment of the TE copies of interest, which is achieved as described in Subheading 3.2.2. The alignment is then used for tree reconstruction. For example, in MEGA, multiple phylogeny algorithms are available by clicking `Phylogeny>Construct Phylogeny`. Alternatively, for datasets with low sequence divergence, higher phylogenetic resolution may be reached by using network phylogenetic approaches (46, 47). Several programs for reconstructing networks, such as NETWORK, are available (48) (see Note 20).

4. Notes

1. While UNIX is typically stated as a requirement, many of these tools also work under the UNIX-based Macintosh OS X operating system, and also under Microsoft Windows with environments like Cygwin or MSYS.

2. The human genome can be in theory replaced by any other genome. If working with a genome for which a library does not exist and no analysis of TEs in a closely related species has been performed, de novo identification of TEs needs to be performed first to create a personal library for the species (see Subheading 3.1.2). Alternately, an analysis on the basis of protein similarities can be performed (e.g., see http://www.repeatmasker.org/cgi-bin/RepeatProteinMaskRequest). However, the latter approach does not detect TEs that lack typical protein structures, e.g., SINEs are not identified.
3. The classic Repbase library is modified for RepeatMasker, in particular to improve the annotation of long TEs.
4. Also required are: (1) a UNIX-based system with perl 5.8.0 or higher, (2) either Cross_Match (obtained from http://www.phrap.org, select "Phred/Phrap/Consed") or WU Blast (available from http://blast.wustl.edu/licensing/), and (3) a TE library downloadable from http://www.girinst.org.
5. FASTA is a text-based file format that represents nucleic acid or protein sequences and is characterized by a text description line beginning with > (no space between > and the text), followed by sequence in the next text line.
6. Cross_match is described as more sensitive in identifying TEs compared to WU Blast.
7. We also suggest that readers familiarize themselves with other options for possible integration within their analysis. These options are largely self-explanatory. In addition, the RepeatMasker documentation provides further detailed information.
8. In principle, the same analysis can be performed with a local installation of RepeatMasker. The corresponding parameters can be selected from the command line.
9. The ID information is important because long elements are particularly disposed to have multiple Ns (i.e., ambiguous or unsequenced bases) within their sequence boundaries (depending on the quality of the genome assembly), and many TEs may also be nested within other TEs. Using ID information, it can often be distinguished if the fragments of the TE belong to one or two separate insertions. While the ID information in most cases is accurate, we recommend checking this information manually if this information is of particular interest for the performed analysis.
10. RepeatScout requires assembled sequences, or at least scaffolds of a genome for the annotation of repeats. The assembly of new genomes, especially without general knowledge of the repeat composition, is challenging and may result in loss of

repetitive sequences. ReAS is one of the few programs for the de novo identification of TEs that requires whole shotgun reads and not assembled genomes.

11. For many users, an analysis of a single or fractional chromosome per-run may be a present-day limit, given common RAM configurations and the RepeatScout v1 software itself. RepeatScout v1 does not provide intrinsic support for multiple CPUs; and its internal use of signed 4-byte integers limits runs to FASTA files with a maximum of 2 Gbp.
12. A list of modifiable parameters, which usually do not need to be adjusted, can be found in the help file (--h) for RepeatScout.
13. Alternately, sequence traces can be used with some procedural modifications; we highlight these in the notes of the appropriate sections.
14. Genomes of other species are also available from ftp://ftp.ncbi.nih.gov/genomes. Different versions of assembled reference genomes can be downloaded from UCSC (http://genome.ucsc.edu). To our knowledge the ref_genome is not available from UCSC.
15. Depending on the TE of interest, an approximately 50 bp-long conserved region of the TE may be used as a query sequence.
16. ClustalW is available as a command line interface or as a graphical user interface (ClustalX), downloadable at ftp://ftp.ebi.ac.uk/pub/software/clustalw2/. It is also implemented in biological sequence analyses software, such as BioEdit.
17. For subfamilies with relatively recent periods of activity, individual copies will be similar to the consensus sequence; however, for older repeats individual members are usually far more divergent, and a well-constructed subfamily consensus sequence is the only suitable query for computational data mining.
18. Freely available for download at http://www.megasoftware.net/
19. Alternatively, the age of a subfamily can be estimated without reconstructing a subfamily consensus sequence. This can be achieved by calculating the average divergence between any two copies of the subfamily (in MEGA, open a .meg file containing an alignment of individual TE copies of interest and click `Distances>Compute Overall Mean`). Assuming that divergence has accumulated at the same rate among copies, approximate subfamily age can be estimated as half the average divergence divided by the substitution rate.
20. Freely available for download at http://www.fluxus-engineering.com/netwinfo.htm. NETWORK requires a specific file format

(containing an .rdf extension) that can be created manually using a text editor or automatically by converting a FASTA file into .rdf format using a program available for purchase from the NETWORK website.

Acknowledgments

Our research is supported by National Science Foundation BCS-0218338 (MAB) and EPS-0346411 (MAB), National Institutes of Health RO1 GM59290 (MAB) and PO1 AG022064 (MAB), and the State of Louisiana Board of Regents Support Fund (MAB). RC is supported by a Young Investigator ATIP award from the Centre National de la Recherche Scientifique (CNRS).

References

1. Lander, E.S., Linton, L.M., Birren, B., Nusbaum, C., Zody, M.C., Baldwin, J., et al. (2001) Initial sequencing and analysis of the human genome. *Nature* **409**, 860–921.
2. Chimpanzee Sequencing and Analysis Consortium (2005) Initial sequence of the chimpanzee genome and comparison with the human genome. *Nature* **437**, 69–87.
3. Gibbs, R.A., Rogers, J., Katze, M.G., Bumgarner, R., Weinstock, G.M., Mardis, E.R., et al. (2007) Evolutionary and biomedical insights from the rhesus macaque genome. *Science* **316**, 222–234.
4. Hedges, D.J. and Deininger, P.L. (2007) Inviting instability: Transposable elements, double-strand breaks, and the maintenance of genome integrity. *Mutat Res* **616**, 46–59.
5. Callinan, P.A., Wang, J., Herke, S.W., Garber, R.K., Liang, P. and Batzer, M.A. (2005) Alu Retrotransposition-mediated deletion. *J Mol Biol* **348**, 791–800.
6. Han, K., Sen, S.K., Wang, J., Callinan, P.A., Lee, J., Cordaux, R., et al. (2005) Genomic rearrangements by LINE-1 insertion-mediated deletion in the human and chimpanzee lineages. *Nucleic Acids Res* **33**, 4040–4052.
7. Sen, S.K., Han, K., Wang, J., Lee, J., Wang, H., Callinan, P.A., et al. (2006) Human genomic deletions mediated by recombination between Alu elements. *Am J Hum Genet* **79**, 41–53.
8. Han, K., Lee, J., Meyer, T.J., Wang, J., Sen, S.K., Srikanta, D., et al. (2007) Alu recombination-mediated structural deletions in the chimpanzee genome. *PLoS Genet* **3**, 1939–1949.
9. Bailey, J.A., Liu, G. and Eichler, E.E. (2003) An Alu transposition model for the origin and expansion of human segmental duplications. *Am J Hum Genet* **73**, 823–834.
10. Jurka, J., Kohany, O., Pavlicek, A., Kapitonov, V.V. and Jurka, M.V. (2004) Duplication, coclustering, and selection of human Alu retrotransposons. *Proc Natl Acad Sci U S A* **101**, 1268–1272.
11. Lobachev, K.S., Stenger, J.E., Kozyreva, O.G., Jurka, J., Gordenin, D.A. and Resnick, M.A. (2000) Inverted Alu repeats unstable in yeast are excluded from the human genome. *Embo J* **19**, 3822–3830.
12. Stenger, J.E., Lobachev, K.S., Gordenin, D., Darden, T.A., Jurka, J. and Resnick, M.A. (2001) Biased distribution of inverted and direct Alus in the human genome: implications for insertion, exclusion, and genome stability. *Genome Res* **11**, 12–27.
13. Pickeral, O.K., Makalowski, W., Boguski, M.S. and Boeke, J.D. (2000) Frequent human genomic DNA transduction driven by LINE-1 retrotransposition. *Genome Res* **10**, 411–415.
14. Xing, J., Wang, H., Belancio, V.P., Cordaux, R., Deininger, P.L. and Batzer, M.A. (2006) Emergence of primate genes by retrotransposon-mediated sequence transduction. *Proc Natl Acad Sci U S A* **103**, 17608–17613.
15. Morrish, T.A., Gilbert, N., Myers, J.S., Vincent, B.J., Stamato, T.D., Taccioli, G.E., et al. (2002) DNA repair mediated by endonuclease-independent LINE-1 retrotransposition. *Nat Genet* **31**, 159–165.

16. Sen, S.K., Huang, C.T., Han, K. and Batzer, M.A. (2007) Endonuclease-independent insertion provides an alternative pathway for L1 retrotransposition in the human genome. *Nucleic Acids Res* **35**, 3741–3751.
17. Mi, S., Lee, X., Li, X., Veldman, G.M., Finnerty, H., Racie, L., et al. (2000) Syncytin is a captive retroviral envelope protein involved in human placental morphogenesis. *Nature* **403**, 785–789.
18. Cordaux, R., Udit, S., Batzer, M.A. and Feschotte, C. (2006) Birth of a chimeric primate gene by capture of the transposase gene from a mobile element. *Proc Natl Acad Sci U S A* **103**, 8101–8106.
19. Boissinot, S., Entezam, A. and Furano, A.V. (2001) Selection against deleterious LINE-1-containing loci in the human lineage. *Mol Biol Evol* **18**, 926–935.
20. Cordaux, R., Lee, J., Dinoso, L. and Batzer, M.A. (2006) Recently integrated Alu retrotransposons are essentially neutral residents of the human genome. *Gene* **373**, 138–144.
21. Schmid, C.W. (2003) Alu: A parasite's parasite? *Nat Genet* **35**, 15–16.
22. Brosius, J. and Gould, S.J. (1992) On "genomenclature": A comprehensive (and respectful) taxonomy for pseudogenes and other "junk DNA". *Proc Natl Acad Sci U S A* **89**, 10706–10710.
23. Liu, W.M., Chu, W.M., Choudary, P.V. and Schmid, C.W. (1995) Cell stress and translational inhibitors transiently increase the abundance of mammalian SINE transcripts. *Nucleic Acids Res* **23**, 1758–1765.
24. Schmid, C.W. (1998) Does SINE evolution preclude Alu function? *Nucleic Acids Res* **26**, 4541–4550.
25. Brookfield, J.F. (2005) The ecology of the genome – mobile DNA elements and their hosts. *Nat Rev Genet* **6**, 128–136.
26. Le Rouzic, A., Dupas, S. and Capy, P. (2007) Genome ecosystem and transposable elements species. *Gene* **390**, 214–220.
27. Jurka, J., Kapitonov, V.V., Pavlicek, A., Klonowski, P., Kohany, O. and Walichiewicz, J. (2005) Repbase Update, a database of eukaryotic repetitive elements. *Cytogenet Genome Res* **110**, 462–467.
28. Kohany, O., Gentles, A.J., Hankus, L. and Jurka, J. (2006) Annotation, submission and screening of repetitive elements in Repbase: RepbaseSubmitter and Censor. *BMC Bioinformatics* **7**, 474.
29. Edgar, R. C. and Myers, E. W. (2005) PILER: identification and classification of genomic repeats. *Bioinformatics* **21** Suppl. 1, i152–i158.
30. Li, R., Ye, J., Li, S., Wang, J., Han, Y., Ye, C., et al. (2005) ReAS: Recovery of ancestral sequences for transposable elements from the unassembled reads of a whole genome shotgun. *PLoS Comput Biol* **1**, e43.
31. Bao, Z. and Eddy, S.R. (2002) Automated de novo identification of repeat sequence families in sequenced genomes. *Genome Res* **12**, 1269–1276.
32. Price, A.L., Jones, N.C. and Pevzner, P.A. (2005) De novo identification of repeat families in large genomes. *Bioinformatics* **21** Suppl. 1, i351–i358.
33. Wang, J., Song, L., Gonder, M.K., Azrak, S., Ray, D.A., Batzer, M.A., et al. (2006) Whole genome computational comparative genomics: A fruitful approach for ascertaining Alu insertion polymorphisms. *Gene* **365**, 11–20.
34. Konkel, M.K., Wang, J., Liang, P. and Batzer, M.A. (2007) Identification and characterization of novel polymorphic LINE-1 insertions through comparison of two human genome sequence assemblies. *Gene* **390**, 28–38.
35. Altschul, S.F., Gish, W., Miller, W., Myers, E.W. and Lipman, D.J. (1990) Basic local alignment search tool. *J Mol Biol* **215**, 403–410.
36. Wang, J., Song, L., Grover, D., Azrak, S., Batzer, M.A. and Liang, P. (2006) dbRIP: A highly integrated database of retrotransposon insertion polymorphisms in humans. *Hum Mutat* **27**, 323–329.
37. Milosavljevic, A., Haussler, D. and Jurka, J. (1989) Informed parsimonious inference of prototypical genetic sequence. In: *Proceedings of the Second Annual Workshop on Computational Learning Theory* (Rivest, R., Haussler, D. and Warmuth, M.K., eds.), pp. 102–117. Morgan Kaufman, San Mateo.
38. Milosavljevic, A. (1990) Categorization of Macromolecular Sequences by Minimal Length Encoding, University of California at Santa Cruz.
39. Keich, U. and Pevzner, P.A. (2002) Finding motifs in the twilight zone. *Bioinformatics* **18**, 1374–1381.
40. Price, A.L., Eskin, E. and Pevzner, P.A. (2004) Whole-genome analysis of Alu repeat elements reveals complex evolutionary history. *Genome Res* **14**, 2245–2252.
41. Xing, J., Hedges, D.J., Han, K., Wang, H., Cordaux, R. and Batzer, M.A. (2004) Alu element mutation spectra: molecular clocks and the effect of DNA methylation. *J Mol Biol* **344**, 675–682.
42. Jurka, J. (1994) Approaches to identification and analysis of interspersed repetitive DNA sequences. In: *Automated DNA Sequencing and Analysis*

(Adams, M.D., Fields, C. and Venter, J.C., eds.), pp. 294–298. Academic Press, London.

43. Smit, A.F., Toth, G., Riggs, A.D. and Jurka, J. (1995) Ancestral, mammalian-wide subfamilies of LINE-1 repetitive sequences. *J Mol Biol* **246**, 401–417.
44. Pace, J. K., II and Feschotte, C. (2007) The evolutionary history of human DNA transposons: evidence for intense activity in the primate lineage. *Genome Res* **17**, 422–432.
45. Kumar, S., Tamura, K. and Nei, M. (2004) MEGA3: Integrated software for Molecular Evolutionary Genetics Analysis and sequence alignment. *Brief Bioinform* **5**, 150–163.
46. Posada, D. and Crandall, K.A. (2001) Intraspecific gene genealogies: trees grafting into networks. *Trends Eco Evol* **16**, 37–45.
47. Cordaux, R., Hedges, D.J. and Batzer, M.A. (2004) Retrotransposition of Alu elements: how many sources? *Trends Genet* **20**, 464–467.
48. Bandelt, H.J., Forster, P. and Rohl, A. (1999) Median-joining networks for inferring intraspecific phylogenies. *Mol Biol Evol* **16**, 37–48.

Chapter 9

Laboratory Methods for the Analysis of Primate Mobile Elements

David A. Ray, Kyudong Han, Jerilyn A. Walker, and Mark A. Batzer

Abstract

Mobile elements represent a unique and powerful set of tools for understanding the variation in a genome. Methods exist not only to utilize the polymorphisms among and within taxa to various ends but also to investigate the mechanism through which mobilization occurs. The number of methods to accomplish these ends is ever growing. Here, we present several protocols designed to assay mobile element-based variation within and among individual genomes.

Key words: Laboratory methods, Transposable element, Insertion, Identification, Classification, Consensus sequence, Subfamily, Assay, Transpositional activity, Primate, Phylogeny inference

1. Introduction

Mobile elements are interspersed repetitive DNA sequences with the unique ability to spread copies of themselves throughout the genome they occupy. As a result, these sequences can comprise a large proportion of the genomes in which they are found (1, 2). Mobile elements may be divided into two classes depending on how they mobilize and the type of intermediate they use. Class I elements include the retrotransposons, which utilize an RNA intermediate during retrotransposition, while DNA transposons, Class II, utilize a DNA intermediate during mobilization (3). While DNA transposons have had periods of activity early in primate evolution, all major recent activities in the human lineage have been retrotransposon-based (1, 4). Thus, we focus on these in this chapter.

Retrotransposons from the human lineage include L1 (a Long INterspersed Element), *Alu* (a primate-specific Short INterspersed Element), and SVA (a composite retrotransposon).

Michael R. Barnes and Gerome Breen (eds.), *Genetic Variation: Methods and Protocols*, Methods in Molecular Biology, vol. 628, DOI 10.1007/978-1-60327-367-1_9, © Springer Science+Business Media, LLC 2010

Together these elements have had significant impacts on the architecture of primate genomes (5). They comprise over ~40% of the human genome by mass and are the most abundant interspersed elements therein (6). Because of their high copy number, these interspersed repeats have been a significant source of variation as a result of insertion, transduction, and post-integration recombination among elements (6, 7). During retrotransposition, the RNA copy is reverse transcribed by target primed reverse transcription (TPRT) and subsequently integrated into the host genome (8–10). Unable to retrotranspose autonomously, *Alu* and SVA elements are thought to borrow the enzymatic factors required for their mobilization from L1 elements (8, 11–15), which encode a protein complex with endonuclease and reverse transcriptase activity (16, 17).

Over millions of years of primate evolution, retrotransposons have tended to accumulate in a hierarchical manner. This pattern is a direct result of the mechanisms through which they mobilize and insert, a modified version of the master gene model (18–21). Evidence suggests that a subfamily will accumulate copies for a certain period of time and then become quiescent. Other newer subfamilies subsequently become active, and the pattern repeats itself. This pattern is well illustrated by the *Alu* family of SINEs. Over time, the *Alu* element has diversified into a variety of subfamilies, each with its own set of diagnostic sequence characteristics and period of activity. For example, during the early stages of primate evolution, *Alu*J subfamilies were active. The activity of these subfamilies was later reduced (if not extinguished), and the *Alu*S subfamilies, derivatives of *Alu*J, became active. Thus, while *Alu*J elements are found in all primates, *Alu*S elements are found only in anthropoid primates (tarsiers, Platyrrhini, Catarrhini). The *Alu*Y subfamilies (22) are even more taxonomically specific in that they began their expansion in catarrhine primates after the platyrrhine-catarrhine split (23). Thus, each taxon has a unique pattern of insertions of which some are shared with other closely related taxa and others that are unique to that lineage. For example, the most recent *Alu* elements to mobilize in our own genome belong to a series of *Alu*Y subfamilies (*Alu*Yb8, *Alu*Ya5a2, etc.) that are exclusively or primarily specific to the human branch of the primate tree (24–27).

As genetic markers, retrotransposons of all sorts offer certain advantages over more commonly used genetic characters such as microsatellites and sequence data. First and foremost is the observation that these markers are an essentially homoplasy-free set of characters (28–32). Unlike many other genetic markers, they tend to exist as character states for no other reason than inheritance from a common ancestor. Thus, they are almost invariably identical by descent, not just identical by state. As a result, they can be used to provide an extremely accurate picture of evolutionary and population relationships (33–39). We also know that the ancestral state at any locus is the absence of the element and once the element is inserted it typically remains there indefinitely. These characteristics result in a relatively simple evolutionary model to be applied when interpreting the data.

SINEs and other retrotransposons share other desirable characteristics as well. The vast majority of mobile element insertions in the host genome are neutral residents (40). The process of genotyping individuals to determine insertion presence or absence at any given set of loci is a relatively simple task involving easily distinguishable fragments on a simple agarose gel stained with ethidium bromide. Multiplexing of loci is possible (41, 42) and fluorescently labeled primers may also be used if one is interested in automated analysis (43). These features make the analysis of *Alu* elements a robust tool for tracking human geographic origins.

In the following pages, we will describe several techniques that have been used to investigate aspects of primate biology and mobile element biology in the primate lineage. We will not only focus on the human lineage but will also mention some techniques that are widely applicable to other taxa, especially non-human primates. We will describe the laboratory techniques required to investigate questions from the fields of forensics, population biology, phylogenetics, genome evolution, and the biology of the elements themselves. One advantage primate researchers have over many other taxa is the availability of a variety of primate genome sequences to serve as a resource and reference in their work. Many of the laboratory techniques to be described benefit from the availability of these sequences and we suggest the reader reference the companion chapter in this volume dedicated to computational analysis of primate/human mobile elements.

1.1. Forensic Applications

Many of the unique properties of mobile elements make them ideally suited for a variety of forensic applications. This section will focus on the *Alu* element, the most abundant class of SINE in the human genome. Most *Alu* elements have become permanent residents of the human genome and are "fixed present", meaning that all individuals are homozygous for the insertion at a particular locus. The continued expansion of *Alu* elements throughout primate evolution has created several recently integrated "young" subfamilies that are present in the human genome, but largely absent from nonhuman primates (24, 25, 27, 44, 45). Some members of these young *Alu* subfamilies have been inserted in the human genome recently enough that individuals remain polymorphic for the insertion presence/absence. Both fixed and polymorphic *Alu* elements have been utilized successfully as robust forensic tools.

Forensic DNA analysis typically begins with the quantitation of human-specific DNA obtained from the biological sample. This is essential to determine the most appropriate autosomal and Y chromosome analysis strategies to perform (46). Highly sensitive methods for quantitation of human DNA based on *Alu* elements have been reported (46–53). These methods take advantage of the high copy number of fixed *Alu* elements in the human genome to maximize sensitivity. Human DNA quantitation based on *Alu* elements is evolving as the preferred method in the forensic community (50). The method described in this chapter utilizes a subfamily of *Alu* elements, enriched in the human genome as compared to other primate species, to maximize human specificity (52, 53).

Another important forensic use for *Alu* elements is human gender identification (54). Fixed *Alu* insertions on either the X or the Y chromosome provide a simple and reliable system to identify them. *AluSTXa* and *AluSTYa* loci demonstrate 100% accuracy in X and Y chromosome identification. The combination of these two markers provides added assurance that gender identification results are accurate since two completely independent mutations would have to occur to affect the outcome.

When one thinks about forensic DNA analysis what typically comes to mind is obtaining a "match" between a crime scene DNA sample and an alleged criminal suspect, thus "solving the case." Frequently however, tools that narrow the potential pool of suspects are essential precursors to a positive identification. The inferred ancestral origin of a DNA specimen is one type of predictor evidence which can advance a criminal investigation (55). Polymorphic *Alu* insertions have been widely used to study human genetic variation in the world populations (6, 56–60).

1.2. Taxonomic Applications

One of the most productive areas of mobile element application has been in the arena of phylogenetic inference. Numerous difficult questions regarding the evolutionary history of the primate lineage have been successfully addressed using *Alu* elements as tools. For example, Salem et al. (37) confidently resolved the human-chimpanzee-gorilla trichotomy and Ray et al. (38) successfully determined the controversial branching order of three families of platyrrhine (New World) primates. Utilizing retrotransposons as phylogenetic markers has been described a number of times. However, phylogenetic analysis of the primate lineage is unique due to the existence of several "reference" genomes. The human (1), chimpanzee (61), and macaque (62) genomes have been released and the marmoset and orang-utan genomes will likely be released in the near future. These genomes provide a valuable resource in determining potentially informative insertions and primer design.

One important consequence of the hierarchical accumulation of retrotransposons in the genome is the ability to target subfamilies of the retrotransposon family that were active during the evolutionary period of interest. For example, if a researcher's interest is in the recent evolutionary history of tamarins, he or she would want to focus on elements belonging to the *Alu*Ta subfamilies instead of *Alu*Y, *Alu*S, or *Alu*J: the reason being that all the latter families were either inactive during that period or never proliferated in that lineage. *Alu*Ta, on the other hand has been active in the tamarin lineage over the last fifteen to twenty million years and many informative insertions will likely be present. Methods described in the companion chapter on computational analysis can aid researchers in determining the sequences that should be targeted for any particular question.

In laboratories dealing with primate genetics, it is critical that researchers be sure that they are handling DNA from the appropriate taxon. For instance, very often researchers collect or receive DNA that was collected in a "non-invasive" manner (i.e., "divorced" tissues such as hair or feces) (63–65). This is especially true during investigations of the illegal wildlife trade and identification of seized products (64, 66, 67). Even when laboratories produce their own "in-house" genomic DNA via cell culture, cross-contamination can occur among cell cultures and within concurrent large-scale DNA extractions from multiple species. Furthermore, simple mishandling of well documented samples may result in the loss of their labels. Future analyses based on these mistaken identities can be compromised. We will review an *Alu*-based dichotomous key for the resolution of primate sample identity for researchers in this area.

1.3. Structural Impact of Retrotransposons

Among mobile elements, retrotransposons (e.g., L1, *Alu*, and SVA elements) are major endogenous contributors to the creation of structural variation in primate genomes. The tempo and mode of their amplification during the primate radiation have been shown to be lineage-specific events and thus, retrotransposons have had an extensive impact on the evolutionary history of different primate lineages through shaping of their genomic landscape (1, 61, 68–70). Computational analyses of genomic sequence, along with the use of newly developed cell culture assays, suggest that the overall contribution of retrotransposon-mediated genomic variation involves not only the initial integration event but also a variety of recombination events occurring after that integration (e.g., *Alu* retrotransposition-mediated deletions, L1 insertion-mediated deletion, and *Alu* recombination-mediated deletions) (68, 71–73).

Completion of the human and chimpanzee reference genomes allowed whole-genome comparison studies of L1 and *Alu* insertion-mediated variation in these primate lineages.

The results showed that 24 (~1.3%) of the total ~1,800 human-specific L1 insertions are involved in genomic deletions and are directly responsible for the loss of ~18 kb from the human genome (72), whereas, only ~0.2% of human-specific *Alu* insertions are involved in genomic deletions and are responsible for the loss of ~9 kb from the human genome (71). Post-insertion recombination events, however, were shown to have greater genomic impact. Sen et al. (73) identified 492 *Alu* recombination-mediated genomic deletions which resulted in the loss of ~400 kb of human genomic sequence, and ~60% of these deletions are involved in known or predicted genes. Three events actually deleted functional exons from human genes as compared to orthologous chimpanzee genes (73).

Genome alignment studies such as these have helped us to understand the distribution of retrotransposons and provide insight into their impact on host genomes, but tell us little about their mobilization. It has been the development of in vitro cell culture based assays which have allowed us to study the mobilization dynamics of retrotransposons. A companion chapter in this volume is dedicated to computational methods for the analysis of primate/human mobile elements. Therefore, in this section, we will focus on methods which utilize recently developed cell culture assays to study retrotransposition events and consider their genomic impact in cultured human cells.

The transient cultured cell retrotransposition assay was developed by Moran and his colleagues (74, 75). L1.2A was isolated as a potential progenitor of disease-producing L1 insertions into the factor VIII from patient JH-27 (hemophilia A) (76). To investigate whether the L1.2A has the capacity of an autonomous retrotransposon, the sequence was cloned and subcloned into a pCEP4 expression vector including a *mneol* reporter cassette which is comprised of an antisense copy of a *neo* selectable marker, the heterologous SV40 promoter, and a polyadenylation sequence. The *neo* gene is disrupted by an intron in the opposite transcriptional orientation (74). This genetic system could display L1 retrotransposition in cultured cell lines and help to estimate the frequency of L1 autonomous retrotransposition. On the basis of these achievements, 82 out of 89 L1s with intact ORFs that exist in the human genome were cloned, and the retrotranspositional capability of each was predicted in cultured human 143B TK- osteosarcoma cells (77). Moreover, the characterization of new daughter L1 inserts generated by synthetic retrotransposition-competent L1s in cultured human cells demonstrated that L1 retrotransposition events cause genomic instability such as deletions, duplications, translocations, and intra-L1 rearrangements (78–80) and have the potential to provide the host genome with new gene families through L1-mediated transduction (74, 81).

Through the L1-mediated *Alu* retrotransposition assay, the retrotransposed *Alu* elements and their flanking sequences were investigated to confirm the fact that *Alu* elements are indeed mobilized in trans by using the L1 enzymatic machinery. As a result, the new daughter *Alu* inserts derived from a *neo*^{Tet}-marked *Alu* construct were intact without deletion. Their pre-insertion sites were predominantly close to an L1 endonuclease cleavage site consensus (TT^AAAA) and on each side of the *Alu* inserts were the presence of target site duplications (TSDs), one hallmark of authentic *Alu* retrotransposition, generated by the target-site primed reverse transcription process (10, 17, 82). Moreover, it was noteworthy that only ORF2p products (endonuclease and reverse transcriptase domains) of L1-encoded proteins are essential for the *Alu* retrotransposition (14).

The fact that L1 retrotransposition can create genomic deletions in the human genome was revealed by the systems of L1 retrotransposition in cultured human cells and the plasmid-based rescue technique (see Subheading 3.3.3.). It revealed that ~20% of de novo L1 insertions recognized through cultured cell retrotransposition assays caused genomic deletions at the integration site and the size of DNA sequences deleted through these events ranged up to 71 kb (78–80).

The enormous difference in genomic variation observed between in vitro and in vivo forms of investigation could be caused by evolutionary forces (e.g., selection pressure, the number of retrotransposition-competent L1s, and effective population size) and host defense mechanisms (e.g., RNAi, APOBEC, and methylation).

2. Materials

2.1. Forensic Applications

2.1.1. Buffers and Solutions

1. TBE Buffer:
2. TLE Buffer: 10 mM Tris/0.1 mM EDTA

2.1.2. AluSTYa and AluSTXa Loci for Human Gender Identification

1. Oligonucleotide PCR primers for each locus:

   AluSTYa, Forward 5′-CATGTATTTGATGGGGATAGAGG-3′and Reverse 5′-CCTTTTCATCCAACTACCACTGA-3′, Primers for the *Alu* insertion on X chromosomes, *AluSTXa*, Forward 5′-TGAAGAAATTCAGTTCATAGCTTGT-3′and Reverse 5′-CAGGAGATCCTGAGATTATGTGG-3′. For both loci, males are distinguished as having two DNA amplicons (X and Y chromosomes), while females (two X chromosomes) have only a single amplicon (Fig. 1).

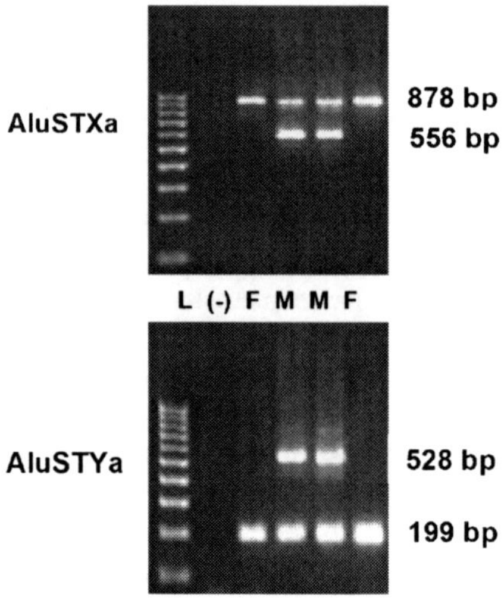

Fig. 1. Mobile element-based human gender determination. An agarose gel chromatograph from the analysis of four individuals using the genetic systems. *AluSTXa* and *AluSTYa* loci are shown. Males are distinguished by the presence of two DNA fragments, while females have a single amplicon. F (female) and M (male) on each lane indicate the gender. L – 100 bp DNA ladder, (–) – negative control consisting of a water template

2. Standard PCR reagents, a thermal-cycler PCR machine, single channel pipettes
3. Horizontal gel electrophoresis unit and power supply; agarose, TBE buffer
4. Ethidium bromide and UV light source to record gel image

2.1.3. Intra-AluYb8 PCR Assay for Human DNA Detection and Quantitation

1. PCR primers: Forward 5′-CTTGCAGTGAGCCGAGATT-3′ and Reverse 5′-GAGACGGAGTCTCGCTCTGTC-3′.
2. TaqMan-MGB probe: 5′FAM or VIC-ACTGCAGTCCGC AGTCCGGCCT-3′-MGBNFQ (Applied Biosystems, Inc.).
3. ABI 7000 Sequence detection system or equivalent and TaqMan PCR core reagents (Kit No. 4304439 or N8080228; Applied Biosystems, Inc).
4. Human Genomic DNA standard (examples: Promega G3041; Novagen #69237)
5. Optical PCR plates and lids (Cat. No. N8010560 and 4360954, respectively; ABI)

2.1.4. Inference of Human Geographic Origins

1. PCR primers for a set of 100 *Alu* insertion polymorphisms and the database of genotypes for 715 individuals of known geographic ancestry from sub-Saharan Africa, East Asia,

Europe, and India (83). These files are available for free download at: (http://batzerlab.lsu.edu; publication #158, Supplementary Data) (55).

2. The program *Structure2.2*: a free software package for using multi-locus genotype data to investigate population structure (84, 85). It is available for free download at http://pritch.bsd.uchicago.edu/software.html.
3. Standard PCR reagents, a thermal cycler PCR machine, single channel pipettes
4. Horizontal gel electrophoresis unit and power supply; agarose, TBE buffer
5. Ethidium bromide and UV light source to record gel image

2.2. Taxonomic Applications

2.2.1. Locus Identification

1. Linker oligonucleotides.
2. *Alu* subfamily-specific oligonucleotides primers.
3. Standard PCR reagents, thermal cycler, single channel pipettes.
4. Restriction enzyme compatible to linker oligonucleotides.
5. Genomic DNA from taxa of interest.
6. Horizontal gel electrophoresis unit and power supply; agarose, TBE buffer.
7. Ethidium bromide and UV light source to analyze PCR products.

2.2.2. Phylogeny Inference

1. Oligonucleotide primers for specific *Alu* insertion loci.
2. Standard PCR reagents, thermal cycler, single channel pipettes.
3. Genomic DNA from taxa of interest.
4. Horizontal gel electrophoresis unit and power supply; agarose, TBE buffer.
5. Ethidium bromide and UV light source to analyze PCR products.

2.2.3. Dichotomous Key Identification

1. Standard set of oligonucleotides primers from Herke et al. (86). Individual researchers must determine if the entire set of oligonucleotides or some subset is required for their particular research.
2. Standard PCR reagents, thermal cycler, single channel pipettes.
3. Genomic DNA from taxa of interest.
4. Horizontal gel electrophoresis unit and power supply; agarose, TBE buffer.
5. Ethidium bromide and UV light source to analyze PCR products.

2.3. Patterns and Processes of Transpositions in Cultured Cells and Within a Genome

2.3.1. Transient Cultured Cell Retrotransposition Assay (70, 71)

1. Obtain *L1.2mneol* expression vector (74): L1.2A, isolated by Dombroski et al., (76), was engineered to create *L1.2mneol* expression vector. To leave a unique *Bam*HI site flanking its 3′ end, the *Bam*HI restriction site at position 4836 of L1.2A was disrupted by site-directed mutagenesis. A *Not*I and a *Sma*I restriction sites were introduced upstream of 5′ UTR and into 3′ UTR at position 5980, respectively. A blunt-ended 2.1 kb *Eco*RI-*Bam*HI fragment bearing the *neo* indicator cassette (87) was ligated to the *Sma*I site resulting in pJCC9 or pJCC8. Thus, the two plasmids contained a tagged L1.2A element, but pJCC8 lacked the L1.2A 5′ UTR. The pJCC9 was restricted with two restriction enzymes of *Not*I and *Bam*HI, generating the 8.1 kb *Not*I-*Bam*HI fragment, which was subcloned into pCEP4 expression vector (Invitrogen) to create pJM101.
2. Neo^{Tet}-marked *Alu* element (14): The *Alu* sequence, integrated into intron 5 of neurofibromatosis type 1 (88), was inserted between 7SL RNA gene enhancer and termination signal using the pDL41–48 plasmid (89). Next, the neo^{Tet} reporter gene (controlled by the SV40 promoter) was inserted upstream of the right monomer poly (A) tail by cleaving the *Alu* sequence-containing plasmid with *Tth*111I (5′-GACNNNGTC-3′) restriction enzyme and ligating with neo^{Tet} PCR product.
3. CMV L1-RP expression vector (14): The cloned L1.2A (76) was inserted, as a blunt-ended *Not*I–*Nsi*I fragment, between the CMV promoter and the SV40 polyadenylation site of pCMVβ (Clontech) to create CMV L1.2 expression vector (90). Next, the L1.2 sequence of the CMV L1.2 expression vector was replaced with $L1_{RP}$ sequence (91) resulting in the CMV L1-RP expression vector.

2.3.2. L1-Mediated Alu Retrotransposition in Cultured Human Cells

1. Liquid N_2 is used to preserve cell lines, either in the vapor phase (–156°C) or in the liquid phase (–196°C). Hela cells (ATCC CCL2) are grown at 37°C in an atmosphere containing 5–7% CO_2 and 100% humidity in high glucose Dulbecco's modified Eagle's medium (DMEM) lacking pyruvate (GIBCO®). DMEM was supplemented with 10% fetal bovine calf serum, 0.4 mM glutamine, and 20 U/mL penicillin-streptomycin (DMEM-complete).
2. Phosphate-buffered saline (PBS) solution: 136.8 mM NaCl, 2.5 mM KCl, 0.8 mM Na_2HPO_4, 1.47 mM KH_2PO_4, 0.9 mM $CaCl_2$, and 0.5 mM $MgCl_2$ ($6H_2O$) in distilled water. The solution is sterilized by using 0.22-µm filter (Millipore) and is stored at room temperature.

3. Geneticin (GIBCO®): Geneticin powder is dissolved in PBS to make a 125 mg/mL stock solution, which is sterilized by using 0.22-μm filter (Millipore) and stored at –20°C.
4. FIX solution: 2% formaldehyde (of a 37% stock solution in ddH_2O) and 0.2% glutaraldehyde (of a 50% stock solution in ddH_2O) in 1 × PBS and is stored at 4°C.

2.3.3. Rescue of L1 Integrants from G418^R Foci (Fig. 5) (74, 92)

1. HeLa genomic DNAs are isolated using the blood and/or cell Midi Prep kit (Qiagen) or the cell and tissue DNA isolation kit (Puregen; Gentra).
2. Plasmid DNAs are purified on Qiagen midi prep columns.
3. For transfection experiments, DNA superhelicity is tested by electrophoresis on 0.6–0.7% agarose-ethidium bromide gels.

3. Methods

3.1. Forensic Applications

3.1.1. Human Gender Identification

1. Dilute *AluSTYa* and *AluSTXa* stock primers to 2 μ*M* in TLE to make a 10× working solution.
2. Obtain DNA from a human male and a human female control (if possible) and dilute all DNA samples to 5 ng/μL for PCR.
3. Set-up PCR reactions with 5 μl (25 ng) of DNA template per 25 μL reaction volume. Prepare a master mix containing PCR reagents per reaction: 1× PCR buffer, 0.2 μ*M* each oligonucleotide primer, 200 u*M* dNTP mix, 1.5 mM $MgCl_2$, and 1 U *Taq* DNA polymerase. Add sterile water for a final volume of 20 μL of mix per reaction. Prepare 20% excess master mix (if you have ten PCR samples, make enough mix for 12 to insure accurate transfer of 20 μL of mix per well).
4. Perform PCR reactions using the following conditions: Initial denaturation for 90 s at 94°C followed by 30–32 cycles of 95°C for 30 s, anneal for 30 s at 58°C (*AluSTYa*) or 60°C (*AluSTXa*), extension at 72°C for 30 s followed by a final extension at 72°C for 2 min.
5. Prepare a 2% agarose gel in 1× TBE buffer containing 0.2 μg/mL ethidium bromide. Load 20 μl of PCR product on gel, size separate PCR amplicons by electrophoresis at 150 V for 1 h, and visualize the genotypes using UV illumination (Fig. 1).

3.1.2. Human DNA Detection and Quantitation

1. Prepare a tenfold serial dilution of human DNA standard from 100 ng to 0.01 pg by first making an aliquot of 20 ng/μL, then diluting a portion of that 1:10 in TLE to make 2 ng/μL, diluting a portion of that 1:10, and so on serially.

2. Use 5 μL of each of the above in duplicate (see Note 1) to prepare a standard curve from 100 ng to 0.01 pg.
3. Use 5 μL of each DNA sample being tested (the unknowns), also in duplicate, in a 50 μL PCR reaction volume.
4. Prepare a master mix using TaqMan PCR core reagents according to the manufacturer's instructions. Each quantitative PCR reaction includes 1× TaqMan PCR buffer, 0.5U AmpErase UNG, 1 μ*M* Intra-*Alu*Yb8 primers from subheading 2.1.3, 100 n*M* TaqMan probe from subheading 2.1.4, 0.5 mM dNTPs, 5.0 mM $MgCl_2$, and 2.5U AmpliTaq Gold DNA polymerase.
5. Add 45 μL master mix to each well containing 5 μL DNA template and carefully seal the optical plate using optical adhesive film (Cat. No. 4360954). Use a plastic sealing spatula or equivalent to avoid touching the optical film with hands.
6. If using an ABI 7000 Prism Sequence Detection System, open a new absolute quantitation file and select the "Setup" icon. Designate each sample well according to standard, unknown, FAM, VIC, etc. Next, select the "Instrument" icon and confirm the PCR cycling conditions listed in the next step, then save the file before starting the run.
7. Perform quantitative PCR using universal PCR cycling conditions as described: 2 min at 50°C for activation of the AmpErase UNG, followed by a denaturation step of 10 min at 95°C to activate the AmpliTaq Gold DNA polymerase, then 40 amplification cycles of denaturation at 95°C for 15 s and 1 min of anneal/extension at 60°C.
8. Following amplification, select the "Results" icon and "amplification plot." Select the wells containing the standards from Step 1 and drag the green threshold bar until it crosses the amplification signal of the standards in the linear phase of amplification (Fig. 2a). Select "Analyze".
9. The ABI Prism 7000 SDS software will calculate the value of each unknown based on the standard curve DNA concentrations.
10. Export the data from the ABI Prism 7000 SDS software into a Microsoft Excel spreadsheet. Calculate the mean and standard deviation for each point on the standard curve and use the Excel "trendline" option to construct the standard curve. Plot each unknown (mean ± SD) along the standard curve to calculate the DNA quantity (Fig. 2b).

3.1.3. Inference of Human Geographic Origins

1. Dilute 100 *Alu* stock primers to 2 μM in TLE to make a 10× working solution for each.

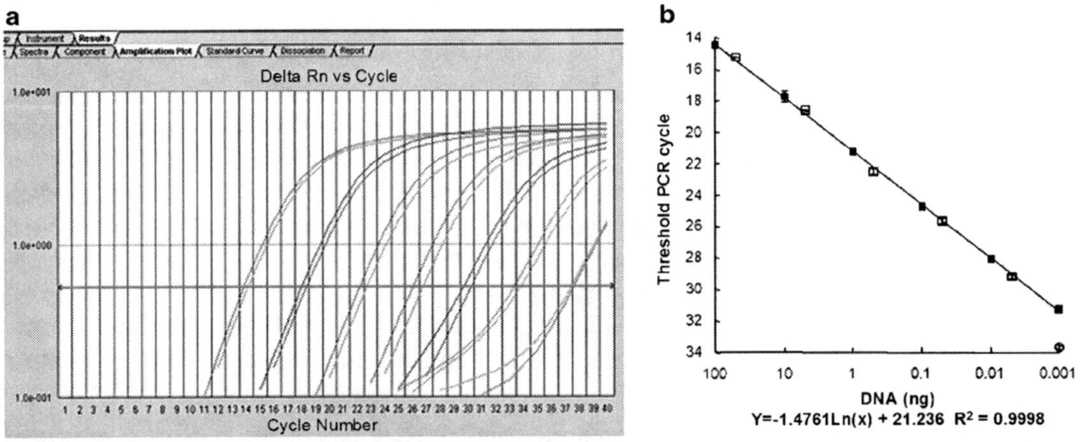

Fig. 2. Quantitative PCR using *Alu* subfamily-specific amplification. Example of a tenfold serial dilution of DNA duplicates using the ABI Prism 7000 Sequence Detection System. (**a**) Fluorescent signal is plotted against PCR cycle number. The threshold cycle (Ct) is defined as the cycle at which the signal crosses the threshold (represented by the horizontal line) during the linear phase of amplification. (**b**) Ct values are exported into a Microsoft Excel spreadsheet where the mean and standard deviation are calculated for each point on the standard curve. Unknown DNA samples are quantified by comparison to the standard curve

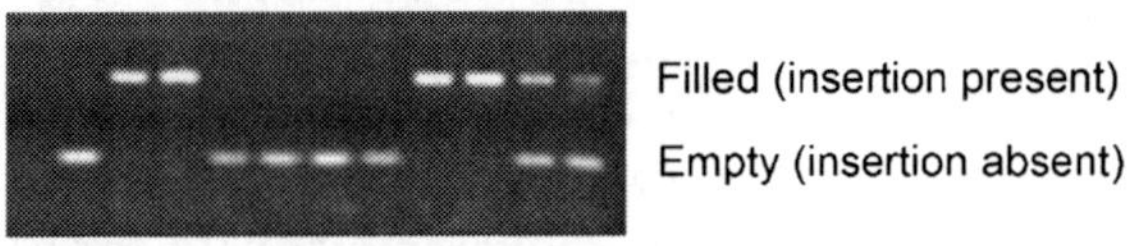

Fig. 3. Gel electrophoresis results using *Alu* insertion loci for human geographic affiliation analysis. The upper band is seen for filled sites and the lower band for empty sites. Individuals exhibiting two bands are presumed to be heterozygous. Individuals for whom only one band is visible indicate a homozygous genotype for either of the alternative states for the locus

2. Perform PCR reactions with at least 10 ng of DNA template per 25 μl reaction volume for 30–32 cycles using the conditions downloaded in subheading 2.1.4.
3. Prepare a 2% agarose gel in 1× TBE buffer containing 0.2 μg/mL ethidium bromide. Load 20 μL of PCR product on gel, size separate PCR amplicons by electrophoresis at 170 V for 1 h, and visualize the genotypes using UV illumination (Fig. 3).
4. Record genotype data in an Excel spreadsheet as: 1, 1 (homozygous present); 1, 0 (heterozygous); 0, 0 (homozygous absent) in rows for each sample as shown in the reference database downloaded in subheading 2.1.4.
5. Prepare the data input file for *Structure* analysis by pasting the genotypes collected for the unknown subjects into a copy of the reference database. Use population code "0" for each unknown.

6. Open *Structure2.2* software by double click on the "gear icon" then select File; New Project (see Note 2).
7. Follow the New Project Wizard steps 1–4: Step 1: Name file and location; Step 2: Number of individuals equals 715 plus the number of your unknowns; Ploidy of data = 2; Number of loci = 100 and Missing data value = –9; Step 3: select box "data file stores data for individuals in a single line"; Step 4: select the following boxes: "individual ID for each individual"; "putative population origin for each individual"; "other extra columns" = 2 for population ID column and continental origin column. Click "Finish" to complete the data input process.
8. Select "Parameter Set"; "New" from the toolbar. Length of Burnin period = 10,000; number of MCMC reps after Burnin = 10,000. On the "Ancestry Model" page select box "Use Population Information." Use the default settings for the remaining tabs and select "OK".
9. Select "Parameter Set"; "Run"; enter number of K populations = 4 at the prompt and hit "OK".
10. The time required to complete the *Structure* analysis run varies depending on your personal computer, the number of individuals being analyzed, and particular parameter settings. Once complete, assess the population assignment of your unknown individuals. Probability of assignment to one of the four pre-defined clusters of at least 80% is a strong indicator of individual ancestry.
11. If the probability of assignment to one of the four pre-defined clusters is less than 80% with significant admixture from one or more of the other clusters, re-run the *Structure* analysis, first assigning the individual to one cluster, and then to each of the other admixed clusters.

3.2. Taxonomic Applications

3.2.1. Locus Identification

Using knowledge of subfamily diagnostic sites, it is a relatively simple task to design primers to experimentally mine a genome with reference to a full genome sequence. The key to this process is to ensure that the diagnostic sites that define the subfamily of interest are well-represented in the primer to be used. Furthermore, it would be advantageous to have the most 3′ base be unique to the subfamily targeted. Several software packages exist to aid in this process.

Identification of potentially informative loci involves the generation of "half-sites" from the genomes of interest. Specifically, a linker ligation protocol first suggested by Munroe et al. (93) and refined by Roy et al. (94) and Ray et al. (38) (Fig. 4) is used to clone the sequences neighboring one side of an insertion. This process involves the digestion of genomic DNA in such a way that

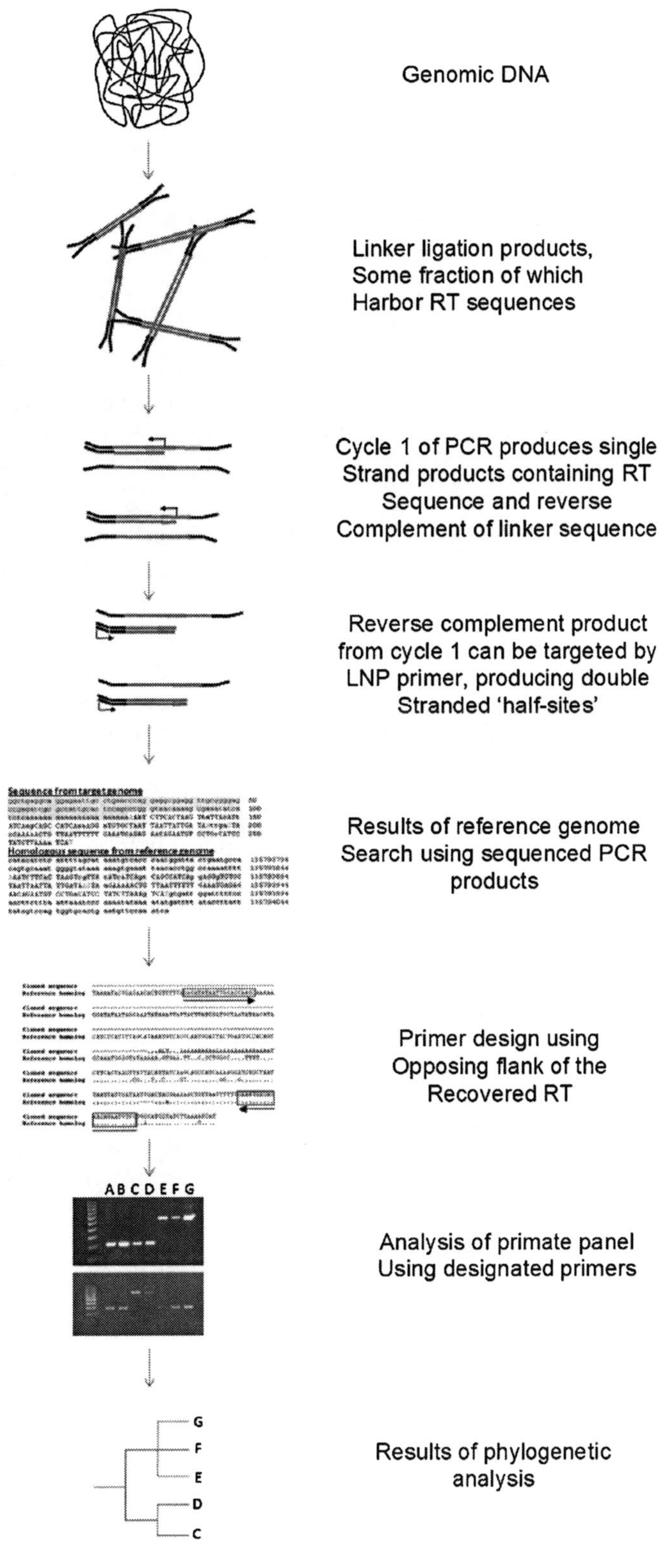

Fig. 4. Schematic representing the steps required to perform a de novo phylogenetic analysis using *Alu* insertion loci

an overhang is produced. That overhang is matched to a set of annealed linkers, which are ligated to the digested genome fragments.

If you have some information on the consensus sequence of the *Alu* subfamily you are targeting, use an alignment of that subfamily consensus and other *Alu* subfamilies to design primers to be used. Primers should be as specific as possible to the subfamily of interest and preferably end with a subfamily specific base. Standard primer design criteria regarding length, GC content, and annealing temperature should be considered (see Notes 3 and 4).

1. Perform a restriction digest of the genome of interest. Five hundred nanograms of genomic DNA from each taxon should be digested using a restriction enzyme leaving an appropriate overhand for the linkers to be ligated to the resulting fragments. For example, we often use the enzyme *Nde*I (CA^TATG), which leaves a 5′ TA overhang. *Nde*I is also a good choice because it does not cut within any known *Alu* subfamily and has a six-base restriction site. This provides an advantage over four-base cutters by producing longer fragments and thus, longer flanking sequences for later computational searches of the reference genome(s). Reactions should be conducted in 120 μl volumes and be followed by heat inactivation of the enzyme at 65°C for 20 min.
2. Produce double stranded linkers by incubating 1 nmol of each linker oligonucleotide (top and bottom) at 94°C for 10 min in a solution of 2× SSC, 10 mM Tris (pH 8). Allow the solution to cool slowly to room temperature. It is important at this point to ensure that your top linker sequences are complementary to the restriction sites that will be produced. For example, when using *Nde*I, we will utilize linkers with the following sequences: *Nde*I_top – TAGAAGGAGAGGACGCTGTCTGTCGAAGG, Universal_bottom – GAGCGAATTCGTCAACATAGCATTTCTGTCCTCTCCTTC. Note the underlined bases in the top linker that will complement the overhang created by the *Nde*I digest.
3. Ligate 12 pmol of the double stranded linkers 0.25 μg of the digested genomic DNA using the ligase manufacturers protocol.
4. Amplify half-sites in 20 μL reactions consisting of the appropriate 1× buffer, 1.5 mM $MgCl_2$, 200 mM dNTPs, 0.25 mM primers (the *Alu*-specific primer and the linker primer, LNP (5′-GAATTCGTCAACATAGCATTTCT-3′)), and 1.5 U Taq polymerase. Amplification conditions that typically work for us follow this temperature regime: 94°C – 2 min, 94°C– 20 s, 62°C – 20 s, 72°C – 1 min, 10 s, for 5 cycles; 94°C – 20 s, 55°C – 20 s, 72°C – 1 min, 10 s, for 25 cycles; 72°C – 3 min.

5. The PCR products will span a range of sizes. Because smaller products (i.e., products with shorter flanking sequences) will be cloned preferentially, we have found it useful to use gel purification to select for fragments of 500–1,000 bp. This ensures that we will obtain enough flanking sequence to increase the probability of finding a single orthologous sequence in the reference genome. We separate the products on a 2% agarose gel and excise the appropriate range. The fragments are then purified using a standard kit such as the Wizard gel purification kit from Promega.
6. Clone the purified PCR products using the TOPO-TA cloning kit for sequencing (Invitrogen) and raise the colonies overnight at 37°C.
7. Select colonies for sequencing by using a sterile toothpick to pick the colonies and incubate in 2–3 ml of Luria Broth (LB) overnight with shaking (200 rpm) at 37°C.
8. Purify the cultures using any of several standard kits. We typically use the Wizard Plus SV Miniprep kit from Promega.
9. Sequencing is performed using any standard method. Our laboratory utilizes the BigDye sequencing reagents from Applied Biosystems and an ABI 3130xl, also from Applied Biosystems. The objective of this step is to obtain enough sequence to verify the presence of the *Alu* insertion and identify the orthologous location in the reference genome.
10. Once the sequence for any given clone has been obtained, the next task is to identify the orthologous sequence in the reference genome. This is typically possible using the web-based Blast-like Alignment Tool (BLAT) hosted at http://genome.ucsc.edu. The search itself is trivial. However, with the expanding number of primate reference genomes available, the choice of genome is important. As of this writing, reference genomes for human, chimpanzee, and macaque might be used. Simply select the reference genome of most closely related to the taxa of interest and input the sequence from the cloned fragment. One of several possible results will be obtained. It is possible that no orthologous sequence will be retrieved. This is not unexpected as there will have been some evolutionary change since the divergence of the two genomes. Often a query will yield a multiplicity of hits. This is often due to the flanking sequence itself being a repetitive region of the genome. For example, a small percentage of the cloned products will likely contain flanking sequences that consist solely of L1 sequence. Unfortunately, these sequences are unlikely to be of much value when designing primers as the resulting primers will have multiple annealing sites in the genome.

The most productive hits are single, high-scoring hits from the genome of interest in which the flanking sequence is unique. When these are encountered, BLAT can be used to expand the coverage of the genome region to determine two important pieces of information. First, you can immediately discover whether the insertion you recovered from the genome of interest is present in the reference genome. This in itself may be a useful information in resolving your phylogeny. Second, you can identify the opposing flanking sequence of the SINE insertion in the reference genome. Using the opposing flank and the flanking sequence from the genome of interest, oligonucleotides primers can be designed using standard methodologies. Primer design should take into account the potential presence of other mobile elements in the flanks. These should be avoided as priming sites for reasons stated earlier.

3.2.2. Phylogeny Inference

1. Prepare a panel of template DNAs to perform your phylogenetic analysis. The panel should include all taxa of interest as well as an appropriate outgroup and negative control (water). Template DNA concentration will be variable depending on the standards of individual laboratories but should be consistent among taxa being examined.
2. Perform amplifications on the panel using appropriate conditions for each primer pair designed using the locus identification protocol mentioned earlier (see Subheading 2.2.1.). Annealing temperatures, $MgCl_2$ concentration, and other factors may differ among primer sets.
3. Use agarose gel electrophoresis to determine insertion patterns for the insertions at each locus. Figure 4 illustrates one pattern obtained from an analysis of New World primate taxa.
4. Each band should be scored as 1 (insertion present) or 0 (insertion absent) for all taxa for which amplification was obtained.
5. While small, there is the possibility that size variation among taxa can be due to some other event that mimics the pattern expected by presence or absence of the *Alu* being assayed. Thus, some method should be used to verify that the source of any size variation is indeed due to the presence or absence of the *Alu* element. DNA sequence analysis provides the most information on each locus but can be cost-prohibitive. One alternative may be to perform hybridization analysis using probes designed from the *Alu* sequence and from the flanking sequences.
6. A matrix of presence/absence of data can be analyzed using any of several available phylogenetic analysis packages including PAUP* (95) and PHYLIP (96). Specific considerations

for phylogeny analysis using SINE insertion data have been previously discussed by Okada and colleagues (28).

3.2.3. An Alu-Based Dichotomous Key

1. Dilute *Alu* stock primers from Herke et al. (86) to 2 μ*M* in TLE to make a 10× working solution for each. The sequences are available for free download at: (http://batzerlab.lsu.edu; publication #190, Supplementary Data).
2. Perform PCR reactions with at least 10 ng of DNA template per 25 μL reaction volume for 30–32 cycles using the conditions downloaded in subheading 2.2.3.
3. Prepare a 2% agarose gel in 1× TBE buffer containing 0.2 μg/mL ethidium bromide. Load 20 μl of PCR product on gel, size separate PCR amplicons by electrophoresis, and visualize the genotypes using UV illumination.

3.3. Patterns and Processes of Transpositions in Cultured Cells and within a Genome

3.3.1. The transient cultured cell retrotransposition assay (75)

1. Plate HeLa cells at 2×10^5 HeLa cells/well in 6-well plates and culture for ~8–14 h at 37°C in an atmosphere containing 5–7% CO_2 and 100% humidity in high glucose DMEM lacking pyruvate (GIBCO®).
2. Transfect cells with 3 μL of FuGene 6 nonliposomal transfection reagents (Roche) and 1 μg of DNA per transfection of HeLa cell in 6-well plates.
3. Co-transfect one set of 6-well plates with equal amount of a reporter plasmid (pGreen Lantern) and an L1 allele tagged with the *mneol* indicator cassette while the others are transfected with only the L1 construct.
4. At three days post-transfection, trypsinize the HeLa cells in the first set of plates and analyze them by using flow cytometry.
   (a) Remove the spent media with a sterile Pasteur pipette.
   (b) Wash the cells with 2 mL PBS (one or two times).
   (c) Remove the PBS and add 0.2–0.3 mL of a modified Versense solution (5 mM EDTA in PBS) which has been pre-warmed to 37°C, and then incubate the plates for 10 min.
   (d) Gently remove the adherent cells from the 6-well plates.
   (e) Transfer the cell suspension to polystyrene tubes by passage through cells snap caps (Falcon) and keep them on ice until flow cytometry analysis.
   (f) Quantify the cells with a Becton Dickinson flow cytometer using a 15 mWatt argon ion laser (488 nm) and fluorescein filter sets (530/30 bandpass).
   (g) Perform data analysis using the CellQuest software.
   (h) The percentage of GFP cells is used to determine the transfection efficiency of each sample.

5. Seed the remaining set at 2×10^5 cells/well in 6-well plates and add 400 μg/mL of G418 to the cells for the selection of L1 retrotransposition.
6. Aspirate the selection media after 12 days (daily re-feeding) and wash the cells in 1 × PBS.
7. Fix the G418^R foci by incubation in FIX solution for 30 min at 4°C.
8. Stain the fixed cells for 30 min with crystal violet (0.2% crystal violet in 5% acetic acid, 2.5% isopropanol) at room temperature and wash with PBS.
9. Determine the retrotransposition efficiency (the number of G418^R foci/the number of transfected cells) using an Oxford Optronics ColCount colony counter.

3.3.2. L1-Mediated Alu Retrotransposition in Cultured Human Cells

1. Plate HeLa cells at 5×10^5 HeLa cell/60-mm dish and grow ~8–14 h at 37°C in an atmosphere containing 5–7% CO_2 and 100% humidity in high glucose DMEM lacking pyruvate (GIBCO®).
2. Co-transfect the dish with 12 μL Lipofectamine and 8 μL Reagent (GIBCO®), 2 μg *neo*^{Tet}-marked *Alu* and 2 μg CMV L1-RP expression vector.
3. For seven days, culture and seed the transfected cells at 5×10^5 cells/100-mm dish.
4. Add 560 μg/mL of Geneticin (GIBCO®) to the cells for G418 selection.
5. After 14 days (daily re-feeding), aspirate the selection media and wash the cells in 1× PBS.
6. Fix the G418^R foci by incubation in FIX solution for 30 min at 4°C.
7. Stain the fixed cells for 30 min with crystal violet (0.2% crystal violet in 5% acetic acid, 2.5% isopropanol) at room temperature and wash with PBS.
8. Determine the retrotransposition efficiency (the number of G418^R foci/the number of transfected cells) using an Oxford Optronics ColCount colony counter.

3.3.3. Rescue of L1 Integrants from G418^R Foci (Fig. 5) (74, 92)

1. Extract HeLa genomic DNA from either a single G418^R focus or small pool (10 to 250) of G418^R foci by using the Puregene cell and tissue DNA isolation kit (Gentra).
2. Perform a restriction digest of the genomic DNA as follow:
   (a) 10 μg of genomic DNA
   (b) 5 μL of 10× buffer (specific to restriction enzyme)

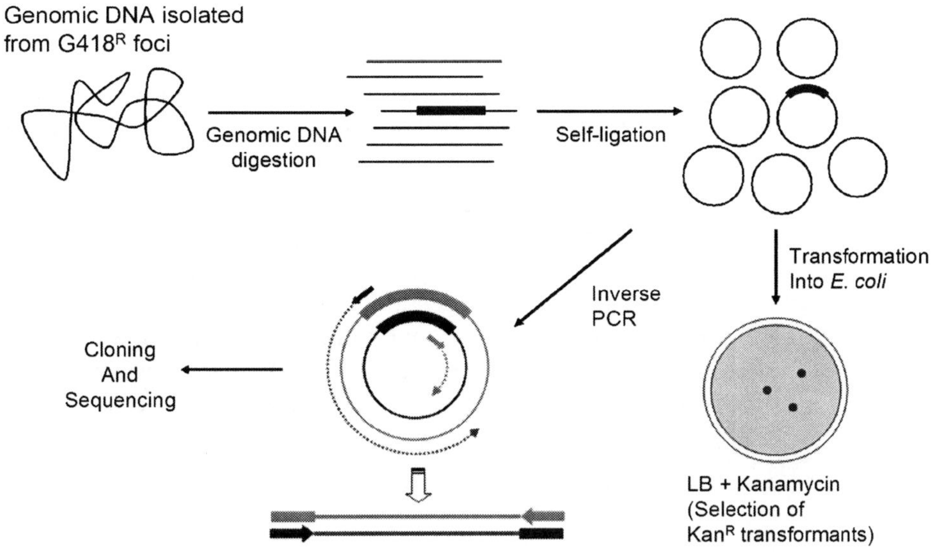

Fig. 5. Rescue of L1 integrants from G418^R foci (74, 92). Genomic DNA is isolated from HeLa cells that have G418-resistance (G418^R) derived from de novo L1 inserts. The new L1 elements (*thick black boxes*) are recovered by either transforming into *E.coli* or performing an inverse PCR

   (c) 20 U of restriction enzyme (HindIII, BglII, BclI, or BamHI (New England Biolabs))
   (d) Add distilled water for a final volume of 50 μL
3. Put the tube in a thermomixer or water bath at 37°C for 2 h (or overnight).
4. Inactivate the restriction enzyme by heating or the Wizard DNA cleanup kit (Promega).
5. Dilute the digested genomic DNA pieces.
6. Prepare the self ligation reaction as follows: (intra-molecule, hopefully)
   (a) 5 μL of the digested genomic DNA (from step 5)
   (b) 50 μL of 10× ligation buffer
   (c) 1 μL of T4 DNA ligase (New England Biolabs)
   (d) Add distilled water for a final volume of 500 μL
7. Put the tube overnight at 14°C.
8. Centrifuge the ligation mixtures through a Microcon-100 at 500×*g* for 14 min.
9. Transform 1 mL of XL1-Blue MRF' $CaCl_2$ competent cells (efficiencies of >1 × 10^8 cfu/μg; Stratagene) with the total concentrated ligation (~1 μg) or perform the inverse PCR using the ligation as a template.

10. Several transformants are visible after overnight growth at 37°C on LB agar plates with 50 µg/mL kanamycin.
11. Extract the plasmid DNA from the resistant clones and then perform restriction mapping, PCR, or DNA sequencing analyses.

The pre-integration site of a de novo L1 insert would be identified by searching BLAT (e.g., hg18; Mar. 2006 freeze) with its each upstream and downstream flanking sequence obtained from above rescue procedure. The acquisition of pre-integration sequences confers the opportunity of additional analyses such as endonuclease cleavage sites, TSD structures, and target sequence alterations derived from the L1 retrotransposition.

4. Notes

1. Quantitative PCR is best prepared using a single channel pipette, electronic repeater style is ideal for duplicates. A typical multi-channel pipette is typically not consistent enough between channels for accurate duplicates.
2. Once *Structure* is open, do not minimize the window. If you leave *Structure* to use another application, minimize that application window when finished and you will return to the *Structure* sub-window to continue the set up. DO NOT click on the *Structure* window – it won't let you continue because the initial sub-window is still open.
3. Primer3: The Primer3 software (http://frodo.wi.mit.edu/) (97) is a useful web interface for designing oligonucleotides primers. The software offers users a variety of options such as the size range of PCR product, primer size, GC content of the primer, and oligonucleotide-melting temperature, which allows users to design optimal primers. In addition, it is linked with BLAST-Like Alignment Tool (BLAT) web browser (http://genome.ucsc.edu/cgi-bin/hgBlat) showing the genomic positions and sequences of PCR products which could be amplified by the primers and thus users are readily able to figure out whether the primers are accurate or specific to their genomic target region.
4. OligoCalc (Oligonucleotide Properties Calculator): The OligoCalc (98) is a web-accessible (http://www.basic.northwestern.edu/biotools/oligocalc.html) and estimates properties for single-stranded DNA and RNA. Features important to consider include self-complementarity, potential hairpin loop formation, and oligonucleotide-melting temperature with and without salt conditions.

Acknowledgments

Our research is supported by National Science Foundation BCS-0218338 (MAB) and EPS-0346411 (MAB), National Institutes of Health RO1 GM59290 (MAB) and PO1 AG022064 (MAB), and the State of Louisiana Board of Regents Support Fund (MAB). DAR is supported by the Eberly College of Arts and Sciences at West Virginia State University.

References

1. Lander, E. S., Linton, L. M., Birren, B., Nusbaum, C., Zody, M. C., Baldwin, J., *et al.* (2001). Initial sequencing and analysis of the human genome. *Nature* **409**, 860–921.
2. Waterston, R. H., Lindblad-Toh, K., Birney, E., Rogers, J., Abril, J. F., Agarwal, P., *et al.* (2002). Initial sequencing and comparative analysis of the mouse genome. *Nature* **420**, 520–62.
3. Berg, D. E. & Howe, M. M., Eds. (1989). Mobile DNA. Washington, DC: American Society for Microbiology.
4. Pace, J. K., 2nd & Feschotte, C. (2007). The evolutionary history of human DNA transposons: evidence for intense activity in the primate lineage. *Genome Research* **17**, 422–32.
5. Deininger, P. L., Batzer, M. A. (1993). Evolution of retroposons. *Evolutionary Biology* **27**, 157–196.
6. Batzer, M. A., Deininger, P. L. (2002). *Alu* repeats and human genomic diversity. *Nature Reviews Genetics* **3**, 370–9.
7. Deininger, P. L, Batzer, M. A. (1999). Alu repeats and human disease. *Mol Genet Metab* **67**, 183–93.
8. Kajikawa, M., Okada, N. (2002). LINEs mobilize SINEs in the eel through a shared 3 sequence. *Cell* **111**, 433–44.
9. Kazazian, H. H., Jr., Moran, J. V. (1998). The impact of L1 retrotransposons on the human genome. *Nature Genetics* **19**, 19–24.
10. Luan, D. D., Korman, M. H., Jakubczak, J. L. & Eickbush, T. H. (1993). Reverse transcription of R2Bm RNA is primed by a nick at the chromosomal target site: a mechanism for non-LTR retrotransposition. *Cell* **72**, 595–605.
11. Wang, W., Kirkness, E. F. (2005). Short interspersed elements (SINEs) are a major source of canine genomic diversity. *Genome Research* **15**, 1798–808.
12. Sinnett, D., Richer, C., Deragon, J. M. & Labuda, D. (1992). *Alu* RNA transcripts in human embryonal carcinoma cells. Model of post-transcriptional selection of master sequences. *Journal of Molecular Biology* **226**, 689–706.
13. Ostertag, E. M., Goodier, J. L., Zhang, Y. & Kazazian, H. H., Jr. (2003). SVA elements are nonautonomous retrotransposons that cause disease in humans. *American Journal of Human Genetics* **73**, 1444–51.
14. Dewannieux, M., Esnault, C. & Heidmann, T. (2003). LINE-mediated retrotransposition of marked Alu sequences. *Nature Genetics* **35**, 41–8.
15. Boeke, J. D. (1997). LINEs and Alus–the polyA connection. *Nature Genetics* **16**, 6–7.
16. Feng, Q., Moran, J. V., Kazazian, H. H., Jr. & Boeke, J. D. (1996). Human L1 retrotransposon encodes a conserved endonuclease required for retrotransposition. *Cell* **87**, 905–16.
17. Jurka, J. (1997). Sequence patterns indicate an enzymatic involvement in integration of mammalian retroposons. *Proceedings of the National Academy of Sciences of the United States of America* **94**, 1872–7.
18. Deininger, P. L., Batzer, M. A., Hutchison, C. A., 3rd, Edgell, M. H. (1992). Master genes in mammalian repetitive DNA amplification. *Trends in Genetics* **8**, 307–11.
19. Cordaux, R., Hedges, D. J. & Batzer, M. A. (2004). Retrotransposition of Alu elements: how many sources? *Trends in Genetics* **20**, 464–7.
20. Matera, A. G., Hellmann, U., Hintz, M. F., Schmid, C. W. (1990). Recently transposed Alu repeats result from multiple source genes. *Nucleic Acids Research* **18**, 6019–23.
21. Shen, M. R., Batzer, M. A. & Deininger, P. L. (1991). Evolution of the master *Alu* gene(s). *Journal of Molecular Evolution* **33**, 311–20.

22. Batzer, M. A., Arcot, S. S., Phinney, J. W., Alegria-Hartman, M., Kass, D. H., Milligan, S. M., *et al.* (1996). Genetic variation of recent *Alu* insertions in human populations. *Journal of Molecular Evolution* **42**, 22–9.

23. Kapitonov, V. & Jurka, J. (1996). The age of *Alu* subfamilies. *Journal of Molecular Evolution* **42**, 59–65.

24. Carroll, M. L., Roy-Engel, A. M., Nguyen, S. V., Salem, A. H., Vogel, E., Vincent, B., *et al.* (2001). Large-scale analysis of the *Alu* Ya5 and Yb8 subfamilies and their contribution to human genomic diversity. *Journal of Molecular Biology* **311**, 17–40.

25. Carter, A. B., Salem, A. H., Hedges, D. J., Keegan, C. N., Kimball, B., Walker, J. A., *et al.* (2004). Genome-wide analysis of the human Alu Yb-lineage. *Human Genomics* **1**, 167–78.

26. Han, K., Xing, J., Wang, H., Hedges, D. J., Garber, R. K., Cordaux, R. & Batzer, M. A. (2005). Under the genomic radar: the stealth model of Alu amplification. *Genome Research* **15**, 655–64.

27. Otieno, A. C., Carter, A. B., Hedges, D. J., Walker, J. A., Ray, D. A., Garber, R. K., *et al.* (2004). Analysis of the human Alu Ya-lineage. *Journal of Molecular Biology* **342**, 109–18.

28. Okada, N., Shedlock, A. M. & Nikaido, M. (2004). Retroposon mapping in molecular systematics. In: *Mobile Genetic Elements: Protocols and Genomic Applications*, Vol. 260, pp. 189–226. Humana Press, Totowa, NJ.

29. Ray, D. A. (2007). SINEs of progress: Mobile element applications to molecular ecology. *Molecular Ecology* **16**, 19–33.

30. Ray, D. A., Xing, J., Salem, A.-H., Batzer, M. A. (2006). SINEs of a *nearly* perfect character. *Systematic Biology* **55**, 928–935.

31. Shedlock, A. M., Okada, N. (2000). SINE insertions: powerful tools for molecular systematics. *Bioessays* **22**, 148–60.

32. Shedlock, A. M., Takahashi, K., Okada, N. (2004). SINEs of speciation: tracking lineages with retroposons. *Trends in Ecology and Evolution* **19**, 545–553.

33. Zietkiewicz, E., Richer, C., Labuda, D. (1999). Phylogenetic affinities of tarsier in the context of primate Alu repeats. *Molecular Phylogenetics and Evolution* **11**, 77–83.

34. Watanabe, M., Nikaido, M., Tsuda, T. T., Inoko, H., Mindell, D. P., Murata, K. & Okada, N. (2006). The rise and fall of the CR1 subfamily in the lineage leading to penguins. *Gene* **365**, 57–66.

35. Takahashi, K., Terai, Y., Nishida, M., Okada, N. (2001). Phylogenetic relationships and ancient incomplete lineage sorting among cichlid fishes in Lake Tanganyika as revealed by analysis of the insertion of retroposons. *Molecular Biology and Evolution* **18**, 2057–66.

36. Sasaki, T., Takahashi, K., Nikaido, M., Miura, S., Yasukawa, Y. & Okada, N. (2004). First application of the SINE (short interspersed repetitive element) method to infer phylogenetic relationships in reptiles: an example from the turtle superfamily Testudinoidea. *Molecular Biology and Evolution* **21**, 705–15.

37. Salem, A. H., Ray, D. A., Xing, J., Callinan, P. A., Myers, J. S., Hedges, D. J., *et al.* (2003). *Alu* elements and hominid phylogenetics. *Proceedings of the National Academy of Sciences of the United States of America* **100**, 12787–91.

38. Ray, D. A., Xing, J., Hedges, D. J., Hall, M. A., Laborde, M. E., Anders, B. A., *et al.* (2005). Alu insertion loci and platyrrhine primate phylogeny. *Molecular Phylogenetics and Evolution* **35**, 117–26.

39. Murata, S., Takasaki, N., Saitoh, M., Okada, N. (1993). Determination of the phylogenetic relationships among Pacific salmonids by using short interspersed elements (SINEs) as temporal landmarks of evolution. *Proceedings of the National Academy of Sciences of the United States of America* **90**, 6995–9.

40. Cordaux, R., Lee, J., Dinoso, L., Batzer, M. A. (2006). Recently integrated Alu retrotransposons are essentially neutral residents of the human genome. *Gene* **373**, 138–44.

41. Kass, D. H. (2003). Generation of human DNA profiles by *Alu*-based multiplex polymerase chain reaction. *Analytical Biochemistry* **321**, 146–9.

42. Thomas, E., Herrera, R. J. (1998). Multiplex polymerase chain reaction of *Alu* polymorphic insertions. *Electrophoresis* **19**, 2373–9.

43. Flavell, A. J., Knox, M. R., Pearce, S. R. & Ellis, T. H. (1998). Retrotransposon-based insertion polymorphisms (RBIP) for high throughput marker analysis. *Plant J* **16**, 643–50.

44. Salem, A. H., Ray, D. A., Hedges, D. J., Jurka, J. & Batzer, M. A. (2005). Analysis of the human Alu Ye lineage. *BMC Evolutionary Biology* **5**, 18.

45. Xing, J., Salem, A. H., Hedges, D. J., Kilroy, G. E., Watkins, W. S., Schienman, J. E., *et al.* (2003). Comprehensive analysis of two *Alu* Yd subfamilies. *Journal of Molecular Evolution* **57**, S76–S89.

46. Shewale, J. G., Schneida, E., Wilson, J., Walker, J. A., Batzer, M. A. & Sinha, S. K. (2007). Human genomic DNA quantitation

system, H-Quant: development and validation for use in forensic casework. *Journal of Forensic Science* **52**, 364–70.

47. Nicklas, J. A. & Buel, E. (2003). Development of an Alu-based, real-time PCR method for quantitation of human DNA in forensic samples. *Journal of Forensic Science* **48**, 936–44.
48. Nicklas, J. A., Buel, E. (2003). Quantification of DNA in forensic samples. *Anal Bioanal Chem* **376**, 1160–7.
49. Nicklas, J. A., Buel, E. (2005). An Alu-based, MGB Eclipse real-time PCR method for quantitation of human DNA in forensic samples. *Journal of Forensic Science* **50**, 1081–90.
50. Nicklas, J. A., Buel, E. (2006). Simultaneous determination of total human and male DNA using a duplex real-time PCR assay. *Journal of Forensic Science* **51**, 1005–15.
51. Sifis, M. E., Both, K., Burgoyne, L. A. (2002). A more sensitive method for the quantitation of genomic DNA by *Alu* amplification. *Journal of Forensic Sciences* **47**, 589–92.
52. Walker, J. A., Kilroy, G. E., Xing, J., Shewale, J., Sinha, S. K. & Batzer, M. A. (2003). Human DNA quantitation using *Alu* element-based polymerase chain reaction. *Analytical Biochemistry* **315**, 122–8.
53. Walker, J. A., Hedges, D. J., Perodeau, B. P., Landry, K. E., Stoilova, N., Laborde, M. E., *et al.* (2005). Multiplex polymerase chain reaction for simultaneous quantitation of human nuclear, mitochondrial, and male Y-chromosome DNA: application in human identification. *Analytical Biochemistry* **337**, 89–97.
54. Hedges, D. J., Walker, J. A., Callinan, P. A., Shewale, J. G., Sinha, S. K., Batzer, M. A. (2003). Mobile element-based assay for human gender determination. *Analytical Biochemistry* **312**, 77–9.
55. Ray, D. A., Walker, J. A., Hall, A., Llewellyn, B., Ballantyne, J., Christian, A. T., *et al.* (2005). Inference of human geographic origins using Alu insertion polymorphisms. *Forensic Science International* **153**, 117–24.
56. Bamshad, M. J., Wooding, S., Watkins, W. S., Ostler, C. T., Batzer, M. A. & Jorde, L. B. (2003). Human population genetic structure and inference of group membership. *American Journal of Human Genetics* **72**, 578–89.
57. Jorde, L. B., Watkins, W. S., Bamshad, M. J., Dixon, M. E., Ricker, C. E., Seielstad, M. T. & Batzer, M. A. (2000). The distribution of human genetic diversity: a comparison of mitochondrial, autosomal, and Y-chromosome data. *American Journal of Human Genetics* **66**, 979–88.
58. Roy-Engel, A. M., Carroll, M. L., Vogel, E., Garber, R. K., Nguyen, S. V., Salem, A. H., *et al.* (2001). Alu insertion polymorphisms for the study of human genomic diversity. *Genetics* **159**, 279–90.
59. Salem, A. H., Kilroy, G. E., Watkins, W. S., Jorde, L. B., Batzer, M. A. (2003). Recently integrated Alu elements and human genomic diversity. *Molecular Biology and Evolution* **20**, 1349–61.
60. Watkins, W. S., Rogers, A. R., Ostler, C. T., Wooding, S., Bamshad, M. J., Brassington, A. M., *et al.* (2003). Genetic variation among world populations: inferences from 100 Alu insertion polymorphisms. *Genome Research* **13**, 1607–18.
61. CSAC. (2005). Initial sequence of the chimpanzee genome and comparison with the human genome. *Nature* **437**, 69–87.
62. Gibbs, R. A., Rogers, J., Katze, M. G., Bumgarner, R., Weinstock, G. M., Mardis, E. R., *et al.* (2007). Evolutionary and biomedical insights from the rhesus macaque genome. *Science* **316**, 222–34.
63. Kohn, M., Knauer, F., Stoffella, A., Schroder, W., Paabo, S. (1995). Conservation genetics of the European brown bear–a study using excremental PCR of nuclear and mitochondrial sequences. *Mol Ecol* **4**, 95–103.
64. Matsubara, M., Basabose, A. K., Omari, I., Kaleme, K., Kizungu, B., Sikubwabo, K., *et al.* (2005). Species and sex identification of western lowland gorillas (Gorilla gorilla gorilla), eastern lowland gorillas (Gorilla beringei graueri) and humans. *Primates* **46**, 199–202.
65. Taberlet, P., Camarra, J. J., Griffin, S., Uhres, E., Hanotte, O., Waits, L. P., *et al.* (1997). Noninvasive genetic tracking of the endangered Pyrenean brown bear population. *Molecular Ecology* **6**, 869–76.
66. Yan, P., Wu, X. B., Shi, Y., Gu, C. M., Wang, R. P., Wang, C. L. (2005). Identification of Chinese alligators (Alligator sinensis) meat by diagnostic PCR of the mitochondrial cytochrome b gene. *Biological Conservation* **121**, 45–51.
67. Domingo-Roura, X., Marmi, J., Ferrando, A., Lopez-Giraldez, F., Macdonald, D. W. & Jansman, H. A. H. (2006). Badger hair in shaving brushes comes from protected Eurasian badgers. *Biological Conservation* **128**, 425–430.
68. Han, K., Lee, J., Meyer, T. J., Wang, J., Sen, S. K., Srikanta, D., Liang, P. & Batzer, M. A. (2007). Alu recombination-mediated structural deletions in the chimpanzee genome. *PLoS Genet* **3**, 1939–49.

69. Khan, H., Smit, A., Boissinot, S. (2006). Molecular evolution and tempo of amplification of human LINE-1 retrotransposons since the origin of primates. *Genome Res* **16**, 78–87.
70. Lee, J., Cordaux, R., Han, K., Wang, J., Hedges, D. J., Liang, P., Batzer, M. A. (2007). Different evolutionary fates of recently integrated human and chimpanzee LINE-1 retrotransposons. *Gene* **390**, 18–27.
71. Callinan, P. A., Wang, J., Herke, S. W., Garber, R. K., Liang, P., Batzer, M. A. (2005). Alu retrotransposition-mediated deletion. *Journal of Molecular Biology* **348**, 791–800.
72. Han, K., Sen, S. K., Wang, J., Callinan, P. A., Lee, J., Cordaux, R., Liang, P. & Batzer, M. A. (2005). Genomic rearrangements by LINE-1 insertion-mediated deletion in the human and chimpanzee lineages. *Nucleic Acids Res* **33**, 4040–52.
73. Sen, S. K., Han, K., Wang, J., Lee, J., Wang, H., Callinan, P. A., *et al.* (2006). Human Genomic Deletions Mediated by Recombination between Alu Elements. *American Journal of Human Genetics* **79**, 41–53.
74. Moran, J. V., Holmes, S. E., Naas, T. P., DeBerardinis, R. J., Boeke, J. D. & Kazazian, H. H., Jr. (1996). High frequency retrotransposition in cultured mammalian cells. *Cell* **87**, 917–27.
75. Wei, W., Morrish, T. A., Alisch, R. S., Moran, J. V. (2000). A transient assay reveals that cultured human cells can accommodate multiple LINE-1 retrotransposition events. *Analytical Biochemistry* **284**, 435–8.
76. Dombroski, B. A., Mathias, S. L., Nanthakumar, E., Scott, A. F., Kazazian, H. H., Jr. (1991). Isolation of an active human transposable element. *Science* **254**, 1805–8.
77. Brouha, B., Schustak, J., Badge, R. M., Lutz-Prigge, S., Farley, A. H., Moran, J. V., Kazazian, H. H., Jr. (2003). Hot L1s account for the bulk of retrotransposition in the human population. *Proceedings of the National Academy of Sciences of the United States of America* **100**, 5280–5.
78. Gilbert, N., Lutz, S., Morrish, T. A. & Moran, J. V. (2005). Multiple fates of L1 retrotransposition intermediates in cultured human cells. *Molecular and Cellular Biology* **25**, 7780–95.
79. Gilbert, N., Lutz-Prigge, S., Moran, J. V. (2002). Genomic deletions created upon LINE-1 retrotransposition. *Cell* **110**, 315–25.
80. Symer, D. E., Connelly, C., Szak, S. T., Caputo, E. M., Cost, G. J., Parmigiani, G., Boeke, J. D. (2002). Human l1 retrotransposition is associated with genetic instability in vivo. *Cell* **110**, 327–38.
81. Moran, J. V., DeBerardinis, R. J., Kazazian, H. H., Jr. (1999). Exon shuffling by L1 retrotransposition. *Science* **283**, 1530–4.
82. Cost, G. J., Boeke, J. D. (1998). Targeting of human retrotransposon integration is directed by the specificity of the L1 endonuclease for regions of unusual DNA structure. *Biochemistry* **37**, 18081–93.
83. Watkins, W. S., Rogers, A. R., Ostler, C. T., Wooding, S., Bamshad, M. J., Brassington, A. M., *et al.* (2003). Genetic variation among world populations: inferences from 100 *Alu* insertion polymorphisms. *Genome Research* **13**, 1607–18.
84. Pritchard, J. K., Stephens, M., Donnelly, P. (2000). Inference of population structure using multilocus genotype data. *Genetics* **155**, 945–59.
85. Falush, D., Stephens, M., Pritchard, J. K. (2003). Inference of population structure using multilocus genotype data: linked loci and correlated allele frequencies. *Genetics* **164**, 1567–87.
86. Herke, S. W., Xing, J., Ray, D. A., Zimmerman, J. W., Cordaux, R., Batzer, M. A. (2007). A SINE-based dichotomous key for primate identification. *Gene* **390**.
87. Holmes, S. E., Dombroski, B. A., Krebs, C. M., Boehm, C. D. & Kazazian, H. H., Jr. (1994). A new retrotransposable human L1 element from the LRE2 locus on chromosome 1q produces a chimaeric insertion. *Nature Genetics* **7**, 143–8.
88. Wallace, M. R., Andersen, L. B., Saulino, A. M., Gregory, P. E., Glover, T. W., Collins, F. S. (1991). A de novo Alu insertion results in neurofibromatosis type 1. *Nature* **353**, 864–6.
89. Ullu, E. & Tschudi, C. (1984). *Alu* sequences are processed 7SL RNA genes. *Nature* **312**, 171–2.
90. Dhellin, O., Maestre, J., Heidmann, T. (1997). Functional differences between the human LINE retrotransposon and retroviral reverse transcriptases for in vivo mRNA reverse transcription. *Embo J* **16**, 6590–602.
91. Kimberland, M. L., Divoky, V., Prchal, J., Schwahn, U., Berger, W. & Kazazian, H. H., Jr. (1999). Full-length human L1 insertions retain the capacity for high frequency retrotransposition in cultured cells. *Human Molecular Genetics* **8**, 1557–60.
92. Morrish, T. A., Gilbert, N., Myers, J. S., Vincent, B. J., Stamato, T. D., Taccioli,

G. E., *et al.* (2002). DNA repair mediated by endonuclease-independent LINE-1 retrotransposition. *Nature Genetics* **31**, 159–65.
93. Munroe, D. J., Haas, M., Bric, E., Whitton, T., Aburatani, H., Hunter, K., Ward, D., Housman, D. E. (1994). IRE-bubble PCR: a rapid method for efficient and representative amplification of human genomic DNA sequences from complex sources. *Genomics* **19**, 506–14.
94. Roy, A. M., Carroll, M. L., Kass, D. H., Nguyen, S. V., Salem, A. H., Batzer, M. A., Deininger, P. L. (1999). Recently integrated human *Alu* repeats: finding needles in the haystack. *Genetica* **107**, 149–61.
95. Swofford, D. L. (2002). PAUP: Phylogenetic analysis using parsimony (*and Other Methods) 4.0b10 edit. Sinauer Associates, Sunderland, Massachusetts.
96. Felsenstein, J. (1989). PHYLIP - Phylogeny Inference Package (Version 3.2). *Cladistics* **5**, 164–166.
97. Rozen, S. & Skaletsky, H. (2000). Primer3 on the WWW for general users and for biologist programmers. *Methods in Molecular Biology* **132**, 365–86.
98. Kibbe, W. A. (2007). OligoCalc: an online oligonucleotide properties calculator. *Nucleic Acids Research* **35**, W43–6.

Chapter 10

Practical Informatics Approaches to Microsatellite and Variable Number Tandem Repeat Analysis

Gerome Breen

Abstract

The second most common source of genetic variation after SNPs is polymorphic tandem repeats, the alleles of which consist of a variable number of repeated units that can be either small (e.g., CA) or large (to >100 nucleotides in length). There are perhaps over half a million of these in the human genome. They have been implicated as functional promoter polymorphisms acting as common genetic risk factors for complex disorders (in diabetes and depression), as pathogenic mutations (Spinocerebellar Ataxias, Huntington's Disease) and in association mapping, linkage and forensics, but while they enjoyed much success and use in early genetic linkage and association studies, they have recently been neglected. While SNPs are markers of great utility in genetic studies, different alleles of a polymorphic tandem repeat represent a very large physical and chemical change to a stretch of DNA sequence. They can act variously as: (a) functional elements binding transcription factors and other proteins that inhibit or promote expression; (b) motif elements affecting the efficiency of mRNA splicing; and (c) elements having physical effects, such as varying the spacing between functional motifs or in altering the structure and melting properties of DNA in their proximity. For these reasons, they are very good a priori functional candidates. Geneticists wishing to work with these polymorphisms need to know how to find them in sequence, use their annotation in genome browsers and online databases, use specialist bioinformatics web-tools for their analysis, and how to go about analyzing them in the lab and for genetic association.

Key words: Tandem repeat, VNTR, Microsatellite, Function, UCSC, TRDB

1. Introduction

The explosion of genetic data availability in the last decade has opened up many new avenues for the application of genetics to the improvement of human health, particularly the common and complex disorders, which previously largely defied researchers seeking to understand their aetiology. However, the utilisation of these data in healthcare and by the pharmaceutical industry is still in its infancy, with the current focus being almost solely on two

Michael R. Barnes and Gerome Breen (eds.), *Genetic Variation: Methods and Protocols*, Methods in Molecular Biology, vol. 628,
DOI 10.1007/978-1-60327-367-1_10,

forms of genetic variation, single nucleotide polymorphisms (SNPs), involving the substitution of one base in DNA for another and Copy Number Variations (CNV), involving the deletion and multiplication of regions >1,000 bp. However, there is currently a relative paucity of funding and general interest in research on a third form of variation, called microsatellites and VNTRs. These highly variable tandem repeats are variously known as microsatellites (1–6 bp motifs), and minisatellites (6–500 bp repeat units) as well as other names such as STRs and VNTRs. To avoid confusion, we shall use the term VNTR (Variable Number Tandem Repeat) to refer to both microsatellites and minisatellites. There are >600,000 candidate VNTRs in the human genome (http://genome.ucsc.edu Simple Repeat Annotation), and many occur in gene regulatory regions (http://discovery.swmed.edu/potion/html/statistics.shtml). This means they are currently second only to SNPs (Six to eight million) in their global frequency in the human genome. Before examining the informatics methods that can be used to find, select, and analyze the polymorphisms, we outline the biological case for their continued importance in genetics.

VNTRs have been reported as being randomly distributed in genomes and not associated with genes (1). In the yeast genome, microsatellites are preferentially located in non-coding DNA (2, 3), except for trinucleotide repeats, which are preferentially located in coding regions. Exonic trinucleotide repeat mutation events avoid frameshifts, and these repeats cluster in regulatory genes where their proneness to mutation and recombination may allow for the rapid evolution of protein sequences (4, 5). However, many mono-and dinucleotide repeats are found in 3 untranslated regions and in introns in yeast, where they may play an evolutionary role in exon shuffling and gene conversion (6). The role of VNTRs in higher eukaryotes is more controversial, with most VNTRs being assumed to be of neutral function, although it has been hypothesized that the overrepresentation of trinucleotides and trinucleotide diseases in humans may be a cost of the increased flexibility of protein structural evolution that in-frame repeats confer on proteins involved in the evolution of the human brain (7). Additional evidence from human–chimpanzee comparisons suggests that humans are relatively deficient in mono-nucleotide repeats but have over representations of larger VNTRs (8). Despite examples such as the Insulin promoter VNTR (9), the overall contribution of VNTRs to the function and expression of genes has not been quantified.

The intrinsic properties and high mutability of tandem repeat sequences also make them good candidates for mapping and function because a change in the number of repeats has a large chemical–physical effect on the DNA sequence, which can lead to changes in gene expression and, in the cases of repeat sequences which code for amino acids, hyper-variability in protein coding sequences. Although the lack of a systematic study of all VNTRs in the genome, analogous to the HAPMAP for SNPs, does not allow us to exactly estimate their functional role, the cumulative evidence, some of which is referenced below, is impressive. The most prominent functional VNTRs are those that have been implicated in monogenic human disease, with microsatellites such as CAG triplet repeats within transcribed regions causing a variety of neurological disorders (10). Larger non-coding VNTRs such as the insulin gene VNTR have been implicated in diabetes (11) and in causing gene-environment interactions in depression (12), attention deficit hyperactivity disorder (13), and cocaine addiction (14). VNTRs affect gene expression (15), transcript splicing (16), and recombination. Thus, multiple strands of evidence suggest that polymorphic tandem repeats are often functional, and that they contribute to disease themselves.

These are frequently argued against in practice, and thus it is perhaps useful to outline in more detail why VNTRs may be functional. First of all, it is obvious that a VNTR encoding a binding site will present an array of recognition sequences to a protein interacting with DNA. Leonid Mirny and colleagues recently outlined new ideas and evidence about how transcription factors find their binding sites in the genome (17). Among other things, they find that after a TF disassociates from its binding site it first slides or hops a distance along the DNA, on average ~660 bp, scanning it for binding sites and if it fails to find another one, then it will leave the DNA strand to embark on a slow genome wide search for a binding site. In this scenario, if a TF is bound to one repeat unit of a VNTR even weakly, once attracted, it is very likely to stay in the same region of DNA even after initial disassociation because of sliding over and attaching to another copy of the binding site alongside the one it has left. The same concept can be generalized to explain why VNTRs may be functional in general – any given sequence motif with weak functional properties repeated multiple times may become a strongly functional element.

This concept has been studied from the reverse angle with the demonstration that identical TF binding sites (TFBSs) cluster in cis-regulatory modules of 100 bp–1 kb in length (18, 19). This arrangement is suggestive of a VNTR type arrangement, and it is possible to find several examples where clusters of conserved TFBSs are encode by a large VNTR. Figure 1 shows tandemly repeated human-rat-mouse conserved TFBSs in a non-repeat masked 7 copy, 19 bp unit. Further evidence was found by Ji et al. (20), who reanalyzed chromatin immunoprecipitation on microarray and other similar (ChIP-CHIP and ChIP-PET) datasets from experiments designed to locate mammalian TFBSs at a

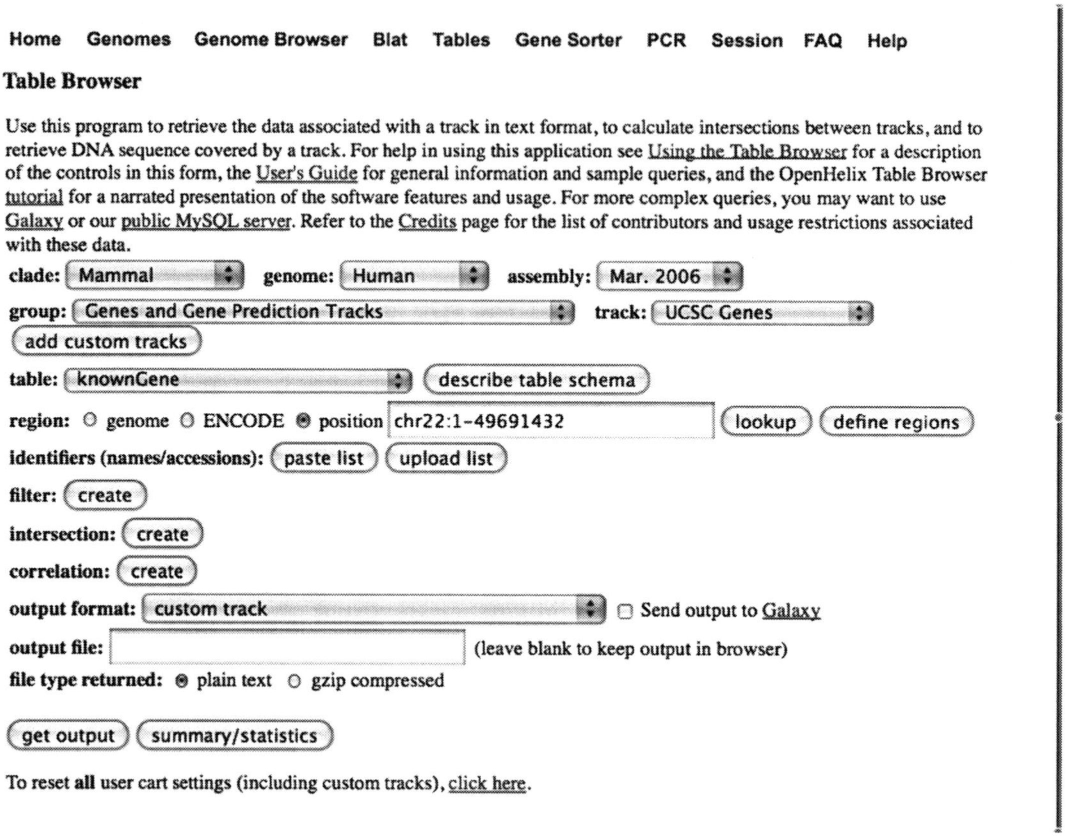

Fig. 1. The UCSC Genome Browser's Table Browser. The Gene annotation can be found in the Group 'Genes and Gene and Prediction Tracks', while the candidate VNTRs can be found in the Simple Repeat track in the Group "Variation and Repeats"

resolution of 0.5–2 kb. Despite repeat masking, they found that the most prominent TFBS sequences found across the multiple unrelated dataset being examined were GGGG(A/C/T)GGGG and TTTTTTT as well as CACACACA, although they urged caution and stated that these patterns were perhaps not of "main interest". This degree of caution is understandable due to problems inherent in the bioinformatics analyses of repeats, such as the fact that some classes are ubiquitous in the genome, that they preferentially association with SINE and LINE elements. However, when such repetitive sequences are more likely to be variable than high-complexity non-repetitive binding sites, it would seem that, when searching for functional variation, these motifs deserve closer attention.

Thus, the biological case for examining VNTR is strong from both a disease and a molecular perspective. There are certain key bioinformatics tasks that geneticists need to carry out in order to use VNTRs in the lab. In the follow sections, we outline some basic bioinformatics methods to find repeats, predict their polymorphism and their function.

2. Materials

All the tools described here are freely available internet web tools which would run on any PC, Mac, or Unix workstation with web access.

3. Methods

3.1. Identifying VNTRs in DNA Sequences

The initial selection of VNTR candidates is relatively straightforward as the presence of tandem repeats in any genome sequence is easy to assess – unlike SNPs which require experimental intervention – by the use of programs which search for sequence repeat motifs such as Tandem Repeat Finder (21), Tandyman (Tandyman: http://biosphere.lanl.gov/tandyman/), the Genequest software program (Dnastar package; LaserGene, Inc., Madison, Wis.), Simple (22), REPuter (23), TRF (24), mreps (25) and ATRHunter (26), etc. The current best solution is to combine the outputs from several of these programs after the input of the DNA sequences in question as each is optimized to find certain types of repeats as most programs will identify only partially overlapping subsets of the actual number of repeats in a given sequence. Exhaustive searching of DNA sequences for sequences with two or more repeats of single nucleotides and larger repeat units is computationally intensive but becomes a more tractable problem when searching for repeats with ten or more copies or, e.g., >20 bp repeats. For example, say we wish to use a program to bioinformatically identify repetitive sequences in a given human gene or region. We can download the current draft human genome sequence data for the gene via the Golden Path Browser at the University of California Santa Cruz (http://genome.ucsc.edu/index.html), by going to http://genome.ucsc.edu/ and downloading the sequences containing and surrounding our candidate gene (+/– 20,000 bp 5′ and 3′) (see chapter in this issue (27) Ref "*Exploring the landscape of the genome.*"; this can also be done via either the Ensembl or NCBI genome browsers). This sequence can be downloaded as FASTA format to a file and saved locally. Then, the sequence can be uploaded to any of several different webservers that will automatically search your sequence files for repeats. One is the MREPs webserver (http://bioinfo.lifl.fr/mreps/mreps.php), which searches for small VNTRs (microsatellites). The options that can be specified for the search are several, but two key concerns are the repeat unit size and number of repeat unit thresholds as well as the number of imperfect repeats tolerated. The MREPS mini-tutorial (http://bioinfo.lifl.fr/mreps/mini_tuto.php) explains some of these concepts and the other filters that can be applied to the searching.

3.2. Predicting Polymorphism Based on the Properties of the Repeated Sequence

Rather than testing each one of the predicted VNTRs in the laboratory, we can filter out those repeat that are unlikely to be polymorphic based on the particular properties of each repeat and it size. General rules for predicting microsatellite polymorphism based on the number of repeats, the repeat unit size and how perfect the repeats are available and several approaches have been published (28–30). For example, for larger VNTRs with >15 bp repeats, we can use the methods of (30) or (29), a tandem repeat history reconstruction algorithm called HistoryR, as well as measures of a number of variables, including the sequence characteristics of unit length, copy number, total length, percent matches, %GC, GC bias, purine/pyrimidine bias, and average entropy. (In practice, it may be easier to use a precomputed list of probable polymorphic repeats such as http://biotools.swmed.edu/repfind/.) This approach is most useful in human studies to eliminate highly unlikely to be polymorphic repeats and to select hypervariable repeats for linkage type approaches. However, in animal studies, this approach is immensely valuable as, in many species, there are no or only very small SNP databases and being able to find several or many highly informative polymorphisms (small polyallelic microsatellites can have heterozygosities exceeding 80%) from even low depth genome sequencing derived from one or a few individuals can be immensely valuable.

3.3. Online Databases of Tandem Repeats and VNTRs

The Simple Repeat annotation in the UCSC human genome browser, in the March 2006/NCBI 31 freeze human sequence's annotation, has 633,715 items. These have been annotated by the Tandem Repeat Finder (TRF) program (21). Additionally, Gary Benson has assembled an excellent set of tools based around his TRF software within TRDB (http://tandem.bu.edu/cgi-bin/trdb/trdb.exe; (31). Skip Garner's group has additionally developed sets of tool and databases funcussed on prediction of polymorphism and identifying VNTRs in genes (http://innovation.swmed.edu/res_inf.htm) one of which is discussed above. In addition, their tool Ereomorph (http://discovery.swmed.edu/eremorph/browse_micro_summary/) has precomputed, almost exhaustive, lists of potential VNTRs for every human gene (http://discovery.swmed.edu/eremorph/browse_micro_summary/), with a focus on microsatellites.

The UCSC and TRDB TRF annotation, however, extends to over 20 species in the with, for example, the Feb 2008 Guinea Pig genome having 652,326 repeats (and no SNPs) annotated. Given its ubiquitous use, it is perhaps worthwhile reviewing the TRF annotation's characteristics. The Simple Repeat annotation's table schema can be accessed from the UCSC Genome Browser table browser and outlines the fields of the table. The most important fields in our opinion are:

- Period – This is the length of the repeat unit with smaller repeats being more likely to be polymorphic;
- copyNum – This is the number of repeats in the tandem repeat. It is dependent on the reading frame of the program (a 4 bp repeat may be also read as an 8 bp repeat or a 12 bp repeat and the program will report all reading frames that exceed its scoring thresholds). The higher this is the more likely the repeat is to be polymorphic, less important for larger repeat units;
- consensusSize – This is the length of the consensus sequence of the repeat and is usually the same as the period;
- perMatch – The Percentage Match, which shows how perfect the total repeat structure is when compared to the repeating unit – the higher this is, the more likely the repeat is to be polymorphic, and less important for larger repeat units;
- perIndel – This is the percentage of insertion and deletions over the total of the repeats units' bases. When this is too high, it will stabilize the repeat such that it does not mutate frequently during meiosis.

3.4. Using a Genome Browser to Find Gene Associated VNTRs

Generally, in order to be functional, VNTRs need to occur within a gene regulatory region within an exon of the gene. Using the tools above, probable polymorphic repeats can be downloaded. There may be a wish to select from our data-set candidate polymorphic TRs occurring in the proximal promoter regions of gene (−500 to +100 from transcription start), those occurring in non-coding and also those occurring in coding exons but to apply different criteria to each one. For example, when including VNTRs in introns, some choices may be to include those with large sequence repeat motifs (>20 bp), those occurring near splice junctions, and those that account for a large% of the sequence of small introns. While repeats from single genes can be found via http://biotools.swmed.edu/repfind/ and other tools like it, the UCSC Genome Browser has Simple Repeat annotation (from TRF) and its Table Browser intersection tool also allows easy generation of region, chromosome or genome wide lists of VNTRs are associated with genes, occurring in their promoters or exons. Say we wish to identify all gene associated VNTRs in promoter regions on chromosome 22. To start off, we can use the Table browser to create a custom tracks the regions 20 kb upstream of each gene (using the guidelines suggested in (32)) or, e.g., the first 500 or 100 bases of the upstream region as preferred. Figure 1 shows the table browser interface and the group and track selection for the UCSC Gene annotation. We can select a region and generate a Custom track Output. When (get output) is clicked, the next webpage allows us to specify the custom track that will be generated (Fig. 2).

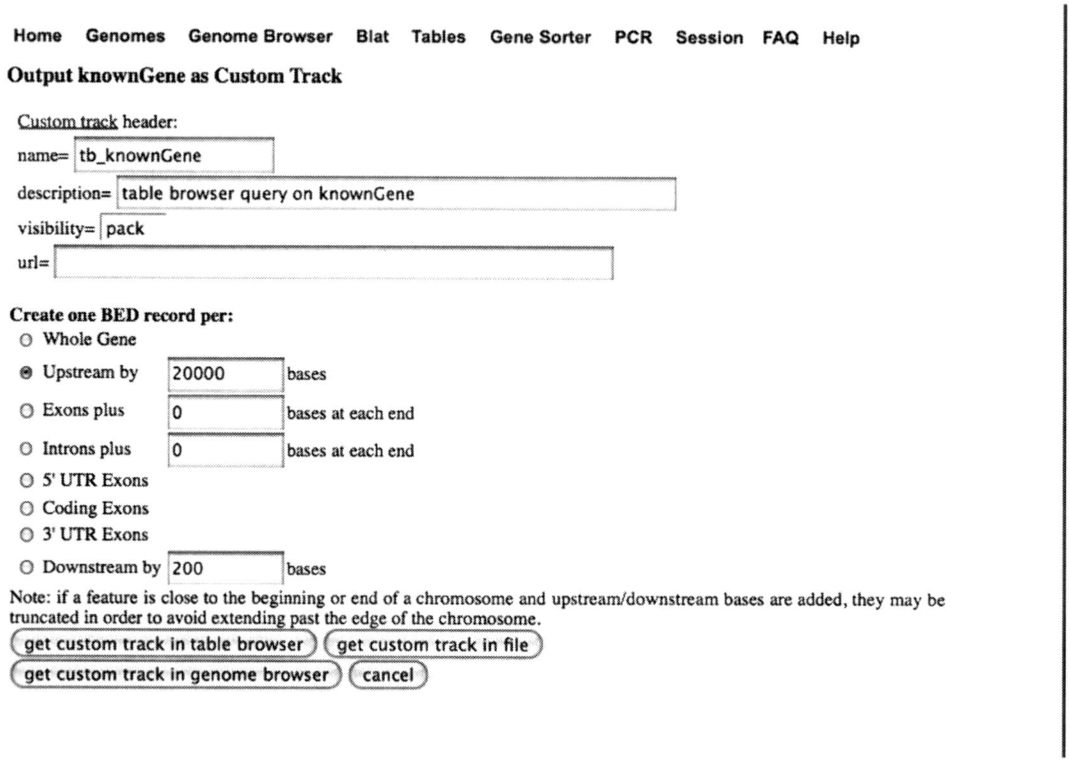

Fig. 2. Custom Track Output Options. By selecting different options on this page, Custome Tracks can be generated containing different items or regions such as Exons, Whole Genes, and Upstream and Downstream regions of defined sizes. The name and description of each track can be entered in the text boxes at the top. The track can then be viewed and/or taken back into the Group "Custom Tracks" for use in the table browser

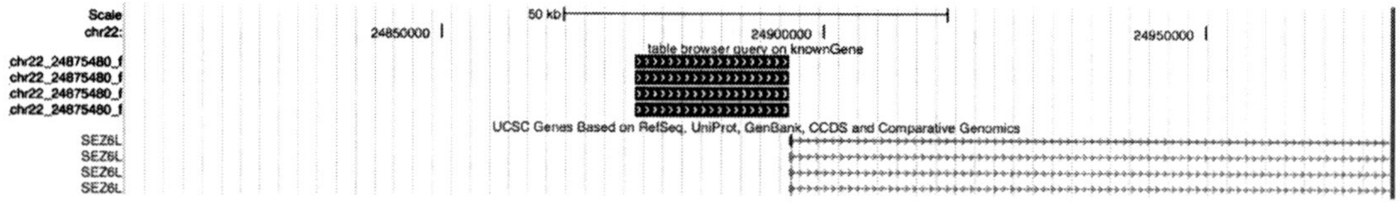

Fig. 3. Custom Track showing Upstream promoter regions of the 4 isoforms of a gene. Duplicate and overlapping gene isoform are a problem for certain types of analyses but even more so for VNTRs. This arises because the finding algorithms may identify multiple overlapping VNTRs where there is just one in the sequence or identify VNTRs that can be equally scored as 4 bp, 8 bp, or 12 bp unit repeats. In either case, this may require filtering of duplicates at a later stage

By querying the table browser, it is possible to generate new custom tracts for the upstream promoter regions of the genes. This yields a custom track that can be displayed in the browser, but more importantly can be queried in the table browser (each isoform of the gene will have its own entry but, as in the case in Fig. 3, (about here) that may share a common promoter, so duplicate entries may occur in some analyses and may need to be filtered later).

3.4.1. Note 1

The intersection of the Simple Repeat track (to be found in the Variation and Repeats group) and this custom track can then be constructed in the table browser. Clicking on the filter button allows selection of potential VNTRs meeting certain properties and shows the various information fields held about each repeat entry, as discussed in subheading 3.3. This filter can then be applied and another custom track can be generated or the position information or DNA sequence can be viewed and downloaded for each repeat. In practice, for large scale tandem repeat analysis, the reader may find it more useful to use UCSC within the Galaxy online tools http://galaxy.psu.edu/ (see chapter 3 in this issue) and many of these analyses can also be performed within TRDB (http://tandem.bu.edu/cgi-bin/trdb/trdb.exe; (31)).

3.5. Examining Potential Functional Roles for VNTRs

In the introduction, we outlined how VNTRs might be functional. Can we examine this question using the online tools we have just described? We might just wish to ask if VNTRs affect transcription factor binding to DNA. It is easy to conduct this sort of analysis using the UCSC Genome browser table, the associated annotations, and the custom track facilities, but there are some caveats.

The Simple Repeat track, in the UCSC March 2006 human annotation, currently has 633,715 items representing 1.95% of the total human sequence (this can be obtained by clicking on summary/statistics for that track in the Table Browser). We might then wish to examine if these potential VNTRs encode transcription factor binding sites. However, there is a problem in that repeat masking (33), which screens DNA sequences for interspersed repeats and low complexity DNA sequences, has been applied to the sequences used to search for transcription factor binding sites (TFBSs) and many other functional annotations effectively excludes many polymorphic VNTRs from being screened by programs which use repeat masked as a filter or in ChIP-seq experiments, where the tiling arrays are designed from repeat masked sequence. Thus to make sure we are comparing like with like, we can intersect the Simple Repeat annotation with the Repeat Masker annotation and take the complement to give the Repeats not found by repeat masking.

When this is done, 104,296 candidate VNTRs survive representing 0.34% of the total human sequence, of which 5,374 are in exons. By contrast, the HAPMAP European SNP panel (3,839,363 SNPs) has 2,567,731 SNPs that are not repeat masked. The Conserved TFBS (cTFBS) track has over three million elements and an interesting comparison to make is how many cTFBSs have SNPs versus how many have VNTRs. Analysis shows that 53,702 SNPs occur in 71,104 binding sites, whereas 3,647 VNTRs overlap 6,829 cTFBSs. This means that VNTRs are overrepresented versus SNPs (3.5% of VNTRs versus 2.1% of SNPs) in cTFBSs but also that each of these VNTRs encodes the TFBS, rather than interrupting it with an average of ~2 sites encode per TFBS-VNTR, with many encoding more (Fig. 4).

This evidence and that outlined above suggests that VNTRs may allow a gene to more easily retain a transcription factor and thus facilitate expression. If this is true, then there should be an enrichment compatible with selection for VNTRs in the core promoter of genes. This can be simply tested by analysis of the first 500 bases immediately upstream of each gene in the March 2006 UCSC Gene annotation shows 9,970 VNTRs (1/3 of which would survive repeat masking) occupying 1,211,357 non-overlapping bases of a total 35,630,944 or 3.4% of bases in >56,000 overlapping promoters, which compares very favourably with the 1.95% VNTRs occupy on a genomewide basis – a 1.7 fold enrichment.

VNTRs can of course encode any type of DNA binding site, including polymerase, splicing factor, and microRNA binding sites as well as other agents of regulation. Notably, in the case of microRNAs, they not only encode tandem arrangements of binding sites (Fig. 5) but also tandem copies of microRNAs themselves (Fig. 6). Overall, it would appear that VNTRs are elements that allow concentration and development of function

Fig. 4. A UCSC genome browser (http://genome.ucsc.edu) view showing multiple copies of a TFBS encoded by a large VNTR

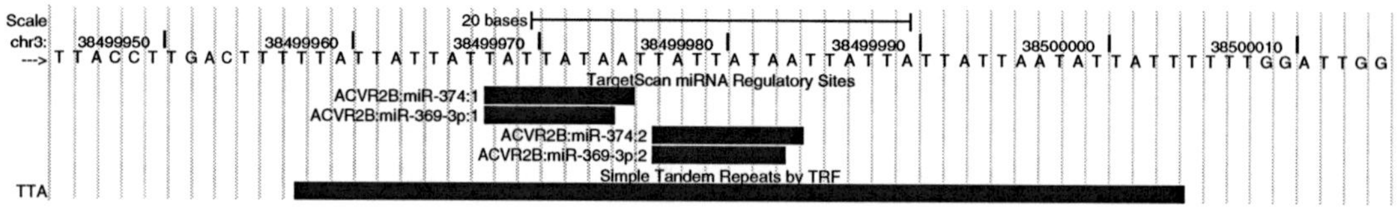

Fig. 5. A UCSC genome browser view shows a VNTR encoding two copies of a TargetScban predicted microRNA binding site in a 3′ untranslated region of a gene. This is one of 67 such examples in the current (March 2006) genome annotation

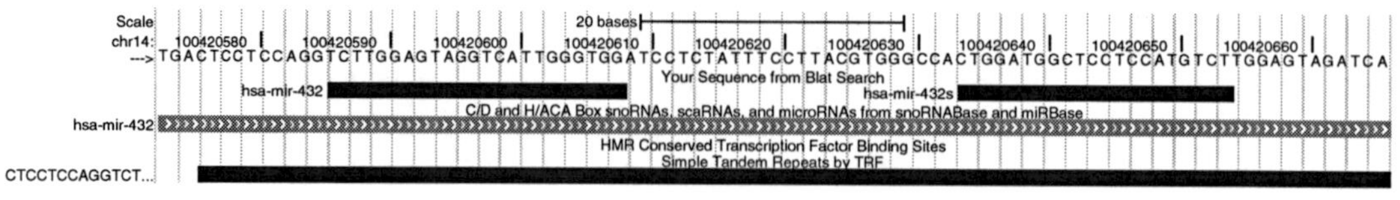

Fig. 6. A microRNA gene, has-mir-432, coded in duplicate by a VNTR. There are 9 such VNTRs encoding 5 microRNAs in the current (March 2006) annotation

and functional variation in the genome. They can encode binding sites and are enriched for in the promoters of genes. All in all, there is a pressing need for a systematic study of the genomewide role of VNTRs in expression, recombination and for non-repeat masked versions of every annotation to be made available as standard.

3.6. Generating VNTR Genotypes in the Next Generation Sequencing Era

VNTRs are currently analyzed by PCR and size separation on capillary electrophoresis machines. The data produced by these are analyzed in software such as GENEMAPPER™, v3.5.1 (Applied Biosystems, CA, USA). This program converts the signals into electropherograms used for visual checking and for automated genotyping calling (Fig. 7). The data generated are then exported as tables to be stored in spreadsheets until statistical analyses were undertaken. These methods are neither highly automatable nor can they be parallelised like SNP microarrays. The current cost of genotyping a VNTR is approximately $0.5 per genotype versus $0.001 for a SNP. The basal problem is that array based technologies cannot be used to genotype VNTRs as the sequence of the VNTR is, in the vast majority of cases, not unique, is frequently too small in many cases for a probe to be designed against it and microarrays do not have the required sensitivity to distinguish reliably different repeat numbers e.g., 10 and 11 copies of a repeat by intensity. However, the advent of next generation sequencing approaches and the 1,000 genomes project means that smaller VNTRs can now be genotyped in a comprehensive genomewide manner. The read lengths of most sequencing technologies are now over 100 bp with some exceeding 400 bp, thus allowing them to capture both the sequence of a VNTR and its (unique) flanks. There are various technical problems to be overcome, such as building calling algorithms for VNTRs into next-generation sequencing analysis pipelines. However, studies have successfully been able to find and call VNTR genotypes in next generation sequencing with the fraction

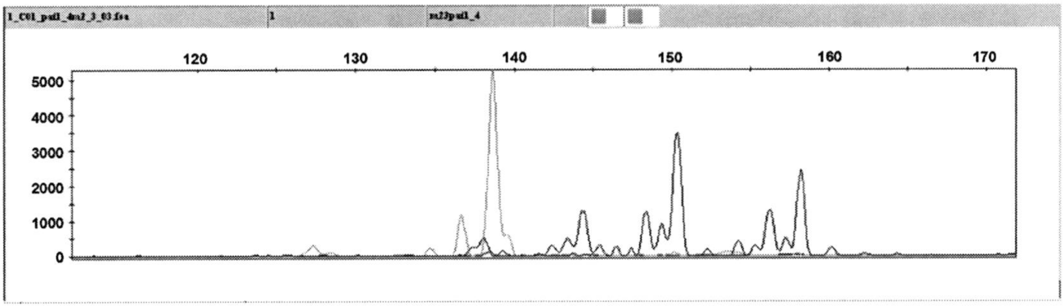

Fig. 7. Illustrating an electropherogram generated by GENEMAPPER™ (v3.5.1). This shows genotypes from one individual genotyped for three different markers using the 5′fluorescent labels FAM, HEX and NED. This technology is labour-intensive and expensive when compared with SNP genotyping

of repeated elements that can be called being proportional to the read length of the sequencing technology. (34). In addition, many of the methods being developed to analyze and call multicopy CNV data may be of great utility in VNTR research; TriTyper is one such program but is currently only useful for triallelic VNTRs (35). Notably, BEAGLE (36) can be used to impute multiallelic VNTRs from SNP data although it does not generate confidence calls for the imputed genotypes (unpublished data, S Cohen, et al.). If large scale genome sequencing such as the 1000 genomes project (http://www.1000genomes.org) can be analyzed to generate a set of VNTR calls analogous to the hapmap for SNPs, then these or further refinements of these programs may be useful in imputing large numbers of VNTRs from SNP genotyping array data.

4. Conclusion

We have reviewed the reasons why VNTRs may be useful genetic markers and the bioinformatic methods and databases that may be used to derive information about them and their potential functional and polymorphic properties. We have seen how these tools can help identify lists of predicted polymorphic VNTRs from unannotated DNA sequences, help us identify the functional properties of VNTRs and, as a case study, we used these tools to make some predictions about their role in the genome with respect to transcription factor binding sites. Lastly, we have looked at methodological developments in next generation sequencing and imputation and their promise to revolutionize the study of VNTRs.

References

1. Epplen JT, Maueler W, Santos EJ. (1998) On GATAGATA and other "junk" in the barren stretch of genomic desert. *Cytogenet Cell Genet.* **80**, 75–82.
2. Richard GF, Dujon B. (1996) Distribution and variability of trinucleotide repeats in the genome of the yeast *Saccharomyces cerevisiae.* Gene. **26**, 165–74.
3. Field D, Wills C. (1998) Abundant microsatellite polymorphism in Saccharomyces cerevisiae, and the different distributions of microsatellites in eight prokaryotes and S. cerevisiae, result from strong mutation pressures and a variety of selective forces. *Proc Natl Acad Sci USA.* **95**, 1647–52.
4. Young ET, Sloan JS, Van Riper K. (2000) Trinucleotide repeats are clustered in regulatory genes in *Saccharomyces cerevisiae. Genetics.* **154**, 1053–68.
5. Hancock JM. (1994) Evolution of sequence repetition and gene duplications in the TATA-binding protein TBP (TFIID). *Nuc Acids Res.* **21**, 2823–2830.
6. Gendrel CG, Boulet A, Dutreix M. (2000) (CA/GT)(n) microsatellites affect homologous recombination during yeast meiosis. *Genes Dev* **14**, 1261–8.
7. Karlin S, Burge C. (1996) Trinucleotide repeats and long homopeptides in genes and proteins associated with nervous system disease and development. *Proc Natl Acad Sci USA* **93**, 1560–5.

8. Webster MT, Smith NG, Ellegren H. (2002) Microsatellite evolution inferred from human-chimpanzee genomic sequence alignments. *Proc Natl Acad Sci USA*. **99**, 8748–53.
9. Vafiadis P, Bennett ST, Colle E, Grabs R, Goodyer CG, Polychronakos C. (1996) Imprinted and genotype-specific expression of genes at the IDDM2 locus in pancreas and leucocytes. *J Autoimmun*. **9**, 397–403.
10. Bowater RP, Wells RD. (2001) The intrinsically unstable life of DNA triplet repeats associated with human hereditary disorders. *Prog Nucleic Acid Res Mol Biol*. **66**, 159–202.
11. Todd JA. (1999) From genome to aetiology in a multifactorial disease, type 1 diabetes. *Bioessays*. **21**:164–74.
12. Caspi A, Sugden K, Moffitt TE, Taylor A, Craig IW, Harrington H, et al. (2003). Influence of life stress on depression, moderation by a polymorphism in the 5-HTT gene. *Science*. **301**, 386–389.
13. Brookes KJ, Mill J, Guindalini C, Curran S, Xu X, Knight J, et al. (2006) A common haplotype of the dopamine transporter gene associated with attention-deficit/hyperactivity disorder and interacting with maternal use of alcohol during pregnancy. *Arch Gen Psychiatry*. **63**, 74–81.
14. Guindalini C, Howard M, Haddley K, Laranjeira R, Collier D, Ammar N, et al. (2006) A dopamine transporter gene functional variant associated with cocaine abuse in a Brazilian sample. *Proc Natl Acad Sci USA*. **103**, 4552–7.
15. Contente A, Dittmer A, Koch MC, Roth J, Dobbelstein M. (2002) A polymorphic microsatellite that mediates induction of PIG3 by p53. *Nat Genet*. **30**, 315–20.
16. Lian Y, Garner HR. (2005) Evidence for the regulation of alternative splicing via complementary DNA sequence repeats. *Bioinformatics*. **21**, 1358–64.
17. Kolesov G, Wunderlich Z, Laikova ON, Gelfand MS, Mirny LA. (2007) How gene order is influenced by the biophysics of transcription regulation. *Proc Natl Acad Sci USA*. **104**, 13948–13953.
18. Gupta M, Liu JS. (2005) De novo cis-regulatory module elicitation for eukaryotic genomes. *Proc Natl Acad Sci USA*. **102**, 7079–7084.
19. Zhou Q, Wong WH. (2004) CisModule, de novo discovery of cis-regulatory modules by hierarchical mixture modeling. *Proc Natl Acad Sci USA*. **101**, 12114–12119.
20. Ji H, Vokes SA, Wong WH. (2006) A comparative analysis of genome-wide chromatin immunoprecipitation data for mammalian transcription factors. *Nucleic Acids Res*. **34**, e146.
21. Benson G. (1999) Tandem repeats finder, a program to analyze DNA sequences. *Nucleic Acids Res* **27**, 573–80.
22. Alba MM, Laskowski RA, Hancock JM. (2002) Detecting cryptically simple protein sequences using the SIMPLE algorithm. *Bioinformatics*. **18**, 672–8.
23. Kurtz S, Choudhuri JV, Ohlebusch E, Schleiermacher C, Stoye J, Giegerich R. (2001) REPuter: the manifold applications of repeat analysis on a genomic scale. *Nucleic Acids Res*. **29**, 4633–42.
24. O'Dushlaine CT, Shields DC. (2006) Tools for the identification of variable and potentially variable tandem repeats. BMC Genomics. **15**, 7:290.
25. Kolpakov R, Bana G, Kucherov G. (2003) mreps, Efficient and flexible detection of tandem repeats in DNA. *Nucleic Acids Res*. **31**, 3672–8.
26. Ydo Wexler, Zohar Yakhini, Yechezkel Kashi, and Dan Geiger (2004) Finding approximate tandem repeats in genomic sequences. *Recomb proceedings*. 223–232.
27. Barnes MR (2009) Exploring the landscape of the genome. Methods in Molecular Biology, (In this issue).
28. Fondon III JW, Mele GM, Brezinschek RI, Cummings D, Pande A, Wren J, et al. (1998) Computerized polymorphic marker identification: experimental validation and a predicted human polymorphism catalog. *Proc Natl Acad Sci USA*, **95**, 7514–9.
29. Näslund K, Saetre P, von Salomé J, Bergström TF, Jareborg N, Jazin E. (2005) Genome-wide prediction of human VNTRs. *Genomics*, **85**, 24–35.
30. Denoeud F, Vergnaud G. (2004) Identification of polymorphic tandem repeats by direct comparison of genome sequence from different bacterial strains, a web-based resource. *BMC Bioinformatics*. **5**, 4.
31. Benson G (2006) TRDB - The tandem repeats database. *Nucleic Acids Research*, **00**, D1–D8.
32. Veyrieras JB, Kudaravalli S, Kim SY, Dermitzakis ET, Gilad Y, Stephens M Pritchard JK. (2008) High-resolution mapping of expression-QTLs yields insight into human gene regulation. *PLoS genetics* **4**, e1000214.
33. Smit AFA, Hubley R, Green P. (1996–2007) RepeatMasker Open-3.0. http.//www.repeatmasker.org.

34. Hillier LW, Marth GT, Quinlan AR, Dooling D, Fewell G, Barnett D, et al. (2008) Whole-genome sequencing and variant discovery in C. elegans. *Nat Methods.* **5**, 183–8.

35. Franke L, de Kovel CG, Aulchenko YS, Trynka G, Zhernakova A, Hunt KA, et al. (2008) Detection, imputation, and association analysis of small deletions and null alleles on oligonucleotide arrays. *Am J Hum Genet* **82**, 1316–33.

36. Browning BL, Browning SR. (2009) A unified approach to genotype imputation and haplotype phase inference for large data sets of trios and unrelated individuals. *Am J Hum Genet* **84**, 210–223.

Chapter 11

Assessing the Impact of Genetic Variation on Transcriptional Regulation In Vitro

Fahad R. Ali, Kate Haddley, and John P. Quinn

Abstract

Two alleles of a gene that contain polymorphic *cis*-regulatory regions can contribute differently to expression levels. Evolutionary changes in such *cis*-regulatory domains are believed to have participated in the cognitive evolution of *H. sapiens* as well as phenotypic diversity. There have been many studies that associate genetic variations to individual's susceptibility to behavioural and affective disorders. *Cis*-acting regulatory polymorphisms can effect gene expression at many levels, such as transcription, mRNA processing efficiency, pre-mRNA splicing, and mRNA stability. *Trans*-acting modulators (such as transcription factors) also play a major role in determining mRNA concentration of a specific allele. Several studies have demonstrated that VNTRs within various genes can support differential gene expression based on copy number and that the function of the VNTR as a transcriptional regulator can be modulated, in part, by transcription factors.

A better understanding of the pathways regulating expression mediated by the VNTRs would complement clinical studies, demonstrating how these domains may be mechanistically involved in the progression of the disorder and may supply more defined targets for pharmaceutical intervention.

Key words: Polymorphism, VNTR, Gene regulation, Behaviour

1. Introduction

The regulation of expression of a given gene can vary between individuals because of epigenetics and polymorphic variation in regulatory domains where, for example, transcription factors may bind. Many studies have shown that *cis*-regulatory loci (e.g., regulatory polymorphisms) in promoters and other non-coding regions of a gene, in addition to transcription factors (*trans*-acting modulators) regulate allele-specific expression (1–6). In the absence of these *cis*-regulatory domains, paternal and maternal alleles of a gene are equally expressed unless one allele is imprinted.

Michael R. Barnes and Gerome Breen (eds.), *Genetic Variation: Methods and Protocols*, Methods in Molecular Biology, vol. 628, DOI 10.1007/978-1-60327-367-1_11, © Springer Science+Business Media, LLC 2010

However, when an individual is heterozygous for a *cis*-regulatory domain, mRNA expression level may vary from each allele, which is termed differential allelic gene expression. Different combinations of these polymorphisms within our genome create the "genetic fingerprint" which contributes to determining phenotypic diversity and can lead in part, to individuality between us. Nearly three decades ago, King and Wilson suggested that changes in the mechanisms controlling gene expression, rather than the DNA sequence itself account for the morphological, behavioural, and cognitive differences between human beings and other primates (7). Therefore, it is suggested that the person we are, is not solely determined by our genes but how their expression is controlled, hence drugs such as cocaine or alcohol vary our behaviour in part by modulating gene expression.

The phenotypic consequences of allelic differential expression will depend on the gene function itself; however, there is growing evidence that genetic variation plays a crucial role in an individual's susceptibility to behavioural and affective disorders. Many alleles that modify disease risk have been identified in various disorders, such as Alzheimer's disease (8), schizophrenia (9, 10), anxiety (11), obsessive compulsive disorder (OCD) (12), unipolar depression (13), bipolar depression (14–16), and Parkinson's disease (17).

Polymorphisms in regulatory regions may also play a major role on how we respond to drugs and our environment. This suggests that individuals with a particular combination of polymorphisms may respond differently to the same medications or environmental stresses. There are several challenges in identifying these *cis*-regulatory regions within the genome, one of which being the tissue-specificity of these regulatory domains as well as the difficulty in identifying the causative regulatory variants that are in linkage disequilibrium with other polymorphic regions "haplotype". Identification of specific polymorphisms which associate with specific disorders, or respond to specific environmental stimuli, could lead to tailored treatments or bespoke medication for individuals based on genomic variation (18).

Various in vitro and in vivo approaches have been carried out to detect differential allelic gene expression. In vitro techniques, such as transient transfection assay, assessing the effect of the polymorphic domain on reporter gene expression, and protein-DNA interaction, such as gel-shift assays and footprinting, are widely used. However, such approaches are difficult to interpret due to the effect of various *trans*-acting factors on allelic expression (e.g., tissue-specific expression) and the lack of the native chromatin configuration of the DNA sequences. Furthermore, the design of the construct is critical; the choice of the fragment size to include in the reporter gene, as polymorphic regions may have a variable effect when assessed separately or when in the context of the full-promoter, in which a naturally occurring haplotype may, exist. An in vivo approach to assess allele-specific *cis*-acting regulatory domains on mRNA expression is allelic expression imbalance (AEI). This is a quantitative method that discriminates expression level when the transcript is heterozygous for a marker (exonic SNPs or intronic SNPs in hnRNA), in which one allele acts as an internal control for the other, eliminating *trans*-acting and other environmental factors that may affect expression (1). Another advantage of such a technique is that the alleles are expressed in their natural chromatin-context.

A subclass of minisatillite polymorphisms which show some degree of degeneration "non-perfect repeats", where one repeat may be slightly different from the next but overall the core consensus sequence is maintained, are termed variable number tandem repeats (VNTRs) (19). The majority of VNTRs are found in non-coding regions of the genome, and many are found at higher density in gene enriched areas when compared to non-genic regions (G. Breen SDGP, IOP, Kings College, London, personal communication), implying their role in transcriptional regulation. VNTRs have the potential to act as transcriptional regulatory domains as they have sufficient DNA sequences in the repeat to act as sequence specific DNA-binding sites for transcription factors. Many of the VNTRs are present in the genome as a feature of an emerging evolution. VNTRs display evolutionary conservation between humans and non-human primates but are often not found in lower mammals (20, 21).

Based on published data by our group, VNTRs can function in both a tissue-specific and stimulus-inducible manner to fine-tune gene expression. This fine tuning could be correlated, mechanistically, not only with normal physiological function and variation between individuals, but also with a predisposition to behavioural disorders by altering neurotransmitter signalling in response to challenges and stress. Furthermore, if stimulus inducible expression varies dependent on a specific polymorphism associated with a disorder, then that may have similar implications in the response of an individual to pharmacological treatment of that disorder (5, 6, 18, 22–24).

2. Materials

2.1. Standard Polymerase Chain Reaction

1. PCR was performed in a PxE 0.2 thermal cycler (Thermo Electron Corporation).
2. 0.2 ml sterile PCR tubes.
3. Reaction Mix.

(a) 10–100 ng DNA template.
(b) 2 mM $MgCl_2$.
(c) 0.2 mM of each dNTP.
(d) 0.2 μM of each primer.
(e) 1 × Diamond reaction buffer.
(f) 1.5 units of Diamond DNA polymerase (Bioline, Cat No. BIO-21059).
(g) 0.5 M Betaine.

2.2. PCR Purification

1. Bench top centrifuge.
2. 1.5 ml Eppendorf tubes.
3. 2 ml collection tubes.
4. QIAquick PCR Purification Kit (Qiagen):
   (a) Buffer PB.
   (b) QIAquick column.
   (c) PE buffer and Ethanol.
   (d) Nuclease-free water.

2.3. Analysis of DNA Using Agarose Gel-Electrophoresis

1. 12 × 14 cm or 20.5 × 10 cm trays and the appropriate combs.
2. Electrophoresis tanks (Hybaid turn and cast submarine gel system, Hybaid, or Savant HG 350 tank).
3. Agarose (multi-purpose agarose, Bioline, Cat. No. BIO-41025).
4. 0.5× TBE buffer.
5. Ethidium bromide (10 mg/ml aqueous solution, Sigma E-5134).
6. Loading buffer.
7. DNA ladders:
   (a) Mass ruler; Fermentas Cat. No. SM0403.
   (b) 100 bp Ladder; Promega Cat. No. G2891.
   (c) 1 Kb Ladder; Promega Cat. No. G7541.
8. MultiImageII Light Cabinet Transluminator (Alpha Innotech Corporation).
9. CCD camera (Alpha Innotech Corporation).

2.4. Recovery of DNA from Agarose-Gels

1. Bench top centrifuge.
2. 1.5 ml Eppendorf tube.
3. 2 ml collection tube.
4. Clean blade.
5. 50°C water-bath.

6. UV translumination.
7. QIAquick Gel Extraction Kit (Qiagen, No. 28706):
   (a) Buffer QG.
   (b) 100% isopropanol.
   (c) QIAquick spin column.
   (d) PE buffer.
   (e) Nuclease-free water.

2.5. Generation of VNTR Reporter Gene Constructs

2.5.1. Reagents for Cloning the VNTRs into Intermediate Vector pT7Blue

1. Heating-block.
2. Intermediate vector pT7Blue (Novagen).
3. End-conversion mix (Novagen).
4. Nuclease-free water.
5. T_4 DNA ligase (New England Biolabs).

2.5.2. Reagents for Cloning the VNTRs into Intermediate Vector pGEM-T

1. Heating-block.
2. Intermediate vector pGEM-T.
3. 0.2 mM of dATP.
4. 1.5 mM $MgCl_2$.
5. 1× Taq reaction buffer.
6. 5 units of *Taq* DNA polymerase.
7. Nuclease-free water.
8. T_4 DNA ligase (New England Biolabs).

2.5.3. Reagents for Cloning the VNTRs into Reporter Gene Vectors

1. Restriction enzymes XbaI, BamHI, NotI, NcoI, BbsI, HindIII, AscI, and Sac I (Promega).
2. AscI linker.
3. pGL3p vector (*Firefly* luciferase driven by a minimal SV40 promoter, Promega).
4. phRL_null renillin expression (Promega).
5. Epstein-Barr virus (EBV) based vector pMep.9
6. T_4 DNA ligase (New England Biolabs).

2.6. Ligation

1. 0.2 ml tubes.
2. 10× ligase buffer.
3. 200 units of T_4 DNA ligase (NEB M0202S).
4. Nuclease-free water.

2.7. Transformation of Chemically Competent E. coli Cells

1. 42°C water-bath.
2. 37°C shaker.

3. Competent *E. coli* cells (DH5-α, Invitrogen-Gibco BRL Cat. No. 18265-017).
4. LB broth.
5. LB agar plates.
6. 100 μg/ml ampicillin.

2.8. Blue/White Screening of Recombinants

1. 37°C incubator.
2. 82 mm LB agar plates.
3. 100 μg/ml ampicillin.
4. 50 mg/ml X-gal (5-bromo-4-chloro-3-indolyl-bD-galactoside).
5. Dimethylformamide (Promega, Cat. No. V3491).
6. 100 mM IPTG (Isopropyl β-D-1-thiogalactopyranoside, BIO-37036).

2.9. Isolation of DNA Constructs from Bacteria

1. Bench top centrifuge (Miniprep) and refrigerated 16,000×*g* centrifuge (Maxiprep).
2. Microcentrifuge tubes (Miniprep) and 50 ml Oakridge tubes (Maxiprep).
3. QIAprep Spin Miniprep Kit (Qiagen, No. 27106) or QIAGEN Plasmid Maxiprep (Qiagen, No. 12263).
4. Resuspension buffer P1.
5. Lysis buffer P2.
6. Neutralization buffer N3 (Miniprep) or buffer P3 (Maxiprep).
7. QIAprep spin column (Miniprep) or QiAGEN tip 500 columns (Maxiprep).
8. PE buffer and Ethanol.
9. Equilibration buffer QBT.
10. Wash buffer QC
11. Elution buffer QF.
12. 100% Isopropanol.
13. 70% ethanol.
14. Nuclease-free water.

2.10. Analytical Restriction Enzyme Digests

Enzymes and buffers for DNA digest were mostly obtained from Promega, or alternatively from New England Biolabs.

2.11. Sequencing

Applied Biosystems model 3730 automated capillary DNA sequencer.

2.12. Ultraviolet Spectroscopy

UV spectrophotometer (Jenway Genova Life Science Analyser Cat. No. 636 031).

2.13. Cell and Tissue Culture

2.13.1. Cell Culture

1. 37°C incubator.
2. T75-culture flasks.
3. JAr cells media:
   (a) RPMI-1640 medium (Bioclear, Autogen).
   (b) 10% heat-inactivated foetal calf serum (Hy-Clone, Logan, UT).
   (c) 2 mg/ml glucose.
   (d) 1 mM sodium pyruvate.
   (e) 2 mM l-glutamine.
   (f) 10 mM HEPES.
   (g) 1% (v/v) 100× penicillin/streptomycin (equates to a final concentration of 100 units penicillin/100 μg streptomycin (Sigma-Aldrich; Cat. No. P0781)).
4. Antibiotic geneticin (GIBCO; Cat. No. 11811-031) for generation of stable cell lines.

2.13.2. Primary Prefrontal Cortical Cultures

1. 37°C incubator.
2. CO_2 gas chamber.
3. Contrast microscope.
4. Bench top centrifuge.
5. Cell counting chamber.
6. T75-culture flasks.
7. 15 ml falcon tubes.
8. 24-well tissue culture plates.
9. Sharp razor blade.
10. Curved forceps (Fisher Scientific Cat. No. DKC-790-D).
11. Micro-scissors (WPI Cat. No 501778).
12. Spatula (Fisher scientific, Cat. No 3006).
13. Petri-dish.
14. Pasteur pipettes.
15. 0.70 μm falcon cell strainer (VWR).
16. Dissection solution: 91 ml Hanks balanced salt solution (HBSS. Invitrogen-Gibco BRL Cat. No. 24020-091) containing 3.5 ml 1 M HEPES, 1 ml 1 M $MgCl_2$, 1 ml 200 mM l-glutamine, 1% (v/v) 100x penicillin/streptomycin (equates to a final concentration of 100 units penicillin/100 μg streptomycin (Sigma-Aldrich; Cat. No. P0781)).
17. Poly-D-lysine (100 μg/ml) (Sigma).
18. Sterile water.
19. 1× PBS.
20. Trypsin/EDTAsolution(Sigma-AldrichLtd,Cat.No.T4049).

21. Culture medium I: DMEM (Bioclear Cat. No. AB2052) supplemented with 10% FCS and penicillin/streptomycin.
22. Culture medium II: Neurobasal-A medium (Invitrogen/Gibco; Cat. No. 10888-022) supplemented with 2% B27 supplement [Invitrogen/Gibco; Cat. No. 17504-044], 2 mM GlutaMAX I and 1% (v/v) gentamycin.

2.14. Cell Treatments

1. Lithium chloride (Sigma-Aldrich; Cat. No. L9650).
2. Cocaine hydrochloride (Sigma-Aldrich; Cat. No. C5776).

2.15. Transfections

1. Bench top centrifuge.
2. Vortex.
3. Humidified 5% CO_2 incubator.
4. ExGen 500 in vitro Transfection Reagent (Fermentas).
5. TransFast Transfection Reagent (Promega).
6. Reporter constructs and/or expression constructs.
7. *Renilla* luciferase (Rluc) cDNA (Novagen) or a modified pMLuc-2 vector containing a minimal TK promoter followed by an optimized *Firefly* luciferase cDNA to normalise for transfections efficiency.
8. 150 mM NaCl.
9. Nuclease-free water.

2.16. Analysis of Transgene Expression by Reporter Gene Assay

1. Glomax 96 microplate luminometer (Promega).
2. Rocking platform
3. Stingray 2.0 software.
4. 96-well plates.
5. Dual Luciferase Assay kit (Promega, Madison Cat. No E1500).
6. 1× PBS.
7. 1× passive lysis buffer.
8. Rocking platform.
9. Firefly luciferase reagent.
10. Sea pansy luciferase reagent.

3. Methods

3.1. General Cloning Methods

3.1.1. PCR Primer Design

Primers were designed with the aid of a primer design computer programme "net primer" http://www.premierbiosoft.com/netprimer/netprlaunch/netprlaunch.html. In general, primers were designed:

1. Of a size between 20 and 25 bp nucleotides long.
2. Melting temperature of 50–65°C.
3. GC content between 40 and 60%.
4. Self complementarity was avoided to minimize primer secondary structure and primer dimer formation.
5. "BLASTN searches" http://blast.ncbi.nlm.nih.gov/Blast.cgi were used to confirm the total gene specificity of the primer sequences chosen.

3.1.2. Standard Polymerase Chain Reaction

Polymerase chain reaction (PCR) was used as a method to amplify specific DNA fragments for use in molecular cloning (see Note 1). PCR was performed in a PxE 0.2 thermal cycler (Thermo Electron Corporation).

50 μl PCR reactions included:

1. 10–100 ng DNA template.
2. 2 mM $MgCl_2$.
3. 0.2 mM of each dNTP.
4. 0.2 μM of each primer.
5. 1× Diamond buffer.
6. 1.5 units of Diamond DNA polymerase (Bioline, Cat. No. BIO-21059).

PCR programmes were adapted from a standard protocol. The annealing temperature was varied according to the melting temperature of the primer pairs used and the extension time was adapted to the size of the expected product, with 1 min extension time for each 1,000 bp as a guide line. The most commonly used reaction conditions were as follows: initial denaturation: 94°C for 5 min for one cycle, denaturation: 94°C for 1 min, annealing: 60°C for 1 min, extension: 72°C for 1 min for 40 cycles, and completion of strands: 72°C for 10 min.

3.1.3. PCR Purification

To purify double-stranded DNA fragments from PCR (primers, nucleotides or polymerase) and other enzymatic reactions, QIAquick PCR Purification Kit (Qiagen) was used following manufacturer's instructions. Briefly,

1. DNA was diluted in a buffer which contained optimal pH and salt conditions.
2. The DNA was applied for subsequent binding to a silica-membrane.
3. Impurities are removed by washes.
4. DNA is eluted in water or TE buffer.

3.1.4. Analysis of DNA Using Agarose Gel-Electrophoresis

For the analysis of PCR products or fragments generated by restriction digests, agarose gel-electrophoresis was employed:

1. 1–2% agarose (multi-purpose agarose, Bioline, Cat. No. BIO-41025) was melted in 0.5× TBE buffer and supplemented with 5 µl ethidium bromide (10 mg/ml aqueous solution, Sigma E-5134).
2. Gels of 100 ml were cast in 12 × 14 cm or 20.5 × 10 cm trays and the appropriate combs were inserted. 50 ml gels were cast in 7 × 10 cm trays, and appropriate combs inserted.
3. Gels were left to set for 30 min at room temperature, and were then submerged in horizontal gel electrophoresis tanks (Hybaid turn and cast submarine gel system, Hybaid, or Savant HG 350 tank) containing 0.5× TBE buffer.
4. Samples were mixed with loading buffer (1× final concentration) and loaded into the wells.
5. The size of a PCR product or restriction digest fragments was determined by loading a DNA ladder (Mass ruler; Fermentas Cat. No. SM0403, 100 bp Ladder; Promega Cat. No. G2891 or 1 Kb Ladder; Promega Cat. No. G7541).
6. In general, gels were run for 1 h at 120 V (Hybaid); however, to separate different variants of the VNTRs, gels were run at 60 V for 3 h.
7. The electrophoretically separated DNA was then visualized with an Evenscan broadband dual wavelength transluminator in a MultiImageII Light Cabinet (both Alpha Innotech Corporation) at a wavelength of 302 nm.
8. Permanent records were taken with a CCD camera (Alpha Innotech Corporation) and stored electronically.

3.1.5. Recovery of DNA from Agarose-Gels

1. PCR products, or specific fragments of a restriction digest, were isolated by running the PCR reaction or the restriction digest out on agarose gels.
2. The desired band corresponding to products of the predicted size were excised from the agrose gel under long wave UV translumination using a clean blade.
3. The DNA was recovered from the gel slice using the QIAquick Gel Extraction Kit (Qiagen, No. 28706) following manufacturer's instructions:
   (a) The gel slice was dissolved in a buffer which contained optimal pH and salt conditions.
   (b) The DNA was applied for subsequent binding to a silica-membrane.
   (c) Impurities are removed by washes.
   (d) DNA is eluted in water or TE buffer.

3.1.6. Generation of VNTR Reporter Gene Constructs

3.1.6.1. Cloning the VNTRs into Intermediate Vectors

Cloning into the Intermediate Vector pT7Blue

The amplified fragments from the PCR reaction (see Note 2) were cloned into the intermediate vector pT7Blue (Novagen) via blunt cloning into the EcoRV site and positive clones were confirmed by sequencing. Briefly,

1. 2 μl of the PCR reaction was added to 5 μl of end-conversion mix (Novagen) in which insert to vector molar ratio of 2.5:1 was obtained.
2. Nuclease-free water to a total of 10 μl was added, and the reaction was mixed gently by stirring with a pipette tip.
3. The reaction was incubated at 22°C for 15 min
4. The reaction was then inactivated by heating at 75°C for 5 min.
5. The reaction was then cooled on ice for 2 min and briefly centrifuged to collect any condensate material.
6. 1 μl (50 ng) blunt vector (Novagen) and 1 μl T_4 DNA ligase (NEB) were added directly to the end-conversion reaction, which brought the total volume to 12 μl.
7. The reaction was mixed gently by stirring with the pipette tip and incubated at 22°C for 15 min or at 4°C overnight.

Cloning into the Intermediate Vector pGEM-T

Cloning of DNA fragments amplified by proofreading DNA polymerases such as, Diamond DNA polymerase, into TA cloning vectors such as, pGEM-T (Promega) gives very low efficiencies, as these enzymes generates blunt-ended PCR fragments, because of removal of the 3′ A overhang usually generated by the 3′ to 5′ exonuclease activity enzymes such as *Taq* polymerase. To clone blunt-ended PCR fragments into the intermediate vector, pGEM-T modification before ligation was required. The gel-purified fragments containing the VNTR of interest were modified using an A-tailing procedure which creates an overhang of adenine nucleotides at the 3′ end of the fragment, complementary to the thymidine overhang found in the pGEM-T vector. In brief,

1. The reaction included 1–7 μl of the purified PCR product, 0.2 mM of dATP, 1.5 mM $MgCl_2$, 1× Taq reaction buffer, 5 units of *Taq* DNA polymerase and nuclease-free water to a final volume of 10 μl.
2. The reaction was incubated at 72°C for 10 min.
3. After this incubation period, the tubes containing the reaction were placed in ice to halt the reaction.
4. After addition of A-overhangs, the fragments were ligated into an intermediate vector pGEM-T (Promega) applying standard ligation reaction.

3.1.6.2. Cloning the VNTRs into Reporter Gene Vectors

Cloning the VNTR into Firefly Luciferase Reporter Vector

1. The VNTR fragments amplified by PCR were cloned into the intermediate vector pT7_Blue (Novagen) via blunt cloning at the EcoRV site.
2. Positive clones were confirmed by sequencing
3. The VNTR inserts were released by enzymatic digestion with XbaI and BamHI.
4. VNTR fragments were cloned into NheI and BglII sites in the multiple cloning site of the pGL3p vector (see Note 3) which carries a reporter gene (*Firefly* luciferase driven by a minimal SV40 promoter, Promega).

3.1.7. Ligation

For ligations, different insert: vector molar ratios were used, usually ranging from 1:1 to 3:1 ratios of molar ends. The amount of insert required was calculated using the following equation:

$$\frac{\text{ng vector} \times \text{kb size of insert}}{\text{kb size of vector}} \times \text{insert : vector ratio} = \text{ng of insert}$$

In the ligation reaction 1 μl (25 ng) of vector and appropriate volume of insert were added to and 1 μl 10× ligase buffer and 1 μl 200 units of T_4 DNA ligase (NEB M0202S) in a total volume of 10 μl, incubated at room temperature for 4 h then at 4°C overnight.

3.1.8. Transformation of Chemically Competent E. coli Cells

Once the generation of recombinant plasmid DNA was confirmed by enzymatic digest and sequencing, intermediate or reporter plasmids were transformed into strains of competent *E. coli* cells (DH5-α, Invitrogen-Gibco BRL Cat. No. 18265-017). Briefly,

1. 50 μl aliquot of competent cells was defrosted on ice.
2. The ligation reaction (10 μl) or 10 ng of plasmid DNA were added to the defrosted cells and subsequently incubated on ice for 30 minutes.
3. The cells were subjected to heatshock in a water bath for 45 s at 42°C and then incubated on ice for 2 min.
4. 950 μl of pre-warmed LB broth was added to the cells, and the culture incubated at 37°C for 1 h, on a shaker at 225 rpm.
5. 50–200 μl of this culture was spread onto LB agar plates supplemented with 100 μg/ml ampicillin, and grown at 37°C overnight.

3.1.9. Blue/White Screening of Recombinants

Intermediate plasmids such as pT7-Blue and pGEM-T enable blue/white screening of recombinants. The plasmid multiple cloning site is within the open reading frame (ORF) of functional *lacZ* encoding active β-galactosidase that can cleave the chromogenic substrate X-gal to yield a blue colony phenotype. Inserts cloned disrupts this ORF, thereby preventing the production of functional β-galactosidase, which results in the white colony phenotype when plated on X-gal/IPTG indicator plates:

1. 82 mm LB agar plates containing 100 μg/ml ampicillin were evenly pre-spread with 35 μl of 50 mg/ml X-gal (5-bromo-4-chloro-3-indolyl-bD-galactoside dissolved in dimethylformamide, Promega Cat. No. V3491), and 20 μl of 100 mM IPTG (Isopropyl β-D-1-thiogalactopyranoside, BIO-37036).
2. After spreading these solutions, the agar plates were placed in an incubator for 30 min at 37°C prior use.
3. 50–200 μl of the transformation culture was streaked out onto these plates, followed by overnight incubation at 37°C.
4. The following day, white positive colonies were picked and screened in more detail for the correct insert.

3.1.10. Isolation of DNA Constructs from Bacteria

3.1.10.1. Mini-Preparation of Plasmid DNA

A small scale preparation of plasmid DNA, for up to 20 μg, was used for screening plasmids after manipulation for molecular cloning. The QIAprep Spin Miniprep Kit (Qiagen, No. 27106) was used for this purpose.

1. This kit uses a modified alkaline lysis method, and the lysate is neutralized and adjusted to high salt binding conditions.
2. The neutralized lysate is cleared by centrifugation.
3. The neutralized lysate is then applied to a silica-gel membrane which selectively absorbs DNA in high-salt conditions.
4. Endonucleases are removed by a wash with buffer PB.
5. Salts are removed by a wash with buffer PE.
6. The plasmid DNA was eluted in nuclease-free water.

3.1.10.2. Maxi-Preparation of Plasmid DNA

For the isolation of up to 500 μg of plasmid DNA, the QIAGEN Plasmid Maxi Kit (Qiagen, No. 12263) was used:

1. The kit employs a modified alkaline lysis procedure which results in a cell lysate containing plasmid DNA among protein, chromosomal DNA, and other cell debris.
2. Debris are cleared from the lysate in a neutralising potassium acetate buffer.
3. The plasmid DNA contained in the supernatant is bound to an anion-exchange column under high salt and low pH conditions.
4. Medium-salt washes remove RNA, proteins etc.
5. The plasmid DNA is eluted with a high-salt wash.

6. The eluted DNA is then precipitated with isopropanol and washed with 70% ethanol.
7. The DNA is resuspended in nuclease-free water.

3.1.11. Analytical Restriction Enzyme Digests

Restriction enzymes were used for molecular cloning and to verify the insertion and position of the VNTR fragments into the plasmid vectors. Restriction enzyme (~5 unit/1 μg DNA) digests were carried out in 1× restriction enzyme buffer. The digests were carried out at the appropriate temperature for the respective enzyme for a minimum time of 3 h. DNA double digestion sequentially using two restriction endonuclease enzymes was performed when the two enzymes buffer salt concentration were not compatible, in such cases the first enzyme that function in a low salt buffer was used first, followed by digestion with the second enzyme that function in a high salt buffer. The second digest was set up adjusted to the volume of the first reaction. Enzymes were mostly obtained from Promega, or alternatively from New England Biolabs. The fragments generated by restriction enzyme reaction were visualized after gel electrophoresis in a UV light transilluminator.

3.1.12. Sequencing

DNA sequencing was performed by The Sequencing Service (School of Life Sciences, University of Dundee, Scotland), using Applied Biosystems Big-Dye Ver3.1 chemistry on an Applied Biosystems model 3730 automated capillary DNA sequencer.

3.1.13. Measurement of DNA Concentration by Spectrophotometry

DNA concentration was determined using UV spectrophotometer (Jenway Genova Life Science Analyser Cat. No. 636 031). The UV spectrophotometer was calibrated using 100 μl dH_2O as blank. After calibration, 1:100 diluted DNA preparation was placed into a quartz cuvette and placed in the cell holder for the determination of concentration, using the following formula: original concentration = O.D value "X" (at wavelength WL of 260 nm) × 50 ng/ml × dilution factor, where 1 O.D. at 260 nm for double-stranded (ds) DNA equals 50 ng/ml of dsDNA.

3.2. Cell and Tissue Culture

3.2.1. Culture of JAr Cells

JAr cells were maintained as monolayers in RPMI-1640 medium (Bioclear, Autogen) and cultured at 37°C, 5% CO_2. Cells were fed three times a week and were generally split once a week when confluent, or more frequently if necessary.

3.2.2. Primary Prefrontal Cortical Cultures

3.2.2.1. Dissection of Prefrontal Cortex from Neonate Wistar Rats

1. Wistar rat neonates, aged 2-7 days old, were killed in accordance with UK schedule one guidelines by neck dislocation.
2. The head was severed from the body using a sharp razor blade applied in the dorsal aspect of the neck area.
3. Using a blade, the skin was cut and the skull was removed using curved forceps (Fisher Scientific Cat. No. DKC-790-D).

4. With the use of micro scissors (WPI Cat. No 501778), a longitudinal incision starting at the bregma point was made along the sagittal suture of the skull.
5. The skull plates were held by curved forceps.
6. The scoop-end of a spatula (Fisher scientific, Cat. No 3006) was placed between the ventral surface of the brain and the inside of the base of the skull.
7. The spatula was then carefully moved from side to side to cut the underlying optic nerve tracts, releasing the brain.
8. The brain was removed using the spatula and placed in a petri dish filled with pre-chilled dissection solution.
9. Once in the petri dish, the frontal cortex was dissected out using a clean scalpel by inserting the blade with a 50° inclination pointing towards the root of the optical bulb.
10. Finally, the dissected frontal cortex was placed into a 15 ml falcon tube containing dissection solution for preparation of dissociated cultures.

3.2.2.2. Coating Tissue Culture Plastics

24-well tissue culture plates used to grow cortical cultures were coated with poly-D-lysine (100 μg/ml) which was prepared as a stock solution at concentration of 10 mg/ml, aliquoted and stored at –20°C under sterile conditions.

1. Prior to transfection, poly-D-lysine was dissolved in sterile dH_2O and 200 μl (12.5 μg/cm^2) were added per well.
2. The 24-well plates were placed in a 37°C incubator overnight to allow the poly-D-lysine to adhere to the surface of the wells.
3. The following day, prior to plating of the cells, the plates were washed twice with 1x PBS to remove any traces of poly-D-lysine.

3.2.2.3. Preparation of Cortical Cultures

1. Cortex tissue were stored in 15 ml falcon tube containing dissection solution and centrifuged at 1,000 rpm for 5 min at room temperature in a bench top centrifuge.
2. The supernatant was replaced with 3 ml of trypsin/EDTA solution (Sigma-Aldrich Ltd, Cat. No. T4049) and placed in an incubator at 37°C for 20 min.
3. The tissue was then centrifuged at 500 rpm for 5 min at room temperature; after which, the trypsin solution was decanted and replaced by fresh pre-warmed culture medium I containing penicillin/streptomycin.
4. The tissue was then centrifuged at 500 rpm for 3 min at room temperature; this procedure was repeated three times.

5. The resulting pellet was dissociated in 5 ml of culture medium I (penicillin/streptomycin) using two Pasteur pipettes with pores of decreasing diameter until the cell suspension was homogeneous and the solution appeared turbid.
6. The resulting cell suspension was passed through a 0.70 μm falcon cell strainer (VWR) to remove debris, centrifuged at 1,000 rpm for 5 min and resuspended in 5 ml of culture medium I (without antibiotics).
7. The dissociated cells were counted under a contrast microscope using a cell counting chamber (10^5 per well).
8. The cells were then plated into poly-D-lysine coated 24-well plates for transfection 24 h later. Poly-D-Lysine was used as this substance will create a matrix for the better adherence of neuronal cultures to the culture flask.
9. After 7 h, medium I was removed and replaced with 1 ml/well of culture medium II. Medium II was renewed prior to transfections.

3.3. Lithium and Cocaine Cell Treatments

1. Cells were plated into 24-well plates.
2. Prior to cells treatment with lithium or cocaine, the cells were incubated in serum-free medium overnight.
3. Cells were subsequently incubated in serum-free medium supplemented with 1 mM LiCl, 1 μM cocaine, or 10 μM cocaine.
4. Lithium and cocaine treated cells were subsequently used in luciferase assays (described below).

3.4. Delivery of Luciferase Constructs into JAr Cell Lines and Rat Cortical Cultures

Reporter gene plasmids and expression constructs were delivered into either cells or prefrontal cortical cells using either ExGen 500 in vitro Transfection Reagent (Fermentas) or TransFast Transfection Reagent (Promega). To normalize for transfection efficiency, (see Note 4) either pmLuc-2 vector containing a minimal TK promoter followed by an optimized *Renilla* luciferase (Rluc) cDNA (Novagen) or a modified pMLuc-2 vector containing a minimal TK promoter followed by an optimized *Firefly* luciferase cDNA were used as an internal control at a ratio of 50:1; VNTR construct: pmLuc-2 plasmid.

3.4.1. ExGen 500 In Vitro Transfection Reagent

ExGen 500 in vitro Transfection Reagent (Fermentas) was used to transfect rat prefrontal cortical cultures and cell lines. ExGen 500 is a polyethylenimine cationic polymer. ExGen 500 and DNA charge-interact and form small, stable, highly diffusible particles that settle on the cell surface. The ExGen 500/DNA complex is then absorbed into the cell by endocytosis. These endosomes are ruptured in the cytoplasm before lysosomal degradation releasing the

ExGen 500/DNA complex, allowing the DNA to be translocated into the nucleus. Following manufacturer's instructions:

1 μg of reporter constructs and 20 ng of the internal control plasmid pmLuc-2 (*Renilla* luciferase or *FireFly* luciferase) were diluted in 100 μl of 150 mM NaCl, followed by gentle vortexing and brief centrifugation.

1. 3.3 μl of ExGen 500 was added per 1 μg of DNA used.
2. The solution was immediately vortexed for 10 s and incubated for 10 min at room temperature.
3. 100 μl of the ExGen 500/DNA mixture was added to each well (the volume of the ExGen 500/DNA mixture represented 10% of the total volume of the culture medium) and gently rocked to achieve even distribution of the complexes.
4. The plates were then centrifuged for 5 min at 280×*g* at room temperature and finally incubated at 37°C for 48 h in a humidified 5% CO_2 incubator.
5. 48 h post-transfection, the cells were lysed to analyze reporter gene expression.

3.4.2. TransFast Transfection Reagent

The TransFast Transfection Reagent (Promega) was used for transfection of plasmid DNA into cell lines. The TransFast Transfection Reagent is comprised of a synthetic cationic lipid and a neutral lipid (DOPE). The lipid complex associates with the DNA and similarly as for ExGen 500, is introduced into the cell by endocytosis and later released into the cytoplasm allowing its passage to the nucleus (Promega Madison, technical bulletin TB260).

24 h prior to transfection, 10^5 cells were seeded into 24-well plates and the Transfast transfection reagent was resuspended in 400 μl nuclease-free water at room temperature, making the final concentration of the cationic lipid component 1 mM, and it was then frozen at −20°C.

1. Immediately before transfection, the culture medium was replaced by serum-free medium.
2. 1 μg of reporter gene constructs and 20 ng of the internal control plasmid pmLuc-2 (*Renilla* luciferase or *FireFly* luciferase) were diluted in 200 μl of serum-free culture medium in an eppendorf tube and mixed with 3–6 μl Transfast transfection reagent, immediately followed by brief vortexing.
3. The mixture was incubated for 10–15 min at room temperature.
4. The culture medium was removed from the 24-well plate and the transfection mixture (200 μl) was added to the cells.

5. The culture plates were returned to the humidified 37°C, 5% CO_2 incubator for 1 h, after which 800 µl of culture medium containing serum was added to each well.
6. The plates were returned to the incubator for 48 h and after this period, the cells were harvested for luciferase assay.

3.5. Co-transfection Experiments

To assess the potential regulation of the VNTR domains by the transcription factors, the full length human expression constructs for specific transcription factors were transfected into cell lines or primary cultures of cortex simultaneously with the reporter constructs. The constructs were co-transfected using transfection protocols described above. In these co-transfection experiments, the total amount of plasmid DNA transfected into the cells was maintained constant. For this, per every 1 µg of either transcription factors expression vector co-transfected with the reporter constructs, an equal amount of inocous DNA (backbone) was included with the VNTR constructs.

3.6. Analysis of Transgene Expression by Reporter Gene Assay

Analysis of the amount of luciferase protein activity produced by the transfected plasmids was estimated using the Dual Luciferase Assay kit (Promega, Madison Cat. No E1500) on extracts of transfected cells (see Note 5). Briefly, cell extracts were obtained as follows:

1. Culture medium was removed and cells were washed twice with PBS.
2. 70 µl of 1× passive lysis buffer per well was added and incubated for 15 min on a rocking platform.
3. 20 µl of the cell lysate were plated into a 96-well plate and transferred into the Glomax 96 microplate luminometer (Promega).
4. Firefly luciferase reagent (100 µl containing the luciferase assay substrate) and sea pansy luciferase reagent (100 µl, containing the sea pansy luciferase substrate) were automatically injected into each well to calculate luminescence intensity.

The sea-pansy luciferase substrate solution was added to each sample to determine the protein production of the internal control (pmLuc-2) to normalize for transfection efficiency, in case the number of cells or the efficiency of the transfection varied from well to well. *Firefly* (green–yellow light) was detected as wavelength of 550–570 nm, whereas, *Renilla* (blue light) was detected as wavelength of 480 nm. The calculated luminescence was processed by GLOMAX software package (Promega).

4. Notes

1. Because of the high GC content of the VNTR sequences, which contribute to formation of DNA secondary structure, 0.5 M Betaine can be added to the PCR mixture and specific polymerases such as Diamond and KOD are recommended.
2. The design of the construct is critical; the choice of the fragment size to include in the reporter gene may have a variable effect when assessed separately or when in the context of the full-promoter. Tissue specificity of the domain tested should also be considered.
3. Restriction site can be introduced into the primers sequences when required for subsequent cloning and to aid orientation.
4. Transfection efficiency must be optimized for various cell lines; this can be done using a CMV-GFP expression vector, in which percentage of transfected cells can be visualised.
5. As mentioned in the introduction, the lack of the native chromatin configuration of the DNA sequences in reporter gene vectors may influence results, therefore, generation of stable cell lines by cloning DNA fragment of interest into Epstein-Barr virus (EBV) based vector is recommended.

References

1. Yan, H., Yuan, W., Velculescu, V.E., Vogelstein, B. and Kinzler, K.W. (2002) Allelic variation in human gene expression. *Science* **297**, 1143.
2. Bray, N.J., Buckland, P.R., Owen, M.J. and O'Donovan, M.C. (2003) *Cis*-acting variation in the expression of a high proportion of genes in human brain. *Hum Genet* **113**, 149–53.
3. Lo, H.S., Wang, Z., Hu, Y., Yang, H.H., Gere, S., Buetow, K.H. and Lee, M.P. (2003) Allelic variation in gene expression is common in the human genome. *Genome Res* **13**, 1855–62.
4. Heils, A., Teufel, A., Petri, S., Stober, G., Riederer, P., Bengel, D. and Lesch, K.P. (1996) Allelic variation of human serotonin transporter gene expression. *J Neurochem* **66**, 2621–4.
5. Klenova, E., Scott, A.C., Roberts, J., Shamsuddin, S., Lovejoy, E.A., Bergmann, S., Bubb, V.J., Royer, H.D. and Quinn, J.P. (2004) YB-1 and CTCF differentially regulate the 5-HTT polymorphic intron 2 enhancer which predisposes to a variety of neurological disorders. *J Neurosci* **24**, 5966–73.
6. Roberts, J.C., Scott, A.M., Howard, M.R., Breen, G., Bubb, V.J., Klenova, E., and Quinn, J.P. (2007) Differential regulation of the serotonin transporter gene by lithium is mediated by transcription factors, CCTC binding protein and Y-Box binding protein 1, through the polymorphic intron 2 variable number tandem repeat. *J Neurosci* **27**, 2793–2801.
7. King, M.C. and Wilson, A.C. (1975) Evolution at two levels in humans and chimpanzees. *Science* **188**, 107–16.
8. Brookes, A.J. and Prince, J.A. (2005) Genetic association analysis: lessons from the study of Alzheimers disease. *Mutat Res* **573**, 152–9.
9. Liu, W., Gu, N., Feng, G., Li, S., Bai, S., Zhang, J., Shen, T., Xue, H., Breen, G., St Clair, D. and He, L. (1999) Tentative association of the serotonin transporter with schizophrenia and unipolar depression but not with bipolar disorder in Han Chinese. *Pharmacogenetics* **9**, 491–5.

10. Bray, N.J., Buckland, P.R., Williams, N.M., Williams, H.J., Norton, N., Owen, M.J. and O'Donovan, M.C. (2003). A haplotype implicated in schizophrenia susceptibility is associated with reduced COMT expression in human brain. *Am J Hum Genet* **73**, 152–61.
11. Evans, J., Battersby, S., Ogilvie, A.D., Smith, C.A., Harmar, A.J., Nutt, D.J. and Goodwin, G.M. (1997) Association of short alleles of a VNTR of the serotonin transporter gene with anxiety symptoms in patients presenting after deliberate self harm. *Neuropharmacology* **36**, 439–43.
12. Baca-Garcia, E., Vaquero-Lorenzo, C., Diaz-Hernandez, M., Rodriguez-Salgado, B., Dolengevich-Segal, H., Arrojo-Romero, M., Botillo-Martin, C., Ceverino, A., et al. (2006) Association between obsessive-compulsive disorder and a variable number of tandem repeats polymorphism in intron 2 of the serotonin transporter gene. *Prog Neuropsychopharmacol Biol Psychiatry* **31**, 416–20.
13. Ogilvie, A.D., Battersby, S., Bubb, V.J., Fink, G., Harmar, A.J., Goodwim, G.M. and Smith, C.A. (1996) Polymorphism in serotonin transporter gene associated with susceptibility to major depression. *Lancet* **347**, 731–3.
14. Battersby, S., Ogilvie, A.D., Smith, C.A., Blackwood, D.H., Muir, W.J., Quinn, J.P., Fink, G., Goodwin, G.M. and Harmar, A.J. (1996) Structure of a variable number tandem repeat of the serotonin transporter gene and association with affective disorder. *Psychiatr Genet* **6**, 177–81.
15. Collier, D.A., Arranz, M.J., Sham, P., Battersby, S., Vallada, H., Gill, P., Aitchison, K.J., Sodhi, M., et al. (1996) The serotonin transporter is a potential susceptibility factor for bipolar affective disorder. *Neuroreport* **7**, 1675–9.
16. Bellivier, F., Leroux, M., Henry, C., Rayah, F., Rouillon, F., Laplanche, J.L. and Leboyer, M. (2002) Serotonin transporter gene polymorphism influences age at onset in patients with bipolar affective disorder. *Neurosci Lett* **334**, 17–20.
17. Skipper, L., Liu, J.J. and Tan, E.K. (2006) Polymorphisms in candidate genes: implications for the current treatment of Parkinson's disease. *Expert Opin Pharmacother* **7**, 849–55.
18. Haddley, K., Vasiliou, A.S., Ali, F.R., Paredes, U.M., Bubb, V.J. and Quinn, J.P. (2008) Molecular genetics of monoamine transporters: relevance to brain disorders. *Neurochem Res* **33**, 652–67.
19. Jeffreys, A.J., Wilson, V. and Thein, S.L. (1985) Hypervariable 'minisatellite' regions in human DNA. *Nature* **314**, 67–73.
20. Lesch, K.P., Meyer, J., Glatz, K., Flugge, G., Hinney, A., Hebebrand, J., Klauck, S.M., Poustka, A., et al. (1997) The 5-HT transporter gene-linked polymorphic region (5-HTTLPR) in evolutionary perspective: alternative biallelic variation in rhesus monkeys. Rapid communication. *J Neural Transm* **104**, 1259–66.
21. Soeby, K., Larsen, S.A., Olsen, L., Rasmussen, H.B. & Werge, T. (2005) Serotonin transporter: evolution and impact of polymorphic transcriptional regulation. *Am J Med Genet B Neuropsychiatr Genet* **136**, 53–7.
22. Fiskerstrand, C.E., Lovejoy, E.A. and Quinn, J.P. (1999) An intronic polymorphic domain often associated with susceptibility to affective disorders has allele dependent differential enhancer activity in embryonic stem cells. *FEBS Lett* **458**, 171–4.
23. MacKenzie, A. and Quinn, J. A. (1999) Serotonin transporter gene intron 2 polymorphic region, correlated with affective disorders, has allele-dependent differential enhancer-like properties in the mouse embryo. *Proc Natl Acad Sci U S A* **96**, 15251–5.
24. Lovejoy, E.A., Scott, A.C., Fiskerstrand, C.E., Bubb, V.J. and Quinn, J.P. (2003) The serotonin transporter intronic VNTR enhancer correlated with a predisposition to affective disorders has distinct regulatory elements within the domain based on the primary DNA sequence of the repeat unit. *Eur J Neurosci* **17**, 417–20.

Chapter 12

Whole Genome Sequencing

Pauline C. Ng and Ewen F. Kirkness

Abstract

Whole genome sequencing provides the most comprehensive collection of an individual's genetic variation. With the falling costs of sequencing technology, we envision paradigm shift from microarray-based genotyping studies to whole genome sequencing. We review methodologies for whole genome sequencing. There are two approaches for assembling short shotgun sequence reads into longer contiguous genomic sequences. In the de novo assembly approach, sequence reads are compared to each other, and then overlapped to build longer contiguous sequences. The reference-based assembly approach involves mapping each read to a reference genome sequence. We discuss methods for identifying genetic variation (single nucleotide polymorphisms, small indels, and copy number variants) and building haplotypes from genome assemblies, and discuss potential pitfalls. We expect methodologies to evolve rapidly as sequencing technologies improve and more human genomes are sequenced.

Key words: Human, Genome, Sequencing, Assembly

1. Value of Whole Genome Sequencing

Currently, whole genome association studies aim to identify the genetic basis of traits and disease susceptibilities using SNP microarrays that capture most of the common genetic variation in the human population. Risk variants for many diseases have been identified. However, with only a few notable exceptions (e.g., age-related macular degeneration, Type 1 diabetes), the risk variants usually explain only a minor fraction of the genetic risk that is known to exist. There are several factors that are likely to contribute to this observation. Common variants may have only minor effects on a phenotype, or have variable penetrance owing to epistatic or epigenetic influences. Two additional factors are rare variants and copy number variants (CNVs). It is known that these types of genomic variation can have important influences on disease phenotypes (1, 2). However, assay of these variants

Michael R. Barnes and Gerome Breen (eds.), *Genetic Variation: Methods and Protocols*, Methods in Molecular Biology, vol. 628, DOI 10.1007/978-1-60327-367-1_12, © Springer Science+Business Media, LLC 2010

cannot be achieved readily using current genotyping microarray technologies.

Whole genome sequencing offers a potential solution by providing the most comprehensive collection of rare variants and structural variation for sequenced individuals. Although currently this is prohibitively expensive to conduct on a large scale, we envision a paradigm shift in the technology because of falling costs. Consequently, there is a need to develop methodology for comparing whole genomes, and we review what is currently under way.

2. Assembly of Human Genome Sequences from Whole Genome Shotgun Reads

Complete sequencing of the ~6 Gb of DNA that uniquely identifies each human individual requires fragmentation of the DNA, redundant sequencing of millions of DNA fragments in lengths of 25–1,000 bases, then assembly of these reads into large contiguous segments that can be ordered and oriented along each chromosome. Owing to the repetitive content of human genome sequence, the most comprehensive assemblies are derived from paired end reads, where the sequence reads are obtained from both ends of each DNA fragment. There are two approaches for assembling shotgun reads into longer contiguous sequences. The first of these, de novo assembly, is the only option if a closely related genome sequence does not yet exist. Here, the sequence reads are compared to each other, then overlapped to build longer contiguous sequences. An alternative approach, reference-based assembly, involves mapping each read to a reference genome sequence, then building a consensus sequence that is similar but not necessarily identical to the backbone reference.

One obvious limitation of the reference-based approach is that novel sequences (absent from the reference) are not readily apparent unless a subsequent de novo assembly is performed on the residual unmapped reads. In terms of computational complexity, de novo assemblies are orders of magnitude more memory intensive than mapping assemblies. This complexity is determined principally by the length and number of fragments. Levy et al. (3) reported the first sequence of an individual human genome, assembled de novo from 32 million Sanger sequence reads (~700 bases long). However, de novo assemblies are much more challenging when using the shorter reads generated by the newer "flow cell sequencing" technologies (4). Several assemblers have been developed to perform de novo assembly of very short sequence reads (25–50 bases), and most are freely available (Table 1). However, at present, their utility is restricted to the assembly of small, bacterial-sized genomes. Even the most mature of the new sequencing technologies, Roche-454, with read lengths of up to 400 bases, paired end read protocols, and a custom

Table 1
De novo assemblers for short sequence reads

ALLPATHS	(42)	
Edena	(43)	http://www.genomic.ch/edena.php
SHARCGS	(44)	http://sharcgs.molgen.mpg.de/
SHRAP	(45)	
SSAKE	(46)	http://www.bcgsc.ca/platform/bioinfo/software/ssake
VCAKE	(47)	http://sourceforge.net/projects/vcake
Velvet	(48)	http://www.ebi.ac.uk/~zerbino/velvet/

Table 2
Reference-based assemblers for short sequence reads

ELAND	http://bioinfo.cgrb.oregonstate.edu/docs/solexa/	Illumina
EULER	http://euler-assembler.ucsd.edu/portal/	Illumina
MOSAIK	http://bioinformatics.bc.edu/marthlab/Mosaik	Illumina/454
MAQ	http://sourceforge.net/projects/maq/	Illumina/SOLiD
Novocraft	http://www.novocraft.com/index.html	Illumina
RMAP	http://rulai.cshl.edu/rmap/	Illumina
SeqMap	http://biogibbs.stanford.edu/~jiangh/SeqMap/	Illumina
SHRiMP	http://compbio.cs.toronto.edu/shrimp/	Illumina/SOLiD
SOAP	http://soap.genomics.org.cn/	Illumina
SXOligoSearch	http://synasite.mgrc.com.my:8080/sxog/NewSXOligoSearch.php	Illumina

assembler, Newbler, has not yet been demonstrated to perform de novo assemblies for human-sized genomes.

As a consequence, current efforts to sequence multiple human genomes using new sequencing technologies rely upon reference based mapping approaches. These include several older sequence alignment tools such as Exonerate, GMAP, MUMmer, and SSAHA (5–8) as well as many that have been developed recently for specific sequencing platforms (Table 2). The second reported genome sequence of a human individual was sequenced using the Roche-454 platform, with the reads mapped to a reference human genome sequence (NCBI Build 36) using the alignment tool, BLAT (9). Over the next few years, it is likely that several hundred human genomes will be sequenced using short read technologies, and reference-based assemblies (http://www.1000genomes.org/).

Clearly, the quality of a genome assembly, whether de novo or reference based, is the most important factor for the subsequent analyses of sequence variation between individual genomes. However, despite continual development of new assembly and alignment algorithms, there are few tools for evaluating the quality of these assemblies. Although individual base calls can be assigned quality values that indicate their reliability (e.g., Phred scores), these values provide no information on the quality of read placement within an assembly. For that, it is necessary to consider additional parameters such as:

1. Mate-pair information. Most shotgun sequence reads are obtained in pairs from both ends of DNA fragments of known length. Consequently, there are distance and orientation constraints on where pairs of reads can be placed relative to each other in an assembly.
2. Unassembled reads. Unless derived from contaminating DNA, unassembled reads are often indicative of misassemblies, such as when tandem repeats are collapsed into a single element.
3. Read coverage. Reads that are derived from different copies of a same repeat are frequently assembled together if the repeat copies are sufficiently similar. This can result in read coverage that is artificially inflated.

The initial reports of individual human genome assemblies have included dedicated browsers that permit selected loci to be examined for mate–pair relationships and read coverage (http://huref.jcvi.org/ and http://jimwatsonsequence.cshl.edu).

3. Methods for Analyses of Whole Genome Shotgun Assemblies

3.1. SNPs, CNVs and Structural Rearrangements

The methodologies used for the identification of variant loci in genome assemblies are dependent on the type of assembly under consideration (i.e., de novo or reference-based). For variants in a de novo assembly, Levy et al., (3) aligned sequencing reads within the assembly, and then identified regions of difference in a one-to-one comparison with the NCBI human reference genome sequence. The contribution of each sequence read to a single position in the consensus was evaluated after the assembly process to identify positions that contain more than one allele. This process identified heterozygous SNPs and indel polymorphisms, and typically two or more reads were required for the initial identification of an alternate allele. Homozygous SNPs were identified when single loci differed in the one-to-one mapping, and all underlying sequence reads supported one allele. Finally, homozygous insertion or deletion loci were identified by the presence or absence of sequence relative to the NCBI assembly, respectively. In contrast, for the reference-based assembly described by Wheeler et al. (9), all reads were aligned directly to the NCBI human reference genome. Poor-quality alignments were defined as those reads aligning for less than 90% of their length, or with more than four substitutions or insertions/deletions with respect to the reference, or reads that matched two locations with nearly equal match score. Reads that passed the alignment quality criteria were then realigned to reference genome fragments using Cross_match software. An error model was developed to separate sequencing error from true genomic variation, and the location and type of each putative true variant was tabulated.

The approaches used by both Levy et al., (3) and Wheeler et al., (9) led to the identification of more than three million SNPs in each of the individual human genome assemblies. A high fraction (~75–80%) of these variants is found in dbSNP. The remainder is composed of rare mutations in the population, mutations that are private to the individual, or false positive errors from the sequencing technology. One of the challenges in sequencing is to distinguish real variants from false positives.

To increase confidence that rare variants are real, several criteria can be imposed. First, only variants with high-quality scores are considered (3, 9). Furthermore, although variants may be rare in the population, they should be present in one of the two chromosome copies. This means that, if the rare mutations are heterozygous, one would expect the rare alleles and the common alleles to each follow a binomial distribution centered at $p=0.5$. Deviations from the binomial distribution could indicate an error in sequencing (or low-frequency, somatic mutations). Other criteria to consider are requirements for two reads to support each allele, or for each allele to be observed on both DNA orientations, although the latter was found to be too stringent (3). Alternatively, because first-order relatives will have a 50% chance of carrying variants, sequencing of relatives can confirm many variants, assuming that they are not de novo mutations.

Rare variants can be validated by inspection of underlying trace data to confirm if a variant is real, though this can be time-consuming. Of the ~380,000 novel nonsilent tumor variants identified in 120,839 exons (21 Mb) by Sjoblom et al. (10), over 90% were excluded as false positives after visual inspection of sequence traces. Subsequent resequencing of the remaining variants caused further exclusion of 32% of the remaining variants. In our analysis of an individual's exome, 35% of novel nonsynonymous variants were not confirmed by subsequent manual trace inspection (11), and of the remaining novel nonsynonymous variants that passed visual inspection, ~25% failed to be confirmed in PCR.

In addition, one should consider any known sequencing biases. In Sanger sequencing, variants called at the beginning or ends of each read were discarded (3). For Roche-454, variants near homopolymers were determined to be low-quality and discarded (9). Thus, for sequencing technologies, it is important to determine what errors a technology introduces and to filter out potential false positives.

With regard to the false-negative rate of missing heterozygous variants, read coverage is also a critical issue. Levy et al., (3) developed a statistical model based on assembly read coverage and on the filtering criteria used for calling high confidence variants. At a given heterozygous locus, the probability of observing both alleles in at least x reads follows the binomial distribution with $p=0.50$ and $n=$ depth of coverage, where x is defined by the filtering criteria. To calculate the false-negative rate genome wide, a Poisson distribution is also incorporated to estimate sequence depth at different loci, where λ is set to the genome sequence coverage (7.5 for SNPs, 5.5 for insertions, 4.9 for deletions, after read filtering is taken into account). As sequencing costs drop, this will allow deeper coverage and decrease the false negative error rate.

It is now recognized that a major fraction of mammalian genetic variation derives from relatively large (>1 kb) segments that are either deleted or amplified to variable degrees among different genomes (12). These segments are known as Copy Number Variants (CNVs), and have been estimated to compose more than 10% of the human genome (13). Historically, CNVs have been studied most frequently by comparative genomic hybridization (CGH). However, it is likely that whole genome sequencing will have a powerful role to play in future classification of CNVs. Wheeler et al. (9) analyzed the depth of sequence coverage across human chromosome 22, where the content of segmental duplications has been characterized extensively. After filtering for high copy number repeats, the unique regions (28 Mb of sequence) showed a narrow distribution of read coverage (50.4 ± 12.8 reads per 5 kb), while all duplications >10 kb and with >95% similarity had demonstrable increases in read coverage. In contrast to CGH, where only the existence of a CNV is revealed, whole genome sequencing offers the potential to resolve the location of each CNV copy, even when amplified segments have been inserted at novel genomic loci. To do this, the genome assemblies must be carefully examined in regions with unusually high read coverage, and extra copies of a CNV relocated by using paired end sequence data to identify the unique sequence of each CNV flank.

At this early stage of whole genome sequencing for human populations, a significant amount of the acquired sequence data is novel (i.e., absent from the NCBI reference). Insertions of novel sequences pose a challenge for reference-based assemblies, particularly those derived from short reads. Following a reference-based assembly, novel sequence reads remain unmapped, and must be assembled de novo in order to obtain contigs that are long enough to annotate for genes etc. For reads of 200–400 bases (Roche-454 platform), it is possible to build contigs of the modest length (~1 kb) and identify putative genes (9). However, for the shorter reads delivered by the Illumina and SOLiD platforms, de novo assembly of unmapped reads on a genome-wide scale has not yet been demonstrated, and will be a significant limitation until resolved.

3.2. Constructing Long-Range Haplotypes

Haplotypes are more strongly correlated with phenotypes than single markers (14–16). Ideally, it will be possible to resolve two distinct haplotypes for each pair of chromosomes in a human diploid genome. Currently, it is possible to reconstruct phased haplotypes using genotypes obtained from SNP microarrays by applying statistical methods to population data, or by incorporating pedigree information (17, 18). When inferring haplotypes from population data, local phasing is usually limited to SNPs within a haplotype block. Also, it is unreliable for estimation of the haplotypic phase between markers that are separated by more than 100 kb (19), and less accurate for rare haplotypes (20). Long-range haplotypes can be reconstructed from pedigree data if genotypes from related family members are available (17). However, these data may be difficult to obtain on a large-scale. There are several alternative methods for obtaining long-range haplotypes that do not involve SNP genotyping microarrays and may be scalable to large-scale studies (20–23).

Whole genome sequence assemblies can be used to produce haplotypes. Bansal, et al. was able to reconstruct haplotypes from a whole diploid genome sequenced by Sanger mate pair reads (19, 24). Paired ends spanning a region with low LD are especially informative for reconstructing haplotypes. In order for a mate pair to be used in haplotype construction, both paired end reads must contain at least one heterozygous SNP. The chance that a mate pair will have this variation depends on the read length; Sanger reads are ~800 bp in length, while newer technologies like Illumina GA and ABI SOLiD can be as short as 35 bp. The probability of observing at least one heterozygous SNP in a read is $1-e^{-\lambda}$, where λ is the length of the read multiplied by the heterozygosity rate. Then, the number of mate pairs that will have at least one heterozygous SNP in both of the paired end reads is $(1-e^{-\lambda})^2$. If a heterozygous SNP occurs every 1,500 bp (3), then for a read length of 800 bp, ~17% of the mate pairs will

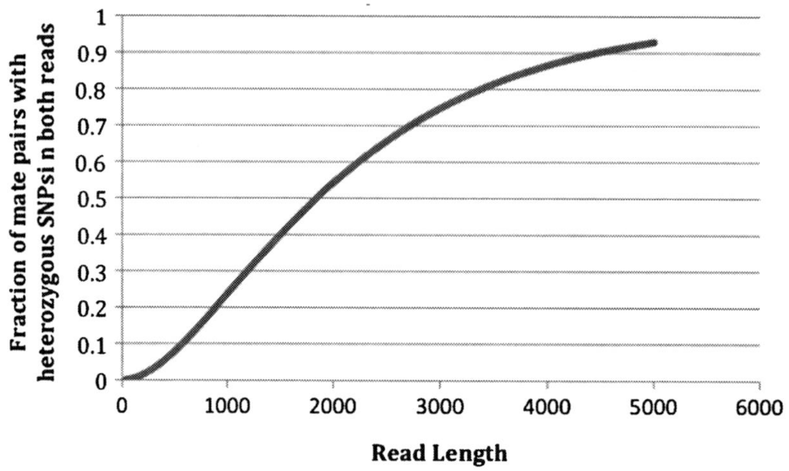

Fig. 1. Mate pairs useful for constructing haplotypes. The fraction of mate pairs that contain at least one heterozygous variant in both mate pair ends is plotted as a function of read length. Mate pairs with short read lengths are less likely to contain heterozygous variation, and hence a smaller fraction would be useful in constructing haplotypes

contain heterozygous SNPs in both of the reads and will be useful in phasing haplotypes (Fig. 1). If a read length is only 35 bp, then ~0.05% of the reads will contain heterozygous SNPs in both reads. Clearly, short-read sequencing technologies are insufficient for construction of haplotypes unless they can generate very high levels of sequence coverage, or until their read lengths can be improved substantially. Furthermore, the distance between the mate pair will determine how long the haplotype can span. Mate pair data from fosmid libraries have 40 kb inserts and these are ideal for constructing long-range haplotypes. Mate pair libraries with 2–10 kb insert sizes may not directly physically link longer haplotypes, but by linking up multiple mate pairs, long-range haplotypes can be inferred (3).

4. Bioinformatic Approaches for Analysis of Rare Variants

When multiple genomes are sequenced, a plethora of variation will be discovered, and one would like to distinguish between the genes containing neutral polymorphisms from those that contain etiological variants. It has been observed that genes involved in disease are enriched in functional variants, namely nonsynonymous mutations that cause amino acid substitutions in the corresponding protein (1, 10, 25–31). For example, Cohen et al. sequenced candidate genes in individuals with low levels of HDL-C who were at risk for coronary atherosclerosis,

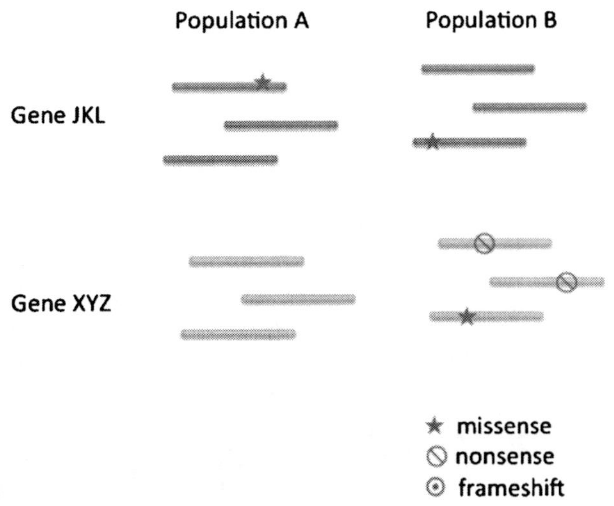

Fig. 2. Comparing the number of functional variants (missense, nonsense, frameshift) to find candidate disease genes. After sequencing many genes in two different populations, one identifies genes enriched in functional variants. For Gene JKL, the number of functional variants in both populations is similar, and therefore this gene is not a candidate disease gene. For Gene XYZ, Population B has more functional variants than Population A, and this gene is considered a candidate disease gene

and compared their variants to individuals with high levels of HDL-C (1). The at-risk cohort had eight times as many nonsynonymous variants when compared to the group not at risk.

In order to distinguish at-risk genes, the number or rate of functional variants in genes is calculated, and those genes with an excess or higher rate of functional variation are further investigated (Fig. 2). Functional coding variants are typically nonsilent changes such as nonsynonymous variants that cause an amino acid change, nonsense changes that introduce a stop codon in the open reading frame, and indels that cause frameshifts in the protein-coding sequence. Those genes that show a higher number or rate of functional variants in an affected population versus a control population are identified as candidate disease genes.

Comparing the number of nonsilent changes requires that similar numbers of cases and controls are sequenced (1). If the number of cases and controls are not similar, then random subsampling can simulate equal numbers of cases and controls (26). When comparing many genes, the nonsynonymous rate can be used (10). By comparing the nonsynonymous mutation rates genes in the affected population to the control population, gene length, sequence composition, and regional mutation rate bias are controlled for (32). Also, one must take into account sequence coverage because variants may be missed in individuals with low coverage, and this could potentially skew the comparison (32).

As sequencing of whole genomes or exomes becomes more common (3, 9, 33–36), more sophisticated statistical methods may be called for. Li and Leal have proposed to collapse the rare variants (37) at a locus, and assess whether the proportions of individuals with rare variants in the cases and controls differ. This method incorporates both common risk variants and rare risk variants, and is robust if functional variants are erroneously excluded or normal variants are erroneously included.

A nonsilent change affects the corresponding protein sequence by altering one or more amino acids. However, this change may or may not be the etiological variant. It is possible that the nonsilent change has no effect on protein function. To be the etiological variant, the nonsilent change has to alter gene function, which then plays some role in the disease. One can further refine the classification of a nonsilent variant by using algorithms that predict whether the resulting amino acid change is likely to affect protein function (1, 30). Correct classification of the truly functional variants increases the power to detect the disease gene (37).

Once candidate genes have been identified, one can try to find a common theme among the genes to support a pathway or function important to the disease. One could examine if certain functional categories (using GO ontology) (10, 38) are enriched in a condition or if the genes occur in certain pathways by using the pathway databases like KEGG, iPATH, BioCarta, sigPathways, Reactome, Panther INOH, MetaCore (25, 29, 38). One could also see if certain protein domains are enriched in the putative disease genes by using Interpro (10, 38). For all these tests, one should take into account the relative representation of different functional classes (39) and the multiple hypotheses being generated with these methods.

The various analyses discussed here can indicate candidate disease genes and their etiological variants. Ideally, one would characterize the variants found in candidate disease genes in subsequent functional studies (40, 41).

5. Conclusion

With the development of new sequencing technologies, whole genome sequencing of human populations is increasingly feasible. These sequencing technologies are now generating terabytes of data on a daily basis. The next challenge will be to analyze these copious amounts of sequence data in the most meaningful manner. In this chapter, we have discussed existing analytic methods, and we expect these to evolve rapidly as more computational biologists gain experience of working with these new types of datasets.

References

1. Cohen, J.C., Kiss, R.S., Pertsemlidis, A., Marcel, Y.L., McPherson, R. and Hobbs, H.H. (2004) Multiple rare alleles contribute to low plasma levels of HDL cholesterol. *Science*, **305**, 869–872.
2. Estivill, X. and Armengol, L. (2007) Copy number variants and common disorders: filling the gaps and exploring complexity in genome-wide association studies. *PLoS Genet*, **3**, 1787–1799.
3. Levy, S., Sutton, G., Ng, P.C., Feuk, L., Halpern, A.L., Walenz, B.P. *et al.* (2007) The diploid genome sequence of an individual human. *PLoS Biol*, **5**, e254.
4. Holt, R.A. and Jones, S.J. (2008) The new paradigm of flow cell sequencing. *Genome Res*, **18**, 839–846.
5. Slater, G.S. and Birney, E. (2005) Automated generation of heuristics for biological sequence comparison. *BMC Bioinformatics*, **6**, 31.
6. Wu, T.D. and Watanabe, C.K. (2005) GMAP: a genomic mapping and alignment program for mRNA and EST sequences. *Bioinformatics*, **21**, 1859–1875.
7. Kurtz, S., Phillippy, A., Delcher, A.L., Smoot, M., Shumway, M., Antonescu, C. and Salzberg, S.L. (2004) Versatile and open software for comparing large genomes. *Genome Biol*, **5**, R12.
8. Ning, Z., Cox, A.J. and Mullikin, J.C. (2001) SSAHA: a fast search method for large DNA databases. *Genome Res*, **11**, 1725–1729.
9. Wheeler, D.A., Srinivasan, M., Egholm, M., Shen, Y., Chen, L., McGuire, A. *et al.* (2008) The complete genome of an individual by massively parallel DNA sequencing. *Nature*, **452**, 872–876.
10. Sjoblom, T., Jones, S., Wood, L.D., Parsons, D.W., Lin, J., Barber, T.D. *et al.* (2006) The consensus coding sequences of human breast and colorectal cancers. *Science*, **314**, 268–274.
11. Ng, P.C., Levy, S., Huang, J., Stockwell, T.B., Walenz, B.P., Li, K. *et al.* (2008) Genetic variation in an individual human exome. *PLoS Genet*, **4**, e1000160.
12. Feuk, L., Carson, A.R. and Scherer, S.W. (2006) Structural variation in the human genome. *Nat Rev Genet*, **7**, 85–97.
13. Redon, R., Ishikawa, S., Fitch, K.R., Feuk, L., Perry, G.H., Andrews, T.D. *et al.* (2006) Global variation in copy number in the human genome. *Nature*, **444**, 444–454.
14. Winkelmann, B.R., Hoffmann, M.M., Nauck, M., Kumar, A.M., Nandabalan, K., Judson, R.S. *et al.* (2003) Haplotypes of the cholesteryl ester transfer protein gene predict lipid-modifying response to statin therapy. *Pharma cogenomics J*, **3**, 284–296.
15. Martin, E.R., Lai, E.H., Gilbert, J.R., Rogala, A.R., Afshari, A.J., Riley, J. *et al.* (2000) SNPing away at complex diseases: analysis of single-nucleotide polymorphisms around APOE in Alzheimer disease. *Am J Hum Genet*, **67**, 383–394.
16. Drysdale, C.M., McGraw, D.W., Stack, C.B., Stephens, J.C., Judson, R.S., Nandabalan, K. *et al.* (2000) Complex promoter and coding region beta 2-adrenergic receptor haplotypes alter receptor expression and predict in vivo responsiveness. *Proc Natl Acad Sci U S A*, **97**, 10483–10488.
17. Kong, A., Masson, G., Frigge, M.L., Gylfason, A., Zusmanovich, P., Thorleifsson, G. *et al.* (2008) Detection of sharing by descent, long-range phasing and haplotype imputation. *Nat Genet*, **40**, 1068–1075.
18. Stephens, M. and Donnelly, P. (2003) A comparison of bayesian methods for haplotype reconstruction from population genotype data. *Am J Hum Genet*, **73**, 1162–1169.
19. Bansal, V., Halpern, A.L., Axelrod, N. and Bafna, V. (2008) An MCMC algorithm for haplotype assembly from whole-genome sequence data. *Genome Res*, **18**, 1336–1346.
20. Zhang, K., Zhu, J., Shendure, J., Porreca, G.J., Aach, J.D., Mitra, R.D. and Church, G.M. (2006) Long-range polony haplotyping of individual human chromosome molecules. *Nat Genet*, **38**, 382–387.
21. Turner, D.J., Tyler-Smith, C. and Hurles, M.E. (2008) Long-range, high-throughput haplotype determination via haplotype-fusion PCR and ligation haplotyping. *Nucleic Acids Res*, **36**, e82.
22. Konfortov, B.A., Bankier, A.T. and Dear, P.H. (2007) An efficient method for multi-locus molecular haplotyping. *Nucleic Acids Res*, **35**, e6.
23. Xiao, M., Gordon, M.P., Phong, A., Ha, C., Chan, T.F., Cai, D. *et al.* (2007) Determination of haplotypes from single DNA molecules: a method for single-molecule barcoding. *Hum Mutat*, **28**, 913–921.
24. Bansal, V. and Bafna, V. (2008) HapCUT: an efficient and accurate algorithm for the haplotype assembly problem. *Bioinformatics*, **24**, i153–i159.
25. Parsons, D.W., Jones, S., Zhang, X., Lin, J.C., Leary, R.J., Angenendt, P. *et al.* (2008) An integrated genomic analysis of human

glioblastoma multiforme. *Science*, **321**, 1807–1812.

26. Romeo, S., Pennacchio, L.A., Fu, Y., Boerwinkle, E., Tybjaerg-Hansen, A., Hobbs, H.H. and Cohen, J.C. (2007) Population-based resequencing of ANGPTL4 uncovers variations that reduce triglycerides and increase HDL. *Nat Genet*, **39**, 513–516.
27. Cohen, J.C., Pertsemlidis, A., Fahmi, S., Esmail, S., Vega, G.L., Grundy, S.M. and Hobbs, H.H. (2006) Multiple rare variants in NPC1L1 associated with reduced sterol absorption and plasma low-density lipoprotein levels. *Proc Natl Acad Sci U S A*, **103**, 1810–1815.
28. Jones, S., Zhang, X., Parsons, D.W., Lin, J.C., Leary, R.J., Angenendt, P. *et al.* (2008) Core signaling pathways in human pancreatic cancers revealed by global genomic analyses. *Science*, **321**, 1801–1806.
29. Greenman, C., Stephens, P., Smith, R., Dalgliesh, G.L., Hunter, C., Bignell, G. *et al.* (2007) Patterns of somatic mutation in human cancer genomes. *Nature*, **446**, 153–158.
30. Wood, L.D., Parsons, D.W., Jones, S., Lin, J., Sjoblom, T., Leary, R.J. *et al.* (2007) The genomic landscapes of human breast and colorectal cancers. *Science*, **318**, 1108–1113.
31. Cancer Genome Atlas Research Network (2008) Comprehensive genomic characterization defines human glioblastoma genes and core pathways. *Nature*, **455**, 1061–1068.
32. Parmigiani, G., Boca, S., Lin, J., Kinzler, K.W., Velculescu, V. and Vogelstein, B. (2009) Design and analysis issues in genome-wide somatic mutation studies of cancer. *Genomics*, **93**(1), 17–21.
33. Albert, T.J., Molla, M.N., Muzny, D.M., Nazareth, L., Wheeler, D., Song, X. *et al.* (2007) Direct selection of human genomic loci by microarray hybridization. *Nat Methods*, **4**, 903–905.
34. Hodges, E., Xuan, Z., Balija, V., Kramer, M., Molla, M.N., Smith, S.W. *et al.* (2007) Genome-wide in situ exon capture for selective resequencing. *Nat Genet*, **39**, 1522–1527.
35. Okou, D.T., Steinberg, K.M., Middle, C., Cutler, D.J., Albert, T.J. and Zwick, M.E. (2007) Microarray-based genomic selection for high-throughput resequencing. *Nat Methods*, **4**, 907–909.
36. Porreca, G.J., Zhang, K., Li, J.B., Xie, B., Austin, D., Vassallo, S.L. *et al.* (2007) Multiplex amplification of large sets of human exons. *Nat Methods*, **4**, 931–936.
37. Li, B. and Leal, S.M. (2008) Methods for detecting associations with rare variants for common diseases: application to analysis of sequence data. *Am J Hum Genet*, **83**, 311–321.
38. Lin, J., Gan, C.M., Zhang, X., Jones, S., Sjoblom, T., Wood, L.D. *et al.* (2007) A multidimensional analysis of genes mutated in breast and colorectal cancers. *Genome Res*, **17**, 1304–1318.
39. Chittenden, T.W., Howe, E.A., Culhane, A.C., Sultana, R., Taylor, J.M., Holmes, C. and Quackenbush, J. (2008) Functional classification analysis of somatically mutated genes in human breast and colorectal cancers. *Genomics*, **91**, 508–511.
40. Marini, N.J., Gin, J., Ziegle, J., Keho, K.H., Ginzinger, D., Gilbert, D.A. and Rine, J. (2008) The prevalence of folate-remedial MTHFR enzyme variants in humans. *Proc Natl Acad Sci U S A*, **105**, 8055–8060.
41. Fahmi, S., Yang, C., Esmail, S., Hobbs, H.H. and Cohen, J.C. (2008) Functional characterization of genetic variants in NPC1L1 supports the sequencing extremes strategy to identify complex trait genes. *Hum Mol Genet*, **17**, 2101–2107.
42. Butler, J., MacCallum, I., Kleber, M., Shlyakhter, I.A., Belmonte, M.K., Lander, E.S. *et al.* (2008) ALLPATHS: de novo assembly of whole-genome shotgun microreads. *Genome Res*, **18**, 810–820.
43. Hernandez, D., Francois, P., Farinelli, L., Osteras, M. and Schrenzel, J. (2008) De novo bacterial genome sequencing: millions of very short reads assembled on a desktop computer. *Genome Res*, **18**, 802–809.
44. Dohm, J.C., Lottaz, C., Borodina, T. and Himmelbauer, H. (2007) SHARCGS, a fast and highly accurate short-read assembly algorithm for de novo genomic sequencing. *Genome Res*, **17**, 1697–1706.
45. Sundquist, A., Ronaghi, M., Tang, H., Pevzner, P. and Batzoglou, S. (2007) Whole-genome sequencing and assembly with high-throughput, short-read technologies. *PLoS ONE*, **2**, e484.
46. Warren, R.L., Sutton, G.G., Jones, S.J. and Holt, R.A. (2007) Assembling millions of short DNA sequences using SSAKE. *Bioinformatics*, **23**, 500–501.
47. Jeck, W.R., Reinhardt, J.A., Baltrus, D.A., Hickenbotham, M.T., Magrini, V., Mardis, E.R. *et al.* (2007) Extending assembly of short DNA sequences to handle error. *Bioinformatics*, **23**, 2942–2944.
48. Zerbino, D.R. and Birney, E. (2008) Velvet: algorithms for de novo short read assembly using de Bruijn graphs. *Genome Res*, **18**, 821–829.

Chapter 13

Detection of Mitochondrial DNA Variation in Human Cells

Kim J. Krishnan, John K. Blackwood, Amy K. Reeve, Douglass M. Turnbull, and Robert W. Taylor

Abstract

The ability to detect mitochondrial DNA (mtDNA) variation within human cells is important not only to identify mutations causing mtDNA disease, but also as mtDNA mutations are being increasingly described in many ageing tissues and in complex diseases such as diabetes, neurodegeneration and cancer. In this review, we discuss the main molecular genetic techniques that can be applied to study the two main types of mtDNA mutation: point mutations and large-scale mtDNA rearrangements. We then describe in detail protocols routinely used within our laboratory to analyse mtDNA mutations in individual human cells such as single muscle fibres and individual neurons to study the relationship between mtDNA mutation load and respiratory chain dysfunction.

Key words: Mitochondrial DNA, Variation, Mutations, mtDNA disease, Polymorphisms, Ageing, Real-time PCR

1. Introduction

The study of mitochondrial DNA (mtDNA) variation within humans is an expanding area of research. Not only are mutations of the mitochondrial genome an important cause of human disease, but they are also becoming frequently described in many other diseases such as cancer, neurodegenerative disease and also normal ageing (1, 2). Mitochondria are ubiquitous organelles found in all nucleated cells, and despite having several cellular functions, their main role is as generators of cellular ATP by oxidative phosphorylation (OXPHOS). Electrons generated by the oxidation of fat and sugars are transferred to oxygen via the redox components of the OXPHOS complexes I–IV found within the inner mitochondrial membrane, forming water. Protons are pumped across the inner membrane from the matrix to the

Michael R. Barnes and Gerome Breen (eds.), *Genetic Variation: Methods and Protocols*, Methods in Molecular Biology, vol. 628, DOI 10.1007/978-1-60327-367-1_13,

inter-membrane space forming an electrochemical gradient, which is used by the fifth and terminal OXPHOS complex, the ATP synthase – to synthesise ATP. The biosynthesis and maintenance of normal respiratory chain complex function is dependent upon the co-ordinated expression and interaction of two DNA molecules: the nuclear and mitochondrial (mtDNA) genomes.

1.1. Mitochondrial Genetics

MtDNA is a circular, double-stranded DNA molecule of size 16.6 kb in humans which encodes 13 essential polypeptides of the OXPHOS system and the necessary RNA machinery (2 rRNAs and 22 tRNAs) for their translation within the organelle, and is able to replicate independently of the nuclear genome (Fig. 1). It is located in the mitochondrial matrix, in close proximity to the inner

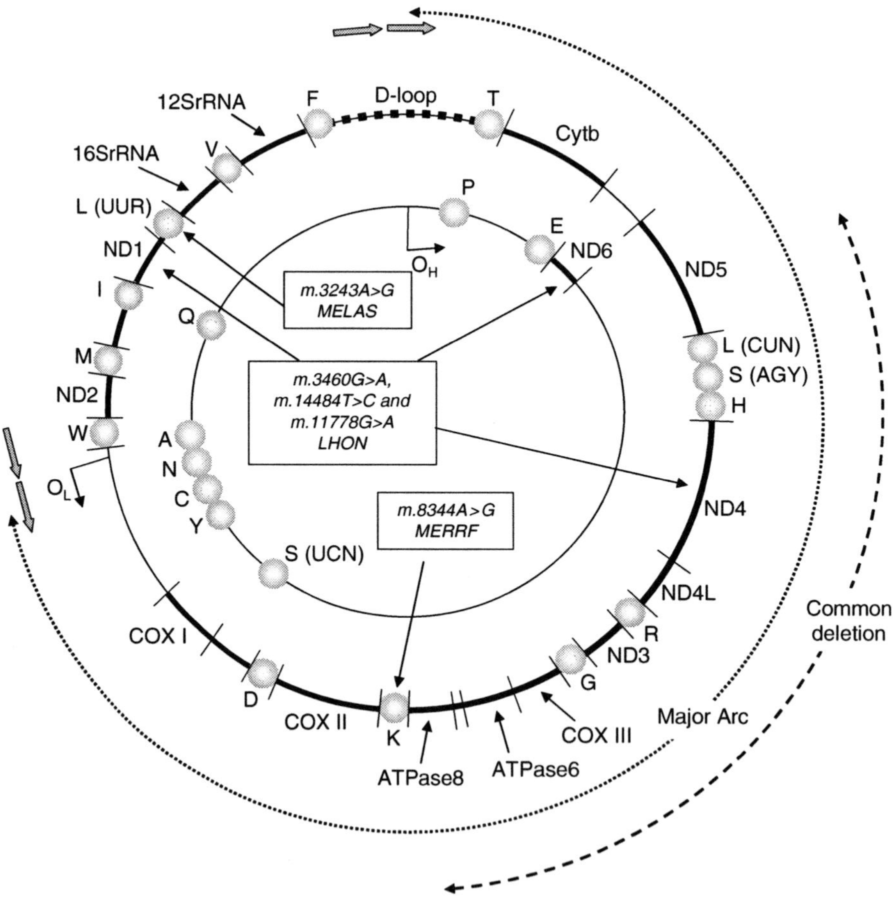

Fig. 1. Schematic diagram of the mitochondrial genome. Displayed are the 13 protein and 2 rRNA encoding genes (solid black lines), 22 tRNAs (*circles*, single letter abbreviated) along with the non-coding region (D-loop). O_H and O_L refer to the origin of H-strand and L-strand replication, respectively. In *boxes* are some of the most common mtDNA point mutations for the mtDNA disorders MELAS, MERRF and LHON and *arrows* pointing to where they are mutated on the genome. Also displayed is the common deletion. Depicted in *block arrows* are the primer positions for the two-round long extension PCR of the major arc

membrane where reactive oxygen species (ROS) are continually produced as a by-product of the electron-transferring reactions of the OXPHOS complexes. For this reason, mtDNA is a prime target for oxidative damage, which can lead to a mutation. The mitochondrial genome can undergo two main types of mutation: point mutations and large-scale deletions. Point mutations can be single nucleotide changes, insertions or deletions but not all are disease-causing. The mitochondrial genome is highly polymorphic. The evolution of human mtDNA has been characterised by the emergence of distinct haplogroups defined by specific mtDNA polymorphic variants, with different haplogroups associated with global ethnic lineages (3). On account of its high mutation rate and strict pattern of clonal, maternal inheritance (2), mtDNA has thus been widely used to make inference about the history of our species and is a favoured genetic tool for evolutionary biologists.

MtDNA exists in multiple copies within cells, the number of mtDNA molecules often reflecting the requirement of that particular tissue for ATP. Consequently, many mtDNA mutations may only be seen in a population of the mtDNA molecules within a cell, with this mixture of wild-type and mutated copies of the mitochondrial genome being referred to as "*heteroplasmy*". The multi-copy nature of the mitochondrial genome means that if a mutation occurs on the genome, a biochemical phenotype is not observed until the level of heteroplasmy reaches a certain threshold level of mutated mtDNA (2). The mechanism by which a particular mtDNA mutation within a cell can expand from the original mutation event to eventually populate the majority of mtDNA molecules has been termed "clonal" (4). Clonal expansion of an mtDNA mutation is dependent on a number of factors such as mtDNA repair, the rate of mtDNA replication, mitochondrial turnover and degradation. It is unknown at present how mtDNA mutations clonally expand, but their presence at high levels within individual cells is an evidence that this process does occur (5–7). Whether clonal expansion is a random or selective event remains to be elucidated. Nevertheless, the analysis of mtDNA mutations must rely on methods that can detect the clonally expanded mtDNA mutation within an individual cell, and ideally quantitate the level at which the mutation occurs. When the mutant copy of mtDNA does exceed the threshold to cause a biochemical defect, the cell becomes respiratory chain deficient, as there are no longer sufficient wild-type mtDNA copies to support biosynthesis of functioning respiratory chain proteins. For a large number of mtDNA mutations, this phenomenon can be observed histochemically using two sequential stains for the mitochondrial respiratory chain enzyme complexes cytochrome *c* oxidase (COX) and succinate dehydrogenase (SDH). Subunits making up the COX complex are encoded by both the nuclear genome and the mitochondrial genome.

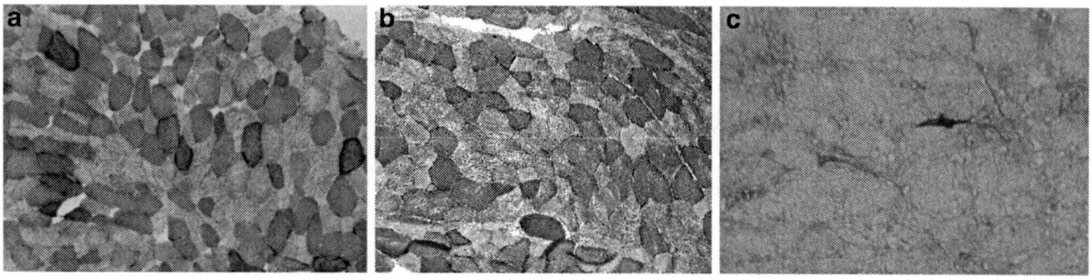

Fig. 2. COX/SDH sequential histochemistry reveals respiratory-deficient cells harbouring clonally expanded mtDNA mutations (**a**) Transverse section through quadriceps muscle from a patient with mitochondrial myopathy due to a mitochondrial tRNA mutation. The muscle fibres would display a characteristic mosaic of colours ranging from *brown* (COX normal) to *dark blue* (COX-deficient), including some fibres which show increased subsarcolemmal accumulation of abnormal mitochondria around the fibre periphery. (**b**) Transverse section through quadriceps muscle of an aged individual showing while there is COX-deficiency it is much less than that seen in the mitochondrial patient. (**c**) A COX-deficient (*blue*) and a COX-positive normal (*brown*) neuron in the midbrain near the red nucleus from a patient with a multiple mtDNA deletion disorder

When an mtDNA mutation affects genes encoding for or required to assemble COX subunits, the cell becomes COX-deficient, and the subsequent (brown) histochemical reaction product is absent. However, SDH is encoded entirely by the nuclear genome and is unaffected by the mtDNA mutation and so by sequential staining with SDH, the COX-deficient cell will exhibit the blue SDH reaction product (Fig. 2) (8). Respiratory chain deficient cells are a classical pathological hallmark of mtDNA involvement in patients with mitochondrial disease and as such the COX/SDH histochemical assay in tissue sections is one of the most important indications of an underlying mitochondrial problem.

1.2. Mitochondrial DNA Mutations and Human Disease

Disease-related mtDNA mutations can be divided into two broad groups: maternally inherited point mutations which are often but not exclusively heteroplasmic and affect protein, tRNA or rRNA genes and large-scale mtDNA rearrangements (mainly single deletions but the molecule may be duplicated) which span several genes. More than 250 pathogenic mutations have now been described that are associated with a remarkable collection of clinical phenotypes, mainly with muscle and brain involvement, and sometimes associated with distinct mitochondrial syndromes (2). The widespread genetic heterogeneity that characterises these disorders is best highlighted by the MELAS (mitochondrial myopathy, encephalopathy, lactic acidosis and stroke-like episodes) syndrome, characterised by recurrent encephalopathy and stroke-like episodes. More than 85% of MELAS cases are caused by a specific mitochondrial tRNA gene mutation (m.3243A>G in $tRNA^{Leu(UUR)}$) yielding a range of biochemical defects (9). However, many other point mutations in this gene, other tRNA genes or even protein-encoding genes such as *MTND5* (10) and *MTND1* (11) can also cause MELAS. In cases where mtDNA disease is suspected, a diagnosis is often only made following an approach that integrates several lines of investigation including clinical, histochemical, biochemical and ultimately molecular genetic tests. Such investigations continue to rely heavily on the study of a clinically affected tissue, often skeletal muscle, in which the characteristic COX-deficient muscle fibres are seen. Whole genome sequencing is now commonplace in many diagnostic centres, with novel mtDNA mutations still regularly being discovered, ever-widening the clinical spectrum of disease phenotypes and allowing accurate estimates of disease prevalence – up to 1:3,500 in some populations may have or be at risk of developing mtDNA disease – to be made (12).

1.3. Acquired mtDNA Mutations

MtDNA mutations have been widely described in ageing human tissues for many years (5, 7, 13–16). Initially as levels of these mutations in ageing tissues rarely exceeded ~1%, their role in the ageing process was doubtful. However, as these original studies were performed mainly on tissue homogenates, it was suggested that the mtDNA mutations could accumulate to high levels in a small subset of cells causing a respiratory chain deficiency and therefore may affect on a focal area of the tissue (17). In fact this has been shown to be the case, as newer techniques now allow the study of individual cells, which can be analysed for mtDNA mutations. High levels of mtDNA mutations leading to respiratory chain deficiency have been shown to occur in many cells such as muscle fibres and neurons (5, 14, 18). Whether or not this does contribute to the ageing process is still unknown, however the most vital functions within a cell are reliant on energy and therefore the observed respiratory chain deficiency will undoubtedly be detrimental for meeting the energy requirements of the cell and ultimately must lead to cell failure and possibly cell death.

1.4. Molecular Genetic Investigation of mtDNA Mutations

The molecular investigation of patients with suspected mtDNA disease includes tests to exclude large-scale mtDNA deletions and common mtDNA point mutations, before proceeding to screening the entire mitochondrial genome for potentially novel mtDNA mutations (19). In this paper, we will discuss the main methods used within our laboratory to analyse these different forms of mtDNA variation, describing many of these specific tests in detail. Within the scope of this paper, we have not discussed the analysis of mtDNA copy number, which is also an important factor for the viability of the cell with mitochondrial disease being caused by mtDNA depletion due to defects in nuclear encoded genes for mitochondrial maintenance. MtDNA copy number naturally varies between different tissues and cell types which makes accurately quantitating mtDNA copy number problematic. However, more recent techniques are sensitive enough to study single cells by real-time PCR which now forms a valuable part of our diagnostic service screening process (20–22).

1.5. Investigation of mtDNA Rearrangements

1.5.1. Southern Blotting

The quantification of mtDNA deletions by Southern blotting still remains the "gold standard" and is an invaluable tool in many mitochondrial disease diagnostic laboratories. The technique was used to identify the first mtDNA deletions in patients with a mitochondrial myopathy (23). The technique requires 2–3 μg of genomic DNA extracted from tissue, and is based upon linearising the circular, mitochondrial genome with a chosen restriction enzyme which cuts the genome at a single site (usually *PvuII* or *BamHI*). The DNA is separated on an agarose gel alongside a DNA ladder, transferred to a membrane, denatured and then hybridised to a radioactively labelled complimentary DNA probe. The result in a control sample is the presence of a single, wild-type band at 16.6 kb, however when smaller bands are observed this indicates the presence of mtDNA deletion(s). The level of deletion present can be quantitated by calculating the intensity of the delete bands relative to the wild-type band. Southern blotting is sensitive enough to detect the levels of deletion ~>5% and is very useful to investigate complex mtDNA rearrangements including mtDNA duplications and triplications (24). To further characterise where mtDNA deletions occur on the genome, different restriction enzymes around the mitochondrial genome can be used to observe whether the deletion removes the restriction enzyme site. Although Southern blotting is undoubtedly the most reliable technique for accurate quantification of mtDNA deletions, for studying single cells where low levels of DNA are obtained, more sensitive techniques are required.

1.5.2. Long Extension PCR

Long extension PCR is also a useful technique for identifying whether a sample contains mtDNA deletions. The advantage of this method over Southern blotting is that small amounts of DNA can be used, in fact more recently it has been used to screen single cells when a two-round amplification is performed (25). The assay uses two primers that span a large region of the mitochondrial genome, typically the major arc which is where the majority of deletions are reported (Fig. 1) (26). On the resulting agarose gel, single or multiple mtDNA deletions appear lower than the wild-type band (Fig. 3). Potential deletion bands can be gel extracted and sequenced to identify the deletions breakpoints. Although Long extension PCR is useful in determining whether a particular DNA sample contains deletions, it is not quantitative as it preferentially amplifies smaller species. However, the convenience of faster results compared to Southern blotting (typically 1 day compared with several days for the Southern blot) along with the low amount of starting template required, ensures this technique is still used for the initial screening process in many mitochondrial diagnostic and research laboratories.

1.5.3. Serial Dilution PCR

Before the advent of newer quantitative systems such as real-time PCR, serial dilution PCR was a popular choice for quantitating

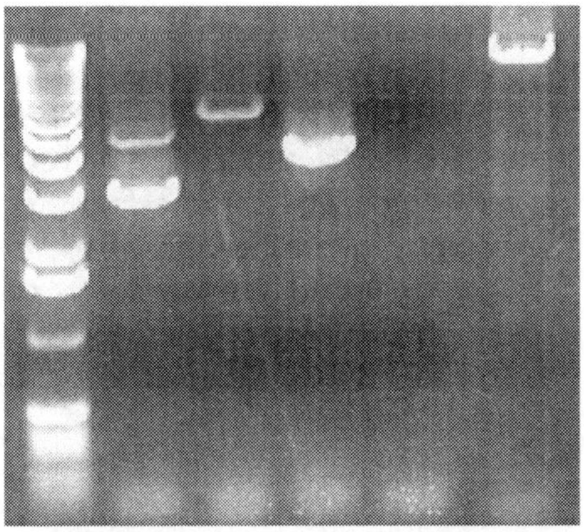

Fig. 3. An example of a typical agarose gel following long extension PCR on an 11 kb region of mtDNA in single cells from a patient with multiple mtDNA deletions. Wild-type molecules amplify the 11 kb region (*lane 6*) however molecules harbouring deletions produce smaller products as shown in *lanes 2–4*. *Lane 1* is a 1 kb ladder and *lane 5* a negative control

mtDNA deletion levels in samples where not enough DNA template was available or deletion levels were below the sensitivity for Southern blotting. Serial dilution PCR were shown to be reliable down to 10^{-5}% (27, 28). The technique works by amplifying a wild-type region of mtDNA and also a region of mtDNA that is removed by the deletion. By serially diluting the starting DNA template and observing at what point the two amplicons are no longer amplified gives an estimation as to the level of mtDNA deletion within a sample. A major disadvantage to this technique is its laborious nature and is therefore rarely used in more recent studies.

1.5.4. Three Primer PCR

The most commonly reported mtDNA deletion in human tissues is a 4,977 bp deletion or the so called "common deletion" (CD) which is located within the major arc between two, 13-bp direct repeats. The CD was first detected in a patient Kearns Sayre syndrome patient and has since been found in a number of tissues in many disease types as well as normal ageing (13, 18, 23, 25, 27, 29). Due to its reported frequent detection, there does exist a certain amount of bias to study this deletion and has potentially resulted in this deletion being reported more frequently than other deletions. To quantitate the levels of the CD, a three primer PCR approach was designed which allows the simultaneous amplification of wild-type and deletion-containing molecules in the same assay (30). The assay uses three primers, two which flank the deletion and one which sits within the deletion site. The PCR conditions are set to amplify only short fragments, thus preventing amplification of the large product between the two primers flanking the deletion in wild type molecules, as this product would be too large.

However, if a deletion is present, this brings the two primers sites close enough together to allow amplification. The wild-type band is amplified from one of the flanking primers and the primer that resides within the deletion (which is removed in the presence of a deletion). The PCR can be performed radioactively or by using a DNA intercalating dye such as SYBR Green to allow accurate quantification of the delete species to levels >1%. The assay is useful for low levels of starting DNA template, however as the assay is deletion specific it can only quantitate the CD and further assays need to be designed to be able to quantitate other deletions.

1.5.5. Real-Time PCR

The development of real-time PCR technology has significantly advanced the capabilities of analysing mtDNA variation in human cells. The high sensitivity of this technique allows rapid and accurate quantification of mtDNA molecules down to as low as 20 copies (21). There are several methods available for use on the real-time PCR system such as SYBR Green, Taqman DNA probes and fluorescent primers. SYBR Green is very sensitive and intercalates any double stranded DNA, while Taqman probes and fluorescent primers can be designed at specific regions of interest. Within our laboratory, we developed a real-time PCR assay using Taqman DNA probes capable of quantitating total mtDNA deletion load within a sample, based on the assumption that the majority of deletions occur in the major arc of mtDNA (31). Two genes, *MTND1* and *MTND4* located within the minor and major arcs of mtDNA, respectively are amplified. The assay relies on the premise that if a deletion is present it will delete the *MTND4* gene and will not generate fluorescence, whereas the *MTND1* gene is unaffected. A ratio of the fluorescence of the two genes allows quantification of the total level of deletion within a given sample. The assay was shown to calculate mtDNA deletion levels comparable to those calculated by Southern blotting (31) and has been used to assess changes in mtDNA deletion level within single muscle fibres in patients subjected to endurance exercise training protocols (32). We recently further enhanced the assay by allowing the simultaneous detection of the two genes within the same reaction by the use of different fluorescent dyes for each gene (33). This improvement allows more accurate quantification, is more cost effective and requires less starting DNA template than the original assay. The *MTND1/MTND4* real-time PCR assay is unable to determine what types of mtDNA deletions are present. However, the assay allows the rapid throughput of DNA samples resulting in fast (~2 h) identification of samples with high levels of mtDNA deletions which can then be further characterised. This assay has also recently been used to identify high levels of mtDNA deletions in substantia nigra neurons in ageing controls and patients with Parkinsons disease (5).

Other studies have designed assays capable of quantitating specific mtDNA deletions (34–37). We have successfully used such an approach to investigate the status of a single, large-scale mtDNA deletion in "identical twins", one of whom was asymptomatic. By designing a real-time PCR assay based upon the specific deletion breakpoint sequence, we were able to show that the unaffected individual harboured mtDNA deletion levels of <0.1% in skeletal muscle, suggesting the mtDNA deletion was present in the oocyte (34).

1.6. Investigation of mtDNA Point Mutations

1.6.1. Whole Genome Sequencing

Several technologies are available to identify mtDNA sequence variation through screening the entire genome, including denaturing high performance liquid chromatography (DHPLC) (38), surveyor nuclease (39, 40) and conventional cycle sequencing. To identify where clonally expanded mtDNA point mutations are present on the mitochondrial genome within individual respiratory deficient cells, it is necessary to be able to analyse the whole mitochondrial genome. Our group have opted to pursue such investigations by determining the entire mtDNA sequence from a single cell, and have described a PCR approach capable of amplifying the whole mitochondrial genome in a series of overlapping primers that span the genome. By using a two-round PCR approach, the whole mitochondrial genome can be analysed from DNA extracted from a single cell (41). Direct sequencing of the PCR products allows identification of any polymorphisms or point mutations. This approach has since been used to identify clonally expanded point mutations in cells from colonic crypts, gut and the stomach (6, 7, 42, 43). Limitations of this assay lie in the sensitivity of direct sequencing which is only sensitive enough to detect levels >25% mutant mtDNA. Most clonally expanded pathogenic mtDNA mutations will be detected at levels >60% but the recent description of the first, functionally dominant mitochondrial tRNA mutation highlights that some mutations may escape detection as they are present at levels near the threshold for detection by sequencing (44).

1.6.2. PCR–Restriction Fragment Length Polymorphism (RFLP) Analysis

Due to the polyploid nature of the mitochondrial genome, the proportion of mtDNA fragments harbouring a particular point mutation can be quantified using several techniques including pyrosequencing (45) and, more conventionally, last hot cycle or last fluorescent cycle PCR–RFLP analysis (46, 47). The demonstration of mtDNA heteroplasmy associated with a biochemical defect remains the most compelling piece of evidence of pathogenicity for any mutation in patients, as it implies a recent mutational event.

As an example, we will describe our assay for determining the level of m.3243A>G mutation load which relies on the point mutation introducing a novel restriction site for the restriction endonuclease HaeIII. A ~200 bp region flanking the mutation of interest is amplified using PCR and is subsequently labelled either radioactively, or fluorescently, in a final last round of PCR (22, 48). Subsequent restriction enzyme digests and DNA resolution reveals mutant and wild-type alleles. DNA can either be resolved using non-denaturing polyacrylamide gel electrophoresis (PAGE) (Fig. 4a) or a capillary based genetic analyser (Fig. 4b) when using radioactivity or fluorescence, respectively. Densitometry (e.g., ImageQuant, version 5.0 Molecular Dynamics) or fragment analysis software (e.g., ABI Prism Genemapper version 3.5, Applied Biosystems) is then used to calculate the proportion of mutant and wild-type bands, allowing quantification of the level of m.3243A>G point mutation. PCR products are labelled in a final cycle rather than during the main amplification as this prevents heteroduplex PCR products forming between wild-type and mutated mtDNA molecules, which cannot be cleaved by the restriction enzyme (46) (Figs. 4a and 4b).

2. Materials

2.1. Single Cell DNA Isolation

1. Tissue sectioning: Fresh tissue, for example muscle is frozen in liquid nitrogen and stored at –80°C until required. Fresh frozen sections are cut at 15 μm (see Note 1) onto glass slides (VWR) or PEN membrane slides (Leica) using a Brights OTF cryostat and air dried at room temperature for 1 h. The sections can then be used immediately or stored at –80°C in airtight slide containers.
2. COX/SDH histochemical staining: The COX reaction requires stock solutions of 5 mM 3,3′-diaminobenzidine tetrahydrochloride (light sensitive) (DAKO) in 0.1 M phosphate buffer, pH 7.0 and 500 μM cytochrome c (light sensitive) (Sigma) in 0.1 M phosphate buffer, pH 7.0 and catalase (Sigma). The SDH reaction requires stock solutions of 1.875 mM nitro blue tetrazolium (Sigma), 1.30 M sodium succinate (Sigma), 2 mM phenazine methosulphate (light sensitive) (Sigma) and 100 mM sodium azide (BDH) all in 0.1 M phosphate buffer, pH 7.0. Slides are washed in Phosphate buffered saline (PBS) (OXOID).

taking the area under the peak. *y* axis, peak intensity, *x* axis, fragment size. (*Bottom*) Radioactive PCR–RFLP analysis of mtDNA heteroplasmy in a patient with the pathogenic m.3243A>G transition. (**a**) Diagram depicting PCR amplification of wild type mtDNA (*left*) and a patient with the pathogenic m.3243A>G transition (*right*). HaeIII restriction sites are displayed as are sizes of DNA fragments following digestion with HaeIII. The resulting PCR fragment size is shown below. (**b**) Densitometric analysis of radioactive m.3243A>G RFLP. *Top of gel*: *Lanes 1* and *15* undigested controls; *lanes 2* and *16*, m.3243A>G positive control, *lanes 3* and *17*, wild-type control, *lanes 4–14* and *18–26*, increasing levels of m.3243A>G heteroplasmy, from 0 to 100% in 5% increments, using a plasmid based assay (22). *Bottom of gel*: calculated levels of heteroplasmy using Imagequant software

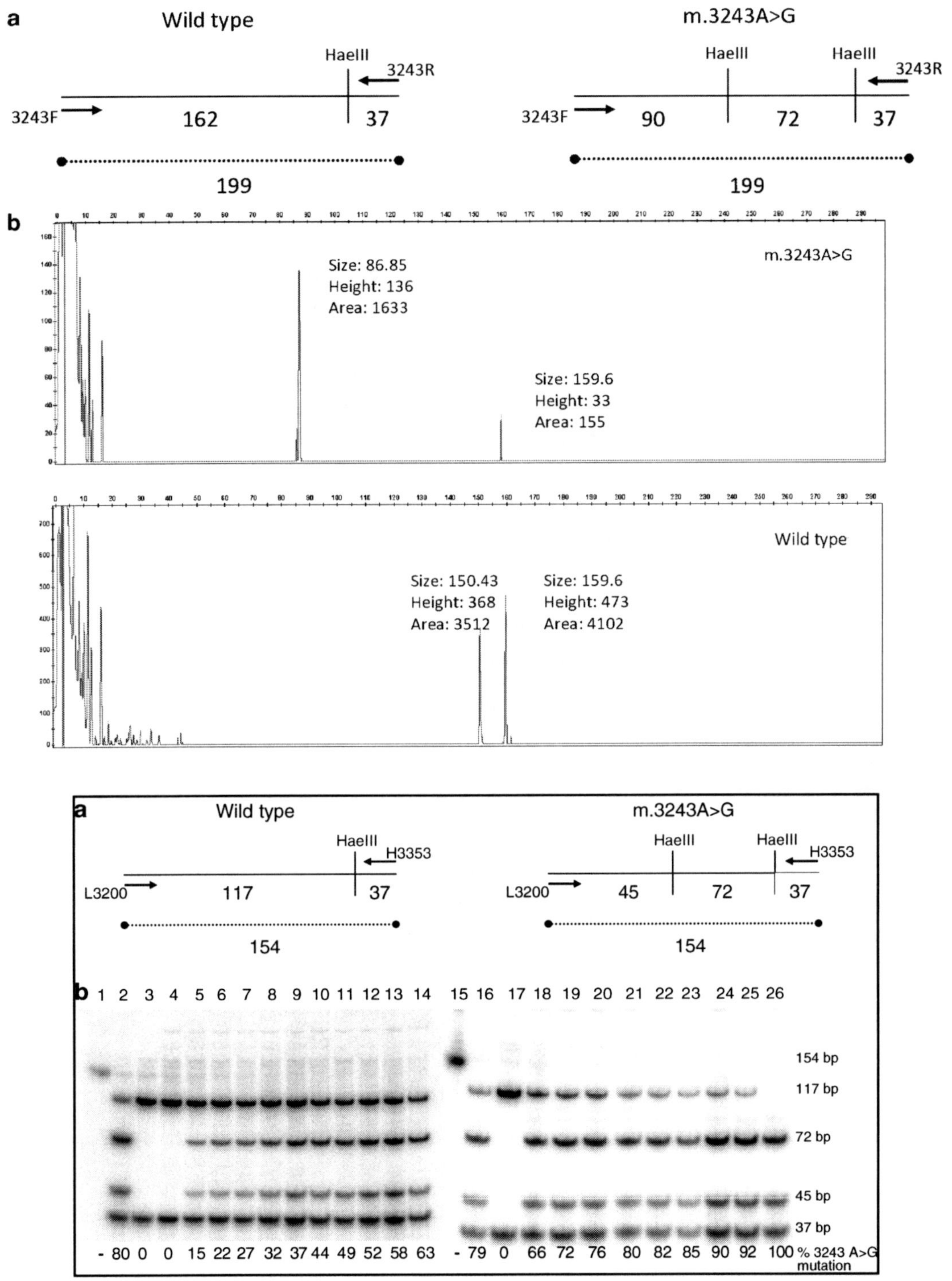

Fig. 4. (*Top*) Fluorescent PCR–RFLP analysis of mtDNA heteroplasmy in a patient with the pathogenic m.3243A>G transition. (**a**) Diagram depicting PCR amplification of wild type mtDNA (*left*) and a patient with the pathogenic m.3243A>G transition (*right*). HaeIII restriction sites are indicated as are sizes of DNA fragments following digestion with HaeIII. The resulting PCR fragment sizes are shown below. (**b**) Fluorescent analysis of m.3243A>G RFLP using fragment analysis software, Genemapper. *Top panel*, m.3243A>G positive control, bottom panel, wild type control. Level of heteroplasmy is derived by

3. Laser-microdissection: Single cells are laser-microdissected into the caps of thin walled 0.5 ml tubes (Eppendorf) using the Leica AS-LMD system.
4. Cell lysis: Stock solutions of 1% Tween-20, pH8.0, 0.5 M Tris–HCl, pH 8.5, and 50 mg/ml proteinase K (NBL Gene Sciences).

2.2. Real-Time PCR for mtDNA Deletions

1. UV hood (Bioair-AURA PCR cabinet).
2. Sterile MicroAmp Optical 96-well reaction plates with adhesive films (both Applied Biosystems).
3. Sterile tubes (e.g., 1.5 ml or 2 ml) for PCR mastermix.
4. Oligonucleotide primers: Two pairs of primers are used to amplify both wild type (mtND1) and deleted (mtND4) mitochondrial genomes simultaneously. These are L3485-ND1/H3532-ND1 and L12087-ND4/H12170-ND4 (Table 1). Stock solutions (10 μM) are stored at −20°C.
5. PCR amplification: 2× TaqMan Universal PCR mastermix from Applied Biosystems. 6 nM TaqMan TAMRA probe for each reaction (labelled VIC for mtND1 and FAM for mtND4). Sterile water.
6. Plate centrifuge (Sigma, Philip Harris).
7. Applied Biosystems 7000 Real-time PCR system.

2.3. Long Extension PCR of the Major Arc

1. The PCR is always set up ice.
2. Sterile 0.2 ml PCR tubes for reactions and 1.5 ml tube for the mastermix.
3. PCR amplification: Expand Long Range dNTPack (Roche), this is supplied with an enzyme mix consisting of thermostable

Table 1
The primers and probes used to identify mtDNA deletions using real-time PCR. Listed are the primer and probes sequences (5′–3′) and position on the mitochondrial genome

Real-time PCR		
L3485-ND1	CCCTAAAACCCGCCACATCT	3,485–3,504
H3532-ND1	GAGCGATGGTGAGAGCTAAGGT	3,532–3,553
ND1 VIC probe	CCATCACCCTCTACATCACCGCCC	3,506–3,529
L12087-ND4	CCATTCTCCTCCTATCCCTCAAC	12,087–12,109
H12170-ND4	CACAATCTGATGTTTTGGTTAAACTATATTT	12,170–12,140
ND4 FAM probe	CCGACATCATTACCGGGTTTTCCTCTTG	12,111–12,138

Taq DNA polymerase and a thermostable DNA polymerase with proofreading activity and three buffers. We use reaction buffer 3 which contains DMSO (20% (v/v)), which prevents DNA depurination and intrastrand secondary structure formation. The enzyme and buffers can be stored at –20°C. Reaction buffer 3 should be checked for the appearance of crystals that may have precipitated before use. Working stock of dNTPs (Roche) is 2.5 mM. Bovine Serum Albumin (BSA) (New England Biolabs) is used at a working stock of 1 mg/ml which is diluted from a stock solution of 10 mg/ml, stored at –20°C, prior to use. Oligonucleotide primers can be designed to amplify any large regions of the mitochondrial genome and we use combinations of the primers used for whole genome sequencing listed in Table 2. For the majority of routine screening, we amplify a 11 kb fragment of the mitochondrial genome using a forward primer 11F and D2R (see Table 2). For single cells, we use a two-round PCR approach using primers 12F and 32R for the first round and 13F and 31R for the second round (see Table 2). Sterile water.

4. DNA template: For standard long range PCR, 10–50 ng of total DNA is required. For best results, we recommend that solutions of DNA are made fresh from concentrated DNA stocks. For single cell long range PCR, 1 μl of the cell lysis is used for the first round PCR and for the second round PCR, 1 μl of a 1/50 dilution of the first round PCR product is used.
5. Thermal cycler: Applied Biosystems 9700.
6. Horizontal gel electrophoresis equipment.
7. Agarose gel containing ethidium bromide, 1× TAE (40 mM Tris-acetate, 1 mM EDTA, pH 8.0) running buffer.
8. 10 μl of 1 kb ladder (Invitrogen).
9. UV transilluminator.

2.4. Whole Mitochondrial Genome Sequencing

1. UV Hood (Bioair-AURA PCR cabinet).
2. Cell lysis.
3. PCR amplification: Oligonucleotide primers are M13 tagged are reconstituted to a 20 μM stock and stored at –20°C (VHBio). Amplitaq gold polymerase, GeneAmp 10× buffer and 25 mM $MgCl_2$ (Applied Biosytems) 10× dNTPS (Roche). Sterile water. ABI GeneAmp Thermal cycler 9700 (Applied biosystems).
4. Gel electrophoresis: Agarose MP (Roche). Hyperladder IV (Bioline).

Table 2
The main oligonucleotides used to detect mtDNA point mutations using RFLP and whole genome sequencing. The list displays the primer sequences (5′–3′) and position on the mitochondrial genome. The list of primers used for whole genome sequencing can also be used in different combinations for long-extension PCR

Fluorescent RFLP		
3243F	CACAAAGCGCCTTCCCC	3,155–3,171
3243R	GCGATTAGAATGGGTACAAT	3,334–3,353
Radioactive RFLP		
H3353	GCGATTAGAATGGGTACAAT	3,334–3,353
L3200	TATACCCACACCCACCCAAG	3,200–3,219
Whole genome sequencing first round primers		
AF	GCTCACATCACCCCATAAAC	627–646
AR	GATTACTCCGGTCTGAACTC	3,087–3,068
BF	ACCAACAAGTCATTATTACCC	2,395–2,415
BR	TGAGGAAATACTTGATGGCAG	4,653–4,633
CF	CCGTCATCTACTCTACCATC	4,489–4,508
CR	GGACGGATCAGACGAAGAG	6,468–6,450
DF	AATACCCATCATAATCGGAGG	6,113–6,133
DR	GGTGATGAGGAATAGTGTAAG	8,437–8,417
EF	AACCACTTTCACCGCTACAC	8,128–8,147
ER	AGTGAGATGGTAAATGCTAG	10,516–10,487
FF	ACTTCACGTCATTATTGGCTC	9,821–9,841
FR	ATAGGAGGAGAATGGGGGATAG	12,101–12,080
GF	ACCCCCCACTATTAACCTACTG	11,866–11,887
GR	GGTAGAATCCGAGTATGTTGG	13,924–13,904
HF	TATTCGCAGGATTTCTCATTAC	13,721–13,742
HR	AGCTTTGGGTGCTAATGGTG	15,997–15,978
IF	CCCATCCTCCATATATCCAAAC	15,659–15,680
IR	GGTTAGTATAGCTTAGTTAAAC	868–847
Whole genome sequencing second round primers		
1F	TGTAAAACGACGGCCAGTTCACCCTCTAAATCACCAG	721–740
1R	CAGGAAACAGCTATGACCGATGGCGGTATATAGGCTGAG	1,268–1,248
2F	TGTAAAACGACGGCCAGTTTAAAACTCAAAGGACCTGGC	1,157–1,177

(continued)

Table 2
(continued)

2R	CAGGAAACAGCTATGACCCTGGTAGTAAGGTGGAGTGGG	1,709–1,689
3F	TGTAAAACGACGGCCAGTAACTTAACTTGACCGCTCTGAG	1,650–1,671
3R	TGTAAAACGACGGCCAGTAACTTAACTTGACCGCTCTGAG	2,193–2,175
4F	TGTAAAACGACGGCCAGTACTGTTAGTCCAAAGAGGAAC	2,091–2,111
4R	CAGGAAACAGCTATGACCTCGTGGAGCCATTCATACAG	2,644–2,625
5F	TGTAAAACGACGGCCAGTCAGTGACACATGTTTAACGGC	2,549–2,569
5R	CAGGAAACAGCTATGACCGATTACTCCGGTCTGAACTC	3,087–3,068
6F	TGTAAAACGACGGCCAGTCAGCCGCTATTAAAGGTTCG	3,017–3,036
6R	CAGGAAACAGCTATGACCGGAGGGGGGTTCATAGTAG	3,374–3,356
7F	TGTAAAACGACGGCCAGTCCTTAGCTCTCACCATCGC	3,533–3,351
7R	CAGGAAACAGCTATGACCAGAGTGCGTCATATGTTGTTC	4,057–4,037
8F	TGTAAAACGACGGCCAGTAATAAACACCCTCACCACTAC	4,005–4,025
8R	CAGGAAACAGCTATGACCGTTTATTTCTAGGCCTACTCAG	4,577–4,556
9F	TGTAAAACGACGGCCAGTACACTCATCACAGCGCTAAG	4,518–4,537
9R	CAGGAAACAGCTATGACCGATTTTGCGTAGCTGGGTTTG	5,003–4,983
10F	TGTAAAACGACGGCCAGTTCCATCATAGCAGGCAGTTG	4,950–4,969
10R	CAGGAAACAGCTATGACCTGTAGGAGTAGCGTGGTAAGG	5,481–5,462
11F	TGTAAAACGACGGCCAGTACCTCAATCACACTACTCCC	5,367–5,386
11R	CAGGAAACAGCTATGACCTAGTCAACGGTCGGCGAAC	5,924–5,906
12F	TGTAAAACGACGGCCAGTCACTCAGCCATTTTACCTCAC	5,875–5,895
12R	CAGGAAACAGCTATGACCATGGCAGGGGGTTTTATATTG	6,430–6,410
13F	TGTAAAACGACGGCCAGTTTAGGGGCCATCAATTTCATC	6,378–6,398
13R	CAGGAAACAGCTATGACCAAGAAAGATGAATCCTAGGGC	6,944–6,924
14F	TGTAAAACGACGGCCAGTATTTAGCTGACTCGCCACAC	6,863–6,882
14R	CAGGAAACAGCTATGACCCATCCATATAGTCACTCCAGG	7,396–7,376
15F	TGTAAAACGACGGCCAGTGGCTCATTCATTTCTCTAACAG	7,272–7,293
15R	CAGGAAACAGCTATGACCGGCAGGATAGTTCAGACGG	7,791–7,773
16F	TGTAAAACGACGGCCAGTTAACATCTCAGACGCTCAGG	7,744–7,763
16R	CAGGAAACAGCTATGACCTACAGTGGGCTCTAGAGGG	8,301–8,283
17F	TGTAAAACGACGGCCAGTACAGTTTCATGCCCATCGTC	8,196–8,215
17R	CAGGAAACAGCTATGACCGTATAAGAGATCAGGTTCGTC	8,740–8,720
18F	TGTAAAACGACGGCCAGTACCACCCAACAATGACTAATC	8,656–8,676

(continued)

Table 2 (continued)

18R	CAGGAAACAGCTATGACCGTTGTCGTGCAGGTAGAGG	9,201–9,183
19F	TGTAAAACGACGGCCAGTATCCTAGAAATCGCTGTCGC	9,127–9,146
19R	CAGGAAACAGCTATGACCATTAGACTATGGTGAGCTCAG	9,661–9,641
20F	TGTAAAACGACGGCCAGTCATCCGTATTACTCGCATCAG	9,607–9,627
20R	CAGGAAACAGCTATGACCTAGCCGTTGAGTTGTGGTAG	10,147–10,128
21F	TGTAAAACGACGGCCAGTCAACACCCTCCTAGCCTTAC	10,085–10,104
21R	CAGGAAACAGCTATGACCAGGCACAATATTGGCTAAGAG	10,649–10,629
22F	TGTAAAACGACGGCCAGTATCGCTCACACCTCATATCC	10,534–10,553
22R	CAGGAAACAGCTATGACCATGATTAGTTCTGTGGCTGTG	11,109–11,089
23F	TGTAAAACGACGGCCAGTCTAATCTCCCTACAAATCTCC	11,054–11,074
23R	CAGGAAACAGCTATGACCTAGGTCTGTTTGTCGTAGGC	11,605–11,586
24F	TGTAAAACGACGGCCAGTTCCTTGTACTATCCCTATGAG	11,541–11,561
24R	CAGGAAACAGCTATGACCCGTGTGAATGAGGGTTTTATG	12,054–12,034
25F	TGTAAAACGACGGCCAGTACAATGGGGCTCACTCACC	12,001–12,019
25R	CAGGAAACAGCTATGACCGTGGCTCAGTGTCAGTTCG	12,545–12,527
26F	TGTAAAACGACGGCCAGTCATGTGCCTAGACCAAGAAG	12,498–12,517
26R	CAGGAAACAGCTATGACCCTGATTTGCCTGCTGCTGC	13,009–12,991
27F	TGTAAAACGACGGCCAGTGCCCTTCTAAACGCTAATCC	12,940–12,959
27R	CAGGAAACAGCTATGACCGGGAGGTTGAAGTGAGAGG	13,453–13,435
28F	TGTAAAACGACGGCCAGTCGGGTCCATCATCCACAAC	13,365–13,383
28R	CAGGAAACAGCTATGACCGTTAGGTAGTTGAGGTCTAGG	13,859–13,839
29F	TGTAAAACGACGGCCAGTACCTAAAACTCACAGCCCTC	13,790–13,809
29R	CAGGAAACAGCTATGACCAGGATTGGTGCTGTGGGTG	14,374–14,356
30F	TGTAAAACGACGGCCAGTCAACCACCACCCCATCATAC	14,331–14,350
30R	CAGGAAACAGCTATGACCAAGGAGTGAGCCGAAGTTTC	14,857–14,838
31F	TGTAAAACGACGGCCAGTATTCATCGACCTCCCCACC	14,797–14,815
31R	CAGGAAACAGCTATGACCGGTTGTTTGATCCCGTTTCG	15,368–15,349
32F	TGTAAAACGACGGCCAGTAGCCCTAGCAACACTCCAC	15,316–15,334
32R	CAGGAAACAGCTATGACCTACAAGGACAGGCCCATTTG	15,896–15,877
D1F	TGTAAAACGACGGCCAGTATCGGAGGACAACCAGTAAG	15,758–15,777
D1R	CAGGAAACAGCTATGACCGTGGGTAGGTTTGTTGGTATC	16,294–16,274
D2F	TGTAAAACGACGGCCAGTCTCAACTATCACACATCAACTG	16,223–16,244

(continued)

Table 2 (continued)

D2R	CAGGAAACAGCTATGACCAGATACTGCGACATAGGGTG	129–110
D3F	TGTAAAACGACGGCCAGTCACCCTATTAACCACTCACG	15–34
D3R	CAGGAAACAGCTATGACCCTGGTTAGGCTGGTGTTAGG	389–370
D4F	TGTAAAACGACGGCCAGTGCCACAGCACTTAAACACATC	323–343
D4R	CAGGAAACAGCTATGACCTGCTGCGTGCTTGATGCTTG	771–752

5. PCR clean up: ExoSAP-IT (GE healthcare). MicroAmp optical 96-well reaction plates (Applied Biosystems).
6. Direct sequencing: BigDye Terminator v3.1 cycle sequencing kit (Applied Biosystems). 125 mM EDTA, 3 M Sodium Acetate (Sigma). Hi-Di Formamide (Applied Biosystems). ABI 3100 Genetic Analyser and ABI Prism 7000 sequence detection system (Applied Biosystems).

2.5. RFLP: Radioactive and Fluorescent

1. Template DNA.
2. PCR reagents: 5 × GoTaq reaction buffer, GoTaq DNA polymerase (5 U/μl) (Promega), 2 mM dNTPS (Boehringer Mannheim).
3. Oligonucleotide primers to amplify the region of interest (Table 2). Five prime fluorescein labelled primers for last fluorescent cycle.
4. Sterile Millipore water.
5. Sterile 0.5 ml PCR tubes.
6. PCR thermal cycler.
7. Horizontal gel electrophoresis equipment.
8. 1% agarose gels containing ethidium bromide and suitable DNA ladder for determining size.
9. UV transilluminator.
10. [α-^{32}P] dCTP (3,000 Ci/nmol). (Amersham Life Science).
11. Pellet paint co-precipitant (Novagen).
12. 3 M Sodium acetate.
13. Ethanol (100 and 75%).
14. Genescan-500 ROX size standards (Applied Biosystems).
15. Cerenkov counter.
16. Appropriate restriction enzyme (e.g., HaeIII for the m.3243A>G transition) (New England Biolabs).
17. Heat block with variable temperature setting.

18. Vertical electrophoresis system for non-denaturing polyacrylamide gels.
19. 5% non-denaturing polyacrylamide gel.
20. 1× TAE and TBE running buffer.
21. Gel dying equipment.
22. Phosphoimager cassette and imaging system such as ImageQuant (Molecular Dynamics).

3. Methods

Our laboratory consists of a mitochondrial diagnostic service as well as a large research group whose primary focus is the study of mtDNA mutations in humans. Therefore, the ability to study mtDNA mutations within individual cells is crucial to our research. Here, we further describe some of the main techniques discussed earlier used routinely within our laboratory to identify and quantitate large-scale mtDNA deletions and point mutations in single cells.

3.1. Single Cell DNA Isolation

1. COX/SDH histochemical staining: Tissue sections that have been cut onto PEN membrane slides and stored at –80°C should be equilibrated at room temperature for 1 h, and then removed from the airtight container and air dried for a further 1 h prior to histochemical analysis. COX activity is detected first using an incubation medium containing 4 mM 3,3′-diaminobenzidine tetrahydrochloride, 100 μM cytochrome c and one flake of catalase. Each section is incubated with 100–200 μl (see Note 2) of incubation medium, depending on the size of the section and incubated at 37°C for up to 50 min (see Note 1) in a humid chamber. Following incubation, any excess medium is discarded and the slides are washed with PBS. SDH activity is detected using an incubation medium containing 1.5 mM nitroblue tetrazolium, 130 mM sodium succinate, 0.2 mM phenazine methosulphate and 1 mM sodium azide. Each section is incubated at 37°C with 100–200 μl (see Note 2) of incubation medium, in a humid chamber for 45 min (see Note 1). Sections are washed in PBS to remove excess medium. COX-deficient cells are easily identified as they do not produce the brown reaction product associated with COX activity, but do react for SDH activity which gives a characteristic blue appearance (8).
2. Following histochemical analysis, slides are air-dried for 1 h and can then be used immediately or stored in an air-tight container at –20°C. Cells of interest are then cut out using the Leica AS-LMD laser microdissection system into thin

walled 0.5 ml tubes. The tubes containing the individual cells are centrifuged at 7,000 × *g* for 10 min. The cells can then be lysed immediately or stored at −20°C.

3. Cell lysis: The cells are lysed with 15 µl of 50 mM Tris–HCl, pH 8.5, 0.5% Tween-20 and 200 µg/ml proteinase K. The cells are incubated for 2 h at 55°C, with agitation every 30 min. This is followed by heat inactivation of the proteinase K at 95°C for 10 min. Alternatively, for Long extension PCR on single cells, we use the QIAamp DNA micro kit (QIAGEN) for DNA extraction.

3.2. Detection of mtDNA Deletions

3.2.1. Long Extension PCR in Single Cells

1. Set up PCR in UV cabinet (see Note 3).
2. Thaw all components stored at −20°C and bring buffer 3 to room temperature (see Note 4). Label some thin walled PCR tubes and place in a rack on ice. Vortex all components prior to use and place on ice. Before use, dilute 1/10 the 10 mg/ml stock of BSA. Combine all components of the mastermix (below) into a sterile eppendorf adding the Expand Taq polymerase until last.

Mastermix	
dNTPs (final concentration 0.35 mM)	8.75 µl
Buffer 3	5 µl
BSA (10 ng/ml)	10 µl
Expand Taq	0.7 µl
dH_2O	21.55 µl
Forward primer	1.5 µl
Reverse primer	1.5 µl
DNA	1 µl[a]

[a]For second round PCR, use 1/50 dilution of first round PCR product

3. Vortex briefly, centrifuge the mix and aliquot 49 µl of the mastermix into each thin walled PCR tube.
4. Add 1 µl DNA template from each sample into a separate PCR tube containing the mastermix. Include a blood sample (wild-type band) and a no template control.
5. Centrifuge the samples briefly, and place back on ice.
6. Using the PCR conditions listed below, place the samples into the block once the heat has reached 93°C. The annealing temperature is dependent upon the oligonucleotide primers used in the PCR. *x* depends on the length of product to be amplified, generally allow 1 min for every 1 kb to be amplified.

Reaction conditions	
Initial denaturation	93°C for 3 min
10 cycles	93°C for 30 s
	Annealing at 55–68°C for 30 s
	68°C for x min
20 cycles	93°C for 30 s
	*55–68°C for 30 s
	68°C for x min + 5 s per cycle
Final extension	68°C for 10–20 min

*primers used and length of PCR product

7. For the second round PCR, dilute 1/50 the first round PCR product and repeat the above protocol with the exception of using a nested pair of primers and the diluted PCR product as the DNA template.
8. When the PCR program is complete, load 10 µl of the PCR product on a 0.7% agarose gel and electrophorese at 50 V for 2–4 h alongside a 1 kb DNA ladder.

3.2.2. Multiplex TaqMan Real-Time PCR to Detect Total mtDNA Deletion Load

1. Set up PCR in a UV cabinet (see Note 3). Thaw components stored at −20°C and keep on ice until use. Vortex all components prior to use then add all components of mastermix (see below) into a sterile eppendorf. Vortex mastermix and centrifuge to remove any drops from the side. Add 24 µl of the mastermix into each well of the 96 well plate.

Mastermix (25 µl volume)	
TaqMan Universal mastermix	12.5 µl
ND1 VIC labelled Probe 100 nM	0.5 µl
ND4 FAM labelled Probe 100 nM	0.5 µl
Primer mtND1 forward 300 nM	0.75 µl
Primer mtND1 reverse 300 nM	0.75 µl
Primer mtND4 forward 300 nM	0.75 µl
Primer mtND4 reverse 300 nM	0.75 µl
dH_2O	7.5 µl
DNA	1 µl

2. Add 1 µl of DNA from each of the samples, perform each sample in triplicate. Include lysis control, no template control and blood sample (see Note 5).
3. Seal the 96-well plate with a film lid ensuring no bubbles. Place a thermal cover on top. Centrifuge the plate briefly to remove any liquid drops from the side of the wells.

4. Load the 96-well plate into the real-time PCR machine.

PCR conditions:	
Pre-incubation steps	2 min at 50°C (Activation of Amperase UNG)
Initial denaturation	10 min at 95°C
40 cycles	15 s at 95°C
	1 min at 60°C

5. After the PCR run has completed, retrieve cycle threshold (Ct) values for each. For data analysis, the equation $1-2^{-\Delta Ct}$ is used to calculate the deletion level in each sample. First, to calculate ΔCt, subtract the Ct value of mtND1 from mtND4, then average the ΔCt values from the triplicates from each sample. Subtract the ΔCt obtained from the blood sample from each of the ΔCt's of the unknown samples. Perform the $1-2^{-\Delta Ct}$ calculation on each of the samples using the calculated ΔCt's and multiply by 100 to calculate the percentage load of mtDNA deletion in each sample.

3.3. Point Mutations

3.3.1. Restriction Fragment Length Polymorphism

We describe in a following section the strategy used to investigate mutation load levels of the pathogenic mitochondrial transition m.3243A>G in the $tRNA^{Leu(UUR)}$ gene responsible for MELAS. Two strategies are described here, one using radioactivity and one fluorescently.

1. PCRs are performed in an UV cabinet. Thaw all PCR reagents, vortex prior to use and place on ice.
2. Primers amplify approximately a 200 bp region flanking the mutation of interest. An example reaction and appropriate primers for radioactive and fluorescent assays are shown in the table below. Suitable quantities of reagents are mixed, depending on how many DNA samples are to be analysed and 24 µl of the mix is aliquoted into labelled PCR tubes. In all assays, a known DNA sample harbouring the mutation of interest (positive control), an unaffected individual (negative control) and a no DNA control are included. Add 1 µl of DNA (typically 50–100 ng/µl) to the master-mix of each PCR tube, vortex and centrifuge.

Mastermix (25 µl volume)	
Go Taq 5× Buffer	5 µl
dNTPs (2 mM)	2.5 µl
Forward Primer (20 µM)	1.5 µl
Reverse Primer (20 µM)	1.5 µl

(continued)

(continued)

Mastermix (25 μl volume)	
dH_2O	13.15 μl
Go Taq polymerase	0.35 μl
DNA	1 μl

3. Place tubes into a thermal cycler and begin the following PCR program. Annealing temperature and extension time are dependent on oligonucleotide sequence and length of PCR product respectively.

PCR conditions	
Initial denaturation	2 min at 95°C
30 cycles	30 s at 95°C
	30 s at xx°C
	Xx s at 72°C
Final extension	5 min at 72°C

xx oligonucleotide sequence and length of PCR product

4. After amplification, PCR products are analysed via agarose gel electrophoresis. A suitable DNA ladder is included to aide sizing of PCR fragments.
5. Following successful amplification, thaw all reagents for the last fluorescent cycle and place on ice. For last radioactive cycle, see step 12.
6. In the last fluorescent cycle, the following reagents are mixed as shown below. The labelling primer is five prime modified with fluorescein (FAM).

Last fluorescent cycle mastermix	
3243 F FAM labelled primer (20 μM)	0.1 μl
3243 R primer (20 μM)	0.1 μl
GoTaq polymerase	0.2 μl

7. To each PCR, 0.4 μl of the fluorescent mix is added, vortexed and centrifuged. Place the PCR tubes into a thermal cycler and begin the following PCR program:

Reaction conditions	
Initial denaturation	45 s at 95°C
Annealing	45 s at 61°C
Extension	1 min at 72°C

8. After labelling, the PCR products are digested with an appropriate restriction enzyme. A restriction enzyme digest mix is created:

Restriction enzyme digest (20 µl total volume)	
MQ	7 µl
Restriction enzyme (HaeIII)	1 µl
10× Buffer 2	2 µl
PCR product	10 µl

9. 10 µl of the restriction digest mix is aliquoted into new labelled PCR tubes and mixed with 10 µl of the labelled PCR and incubated at 37°C overnight. An uncut control is included at this stage.
10. For fragment analysis, suitable size standards are mixed with the PCR samples (see below) and run on a genetic analyser (e.g., ABI PRISM 3100 Genetic Analyzer). It is important that different fluorescent labels are used on PCR samples (e.g., labelled with the fluorophore, FAM) and size standards (e.g., ROX):

Fragment analysis mix (10 µl volume)	
HiDi	8.5 µl
ROX500	0.5 µl
Digested PCR product	1 µl

11. To prepare the samples for loading onto the sequencer mix the HiDi formamide and DNA markers and aliquot 9 µl of the mix into each required well of the 96-well plate. 1 µl of the digested, labelled PCR is then added. Centrifuge the 96-well plate briefly, denature the samples at 95°C for 2 min, then place on ice. Finally, load into the sequencer.
12. Radioactive labelling of PCR products. Following successful amplification, the PCRs are labelled in a last hot cycle:

Last hot cycle mix (2.5 µl volume)	
H3353 primer (20 µM)	1 µl
L3200 primer (20 µM)	1 µl
α^{32}P-dCTP	0.25 µl
Taq polymerase	0.25 µl

13. To each PCR, 2.5 µl of the radioactive mix is added, vortexed and centrifuged. Place the PCR tubes into a thermal cycler and begin the following PCR program:

Reaction conditions	
Initial denaturation	10 min at 95°C
Annealing	2 min at 58°C
Extension	8 min at 72°C

14. The labelled products are precipitated by adding the following reagents and left for 1 h to precipitate at –80°C.

DNA precipitation (54.5 µl volume)	
Pellet paint	2 µl
3 M Sodium acetate (0.1 volume)	2.5 µl
100% ethanol (2 volumes)	50 µl

15. Precipitated DNA is pelleted by centrifugation, washed with 70% ethanol and allowed to air dry.
16. Incorporation of α-^{32}PdCTP is measured using a Cerenkov counter and differences in radioactive incorporation between samples are standardised so that equal amounts (2,000–8,000 cpm) are digested.
17. After labelling, the PCR products are digested with an appropriate restriction enzyme. A restriction enzyme digest mix is created and samples are digested over night at 37°C.

Restriction enzyme digest (20 µl volume)	
Restriction enzyme (HaeIII)	1 µl
10× Buffer	2 2 µl
DNA	17 µl

18. The digested PCR products are resolved through a 5% non-denaturing polyacrylamide gel, which is then dried and exposed to a phosphoimager cassette.
19. To determine the sample level of mutation, the relative levels of mutant and wild-type band must be determined and a ratio derived which then can be converted into a percentage. In the fluorescent analysis described earlier, a 199 bp PCR product was created, which when amplified from wild type mtDNA harbours a single HaeIII recognition site. Digestion with HaeIII generates two products, 162 and 39 bp. The m.3243A>G transition introduces another HaeIII recognition site, which when cleaved generates three products (90, 72 and 37 bp). Using fragment analysis software, such as Genemapper, the mutation level is calculated at the percentage of the area under the mutant (90 bp) allele relative to the combined area under

the 90 and 162 (wild type) allele. The primers used in the fluorescent assay give a larger mutant band than the ones used in the radioactive assay. This is because electropherograms in Genemapper sometimes contain "noise" which can exist up until 40 bps. For radioactive analysis, a 154 bp PCR product was created, which when amplified from wild-type mtDNA harbours a single HaeIII recognition site. Digestion with HaeIII generates two products (117 and 37 bp). The m.3243A>G transition introduces another HaeIII recognition site, which when cleaved generates three products (45, 72 and 37 bp). For quantification, the 117 bp fragment is normalised to the 72 bp fragment for deoxycytosine content, and the mutation level is calculated as a percentage of the amount of radiolabel in the 117 bp fragment relative to the combined amount in the 72 and 117 bp fragments. It is important to determine whether the restriction digest has gone to completion and including an uncut control is especially useful. Ideally, having two restriction sites present in the amplified PCR can aide identification of incomplete digestion. In the above example, the m.3243A>G transition introduces another recognition site, however loss of a restriction site is just as useful. If this is not available, with the mtDNA sequence available, another recognition site can be incorporated into the primer.

3.3.2. Whole Mitochondrial Genome Sequencing from Single Cells

1. DNA extraction and the first round PCR are performed in the UV cabinet. Thaw all components at –20°C, vortex prior to use and place on ice.
2. There are nine first round PCR primer pairs (see Note 6 and Table 2), each giving 2 kb products. Each first round product then acts as the template for four further primer pairs in the second round PCRs (Table 2). For the first round PCR, for each primer pair, aliquot 1.5 µl of forward and reverse primer into the corresponding tube. Then aliquot 1 µl of the DNA lysate into the appropriate tubes. Set up an additional two tubes for the lysis buffer controls, as well as a positive and a negative control. Add all components of the mastermix to a 1.5 ml tube listed below vortex and centrifuge briefly.

First round mastermix	
dH_2O	33.65 µl
10× buffer	5.0 µl
10× dNTPs	5.0 µl
25 mM Mg^{2+} solution	2.0 µl
AmpliTaq Gold	0.35 µl

3. Add 46 μl of the mastermix to each PCR tube, vortex and centrifuge the PCR tubes. Place tubes into a thermal cycler and begin the following PCR program:

First round PCR program		
Initial denaturation	95°C	10 min
38 cycles	94°C	45 s
	58°C	45 s
	72°C	2 min
Final extension	72°C	8 min

4. When the first round PCR is complete, thaw all second round PCR components at –20°C and place on ice. Aliquot 1 μl of each forward and reverse primer (Table 2) into the corresponding PCR tube. Dilute the first round PCR product ¼ (see Note 7) and add 1 μl to the appropriate tubes. Set up two additional tubes for a positive and negative control. Make up a mastermix as follows:

Second round mastermix	
dH_2O	16.87 μl
10× buffer	2.5 μl
10× dNTPs	2.5 μl
AmpliTaq Gold	0.13 μl

5. Add 24 μl of the mastermix to each PCR tube, vortex and centrifuge the PCR tubes. Place tubes into a thermal cycler and begin the following PCR program:

Second round PCR program		
Initial denaturation	95°C	10 min
30 cycles	94°C	45 s
	58°C	45 s
	72°C	1 min
Final extension	72°C	8 min

6. After the PCR program is complete, load 5 μl of each PCR product on a 1.5% agarose gel at 70 V for ~45 min alongside 5 μl of Hyperladder IV.
7. For ExoSAP-IT clean up, place a 96-well plate on ice and add 5 μl of each PCR product to the appropriate wells. Next, transfer the ExoSAP-IT enzyme from the –20°C freezer onto ice and add 2 μl ExoSAP-IT to each well. Place a rubber cover mat onto the 96-well plate, mix briefly and pulse spin down.

8. Incubate the 96-well plate in a thermal cycler, and perform the following program:

ExoSAP-IT program	
37°C	15 min
80°C	15 min

9. After the ExoSAP-IT program has finished, thaw all components for cycle sequencing and place on ice. Vortex all components before use, make up the mastermix as below, vortex and centrifuge. Add 13 μl mastermix to each of the wells, then seal the 96-well plate with caps.

Cycle sequencing mastermix	
Buffer	3 μl
Universal for/rev primer*	1 μl
BigDye 3.1	2 μl
dH_2O	7 μl

*Primers used and length of PCR product

10. Centrifuge the 96-well plate briefly then load into a thermal cycler with following PCR program.

Cycle sequencing PCR program		
Initial denaturation	96°C	1 min
25 cycles	96°C	10 s
	50°C	5 s
	60°C	4 min

11. Once the cycle sequencing program has completed, the samples need to be precipitated using the following procedure:
    (a) Add 2 μl of 125 mM EDTA to each sample in the 96-well plate.
    (b) Add 2 μl 3 M sodium acetate to each sample.
    (c) Briefly centrifuge the 96-well plate.
    (d) Add 70 μl of 100% EtOH to each sample.
    (e) Seal, invert the plate four times to mix and leave for 15 min at room temperature.
    (f) Centrifuge the plate for 30 min at 2,000 g.
    (g) Invert plate on tissue paper and centrifuge up to 50 g.
    (h) Add 70 μl of 70% EtOH to each sample.
    (i) Centrifuge for 15 min at 1,650 g.
    (j) Invert the plate on tissue paper and centrifuge up to 50 g.
    (k) Air dry the plate for at least 20 min in the dark.

12. To prepare the samples for loading onto the sequencer, remove HiDi formamide from the –20°C freezer, thaw and add 10 µl to each well. Centrifuge the 96-well plate briefly, then denature the samples at 95°C for 2 min. Finally, load onto the sequencer.

4. Notes

1. Tissue section sizes vary depending on tissue and use e.g., staining only or LMD.
2. COX/SDH incubation times vary on tissue energetic requirements.
3. For all single cell PCR, UV all plastics e.g., plates, tubes, tips and sterile water for 20 min prior to use.
4. When using the Long Extension PCR Kit, check Reaction Buffer 3 for crystals. If crystals are present leave at room temperature overnight and mix well again, alternatively the buffer can be warmed to ~70°C for 5–10 min.
5. Find a control sample, e.g., typically, blood contains no deletions.
6. Primers are M13-tagged so universal forward and reverse can be used for sequencing any fragment.
7. For most samples, the PCR product from the first round PCRs needs to be diluted. A ¼ dilution is generally adequate, but should be optimised for different cell types if achieving a clean second round PCR product is problematic.

Acknowledgements

KJK has a personal fellowship funded by the Alzheimer's Research Trust. We are grateful for financial support from the Wellcome Trust, SPARKS (Sport Aiding Medical Research for Kids) and the Medical Research Council.

References

1. Krishnan, K.J., Greaves, L.C., Reeve, A.K. and Turnbull, D. (2007) The ageing mitochondrial genome. *Nucleic Acids Research*, **35**, 7399–7405.
2. Taylor, R.W. and Turnbull, D.M. (2005) Mitochondrial DNA mutations in human disease. *Nature Reviews Genetics*, **6**, 389–402.
3. Quintana-Murci, L., Semino, O., Bandelt, H.J., Passarino, G., McElreavey, K. and Santachiara-Benerecetti, A.S. (1999) Genetic evidence of an early exit of Homo sapiens sapiens from Africa through eastern Africa. *Nature Genetics*, **23**, 437–441.
4. Chinnery, P.F., Samuels, D.C., Elson, J. and Turnbull, D.M. (2002) Accumulation of

mitochondrial DNA mutations in ageing, cancer, and mitochondrial disease: is there a common mechanism? *Lancet*, **360**, 1323–1325.

5. Bender, A., Krishnan, K.J., Morris, C.M., Taylor, G.A., Reeve, A.K., Perry, R.H. *et al.* (2006) High levels of mitochondrial DNA deletions in substantia nigra neurons in aging and Parkinson disease. *Nature Genetics*, **38**, 515–517.
6. Greaves, L.C., Preston, S.L., Tadrous, P.J., Taylor, R.W., Barron, M.J., Oukrif, D. *et al.* (2006) Mitochondrial DNA mutations are established in human colonic stem cells, and mutated clones expand by crypt fission. *Proceedings of the National Academy of Sciences of the United States of America*, **103**, 714–719.
7. Taylor, R.W., Barron, M.J., Borthwick, G.M., Gospel, A., Chinnery, P.F., Samuels, D.C. *et al.* (2003) Mitochondrial DNA mutations in human colonic crypt stem cells. *Journal of Clinical Investigation*, **112**, 1351–1360.
8. Johnson, M.A., Bindoff, L.A. and Turnbull, D.M. (1993) Cytochrome c oxidase activity in single muscle fibers: assay techniques and diagnostic applications. *Annals of Neurology*, **33**, 28–35.
9. Goto, Y., Nonaka, I. and Horai, S. (1990) A mutation in the tRNA(Leu)(UUR) gene associated with the MELAS subgroup of mitochondrial encephalomyopathies. *Nature*, **348**, 651–653.
10. Santorelli, F.M., Tanji, K., Kulikova, R., Shanske, S., Vilarinho, L., Hays, A.P. *et al.* (1997) Identification of a novel mutation in the mtDNA ND5 gene associated with MELAS. *Biochemical and Biophysical Research Communications*, **238**, 326–328.
11. Kirby, D.M., McFarland, R., Ohtake, A., Dunning, C., Ryan, M.T., Wilson, C. (2004) Mutations of the mitochondrial ND1 gene as a cause of MELAS. *Journal of Medical Genetics*, **41**, 784–789.
12. Schaefer, A.M., McFarland, R., Blakely, E.L., He, L., Whittaker, R.G., Taylor, R.W. (2008) Prevalence of mitochondrial DNA disease in adults. *Annals of Neurology*, **63**, 35–39.
13. Corral-Debrinski, M., Shoffner, J.M., Lott, M.T. and Wallace, D.C. (1992) Association of mitochondrial DNA damage with aging and coronary atherosclerotic heart disease. *Mutation Research*, **275**, 169–180.
14. Kraytsberg, Y., Kudryavtseva, E., McKee, A.C., Geula, C., Kowall, N.W. and Khrapko, K. (2006) Mitochondrial DNA deletions are abundant and cause functional impairment in aged human substantia nigra neurons. *Nature genetics*, **38**, 518–520.
15. Michikawa, Y., Mazzucchelli, F., Bresolin, N., Scarlato, G. and Attardi, G. (1999) Aging-dependent large accumulation of point mutations in the human mtDNA control region for replication. *Science (New York, NY)*, **286**, 774–779.
16. Muller-Hocker, J. (1989) Cytochrome-c-oxidase deficient cardiomyocytes in the human heart an age-related phenomenon. A histochemical ultracytochemical study. *American Journal of Pathology*, **134**, 1167–1173.
17. Brierley, E.J., Johnson, M.A., Lightowlers, R.N., James, O.F. and Turnbull, D.M. (1998) Role of mitochondrial DNA mutations in human aging: implications for the central nervous system and muscle. *Annals of Neurology*, **43**, 217–223.
18. Bua, E., Johnson, J., Herbst, A., Delong, B., McKenzie, D., Salamat, S. *et al.* (2006) Mitochondrial DNA-deletion mutations accumulate intracellularly to detrimental levels in aged human skeletal muscle fibers. *American Journal of Human Genetics*, **79**, 469–480.
19. Taylor, R.W., Schaefer, A.M., Barron, M.J., McFarland, R. and Turnbull, D.M. (2004) The diagnosis of mitochondrial muscle disease. *Neuromuscular Disorders*, **14**, 237–245.
20. Blakely, E.L., He, L., Gardner J.L., Hudson, G., Walter, J., Hughes, I. *et al.* (2008) Novel mutations in the TK2 gene associated with fatal mitochondrial DNA depletion myopathy. *Neuromuscular Disorders*, **18(7)**, 557–560.
21. Cree, L.M., Samuels, D.C., de Sousa Lopes, S.C., Rajasimha, H.K., Wonnapinij, P., Mann, J.R. *et al.* (2008) A reduction of mitochondrial DNA molecules during embryogenesis explains the rapid segregation of genotypes. *Nature Genetics*, **40**, 249–254.
22. Pyle, A., Taylor, R.W., Durham, S.E., Deschauer, M., Schaefer, A.M., Samuels, D.C. *et al.* (2007) Depletion of mitochondrial DNA in leucocytes harbouring the 3243A- > G mtDNA mutation. *Journal of Medical Genetics*, **44**, 69–74.
23. Holt, I.J., Harding, A.E. and Morgan-Hughes, J.A. (1988) Deletions of muscle mitochondrial DNA in patients with mitochondrial myopathies. *Nature*, **331**, 717–719.
24. Poulton, J., Deadman, M.E. and Gardiner, R.M. (1989) Duplications of mitochondrial DNA in mitochondrial myopathy. *Lancet*, **1**, 236–240.
25. Reeve, A.K., Krishnan, K.J., Elson, J.L., Morris, C.M., Bender, A., Lightowlers, R.N.

et al. (2008) Nature of mitochondrial DNA deletions in substantia nigra neurons. *American Journal of Human Genetics*, **82**, 228–235.

26. MITOMAP. (2006) Centre for Molecular Medicine, Emory University, Atlanta, GA.
27. Corral-Debrinski, M., Horton, T., Lott, M.T., Shoffner, J.M., Beal, M.F. and Wallace, D.C. (1992) Mitochondrial DNA deletions in human brain: regional variability and increase with advanced age. *Nature Genetics*, **2**, 324–329.
28. Corral-Debrinski, M., Shoffner, J.M., Lott, M.T. and Wallace, D.C. (1992) Association of mitochondrial DNA damage with aging and coronary atherosclerotic heart disease. *Mutation Research*, **275**, 169–180.
29. Krishnan, K.J. and Birch-Machin, M.A. (2006) The incidence of both tandem duplications and the common deletion in mtDNA from three distinct categories of sun-exposed human skin and in prolonged culture of fibroblasts. *Journal of Investigative Dermatology*, **126**, 408–415.
30. Sciacco, M., Bonilla, E., Schon, E.A., DiMauro, S. and Moraes, C.T. (1994) Distribution of wild-type and common deletion forms of mtDNA in normal and respiration-deficient muscle fibers from patients with mitochondrial myopathy. *Human Molecular Genetics*, **3**, 13–19.
31. He, L., Chinnery, P.F., Durham, S.E., Blakely, E.L., Wardell, T.M., Borthwick, G.M. *et al.* (2002) Detection and quantification of mitochondrial DNA deletions in individual cells by real-time PCR. *Nucleic Acids Research*, **30**, e68.
32. Taivassalo, T., Gardner, J.L., Taylor, R.W., Schaefer, A.M., Newman, J., Barron, M.J. *et al.* (2006) Endurance training and detraitning in mitochondrial myopathies due to single large-scale mtDNA deletions. *Brain*, **129**, 3391–3401.
33. Krishnan, K.J., Bender, A., Taylor, R.W. and Turnbull, D.M. (2007) A multiplex real-time PCR method to detect and quantify mitochondrial DNA deletions in individual cells. *Analytical Biochemistry*, **370**, 127–129.
34. Blakely, E.L., He, L., Taylor, R.W., Chinnery, P.F., Lightowlers, R.N., Schaefer, A.M. *et al.* (2004) Mitochondrial DNA deletion in "identical" twin brothers. *Journal of Medical Genetics*, **41**, e19.
35. Chabi, B., Mousson de Camaret, B., Duborjal, H., Issartel, J.P. and Stepien, G. (2003) Quantification of mitochondrial DNA deletion, depletion, and overreplication: application to diagnosis. *Clinical Chemistry*, **49**, 1309–1317.
36. Poe, B.G., Navratil, M., Arriaga, E.A. (2007) Absolute quantitation of a heteroplasmic mitochondrial DNA deletion using a multiplex three-primer real-time PCR assay. *Analytical Biochemistry*, **362**, 193–200.
37. Pogozelski, W.K., H.C., Woeller, C.F., Jackson, W.E., Zullo, S.J., Fischel-Ghodsian, N., Blakely, W.F. (2003) Quantification of total mitochondrial DNA and the 4977-bp common deletion in Pearson's syndrome lymphoblasts using a fluorogenic 5′-nuclease (TaqMan) real-time polymerase chain reaction assay and plasmid external calibration standards. *Mitochondrion*, **2**, 415–427.
38. van Den Bosch, B.J., de Coo, R.F., Scholte, H.R., Nijland, J.G., van Den Bogaard, R., de Visser, M. *et al.* (2000) Mutation analysis of the entire mitochondrial genome using denaturing high performance liquid chromatography. *Nucleic Acids Research*, **28**, E89.
39. Bannwarth, S., Procaccio, V. and Paquis-Flucklinger, V. (2006) Rapid identification of unknown heteroplasmic mutations across the entire human mitochondrial genome with mismatch-specific Surveyor Nuclease. *Nature Protocols*, **1**, 2037–2047.
40. Bannwarth, S., Procaccio, V. and Paquis-Flucklinger, V. (2005) Surveyor Nuclease: a new strategy for a rapid identification of heteroplasmic mitochondrial DNA mutations in patients with respiratory chain defects. *Human Mutation*, **25**, 575–582.
41. Taylor, R.W., Taylor, G.A., Durham, S.E. and Turnbull, D.M. (2001) The determination of complete human mitochondrial DNA sequences in single cells: implications for the study of mutations. *Nucleic Acids Research*, **29**, E74-74.
42. McDonald, S.A., Greaves, L.C., Gutierrez-Gonzalez, L., Rodriguez-Justo, M., Deheragoda, M., Leedham, S.J. *et al.* (2008) Mechanisms of field cancerization in the human stomach: the expansion and spread of mutated gastric stem cells. *Gastroenterology*, **134**, 500–510.
43. McDonald, S.A., Preston, S.L., Greaves, L.C., Leedham, S.J., Lovell, M.A., Jankowski, J.A. *et al.* (2006) Clonal expansion in the human gut: mitochondrial DNA mutations show us the way. *Cell Cycle*, **5**, 808–811.
44. Sacconi, S., Salviati, L., Nishigaki, Y., Walker, W.F., Hernandez-Rosa, E., Trevisson, E. *et al.* (2008) A Functionally Dominant Mitochondrial DNA Mutation. *Human Molecular Genetics*, **17(12)**, 1814–1820.

45. White, H.E., Durston, V.J., Seller, A., Fratter, C., Harvey, J.F. and Cross, N.C. (2005) Accurate detection and quantitation of heteroplasmic mitochondrial point mutations by pyrosequencing. *Gene Testing*, **9**, 190–199.

46. Moraes, C.T., Ricci, E., Bonilla, E., DiMauro, S. and Schon, E.A. (1992) The mitochondrial tRNA(Leu(UUR)) mutation in mitochondrial encephalomyopathy, lactic acidosis, and strokelike episodes (MELAS): genetic, biochemical, and morphological correlations in skeletal muscle. *American Journal of Human Genetics*, **50**, 934–949.

47. Tanno, Y., Yoneda, M., Nonaka, I., Tanaka, K., Miyatake, T. and Tsuji, S. (1991) Quantitation of mitochondrial DNA carrying tRNALys mutation in MERRF patients. *Biochemical and Biophysical Research Communications*, **179**, 880–885.

48. McDonnell, M.T., Schaefer, A.M., Blakely, E.L., McFarland, R., Chinnery, P.F., Turnbull, D.M. *et al.* (2004) Noninvasive diagnosis of the 3243A>G mitochondrial DNA mutation using urinary epithelial cells. *European Journal of Human Genetics*, **12**, 778–781.

Chapter 14

An Introduction to Mitochondrial Informatics

Hsueh-Wei Chang, Li-Yeh Chuang, Yu-Huei Cheng, De-Leung Gu, Hurng-Wern Huang, and Cheng-Hong Yang

Abstract

In this chapter, we review the public resources available for human mitochondrial DNA and protein related bioinformatics, with a special focus on mitochondrial single nucleotide polymorphisms (mtSNPs). We also review our own freeware tool V-MitoSNP, giving an overview of its implementation and program workflow. Apart from these, we review several protocols for the graphic input of genes, keywords, gene searching by sequence, mtSNP searching by sequence, restriction enzyme mining, primer design, and virtual electrophoresis for PCR-RFLP genotyping. Some databases with similar function are integrated and compared.

Key words: Mitochondrial genome, Variation, Polymorphism, SNP, Database, BLAST, RFLP, Genotyping, Primer design

1. Introduction

The complete nucleotide sequence of the human mitochondrial (mt) genome has been established (1) and corrected (2). In humans, the mt genome contains a circular dsDNA molecule with 16,569 base pairs (bps) including 37 genes encoding 13 essential polypeptides involved in oxidative phosphorylation and the RNA machinery (2 rRNAs and 22 tRNAs) for their translation within the organelle. The remaining protein subunits that make up the respiratory chain system, together with those required for mtDNA maintenance, are nuclear-encoded, synthesized on cytoplasm, and sorted to the correct location of mitochondria (3).

The role of mitochondrial DNA mutations (polymorphisms) in many human diseases (3), cancers (4), and evolutionary studies (5) has been widely reviewed. Therefore, there is a great need for

Michael R. Barnes and Gerome Breen (eds.), *Genetic Variation: Methods and Protocols*, Methods in Molecular Biology, vol. 628, DOI 10.1007/978-1-60327-367-1_14, © Springer Science+Business Media, LLC 2010

tools that allow the integrated analysis of the mitochondrial genome and its protein products. To date, many resources annotate mitochondrial data with functional and structural information. Many locus or disease-specific databases are beginning to integrate functional information, such as DNA mutation/polymorphism and protein structure, into their annotation sets.

Here, we identify key resources for mitochondrial genome (see Table 1) and protein (see Table 2) annotation and analysis, with a focus on the analysis of mitochondrial variation. Table 1 contains general databases addressing the following: the human mtDNA genome; human mtSNP data; mtDNA sample collections and others; and a database of human nuclear DNA (nuDNA) coding for mitochondrial protein. Table 2 focuses on protein products, with a database of human mitochondrial- and nuclear-encoded proteins.

Among them, an excellent resource for mtSNP locations and other genome annotations is MITOMAP (6), which provides information on mtDNA polymorphisms (includes mini insertions & deletions), mtDNA mutations with reports of disease associations, major rearrangements, nuclear genes involved in mitochondrial disease, and mitochondrial pseudogenes. Another powerful resource for mtSNP analysis is mtDB (7), which provides complete human mitochondrial sequences for many populations, i.e., 1,865 complete sequences and 839 coding region sequences, demonstrating mtSNP sites, allowing the identification of population-specific mt variants. Recently, many other human mitochondrial genome databases have been developed such as HmtDB (8), Mitochondriome, and MitoData (see Table 1).

Other, more general resources also provide mitochondrial variation information in a disease context, including the NCBI Online Mendelian Inheritance in Man (OMIM) database and GOBASE (9). Some databases were developed with a specific focus on mtDNA variations (polymorphisms) such as V-MitoSNP (10) and GiiB-JST mtSNP (11). Some SNP tools provide BLAST function for nuclear- and mtSNPs such as SNP-BLAST (12) in dbSNP of NCBI, SNP500Cancer (13), BLAT (14) of UCSC Genome Browser (15), and our developed V-MitoSNP (10). Other databases have focused on the collection of mitochondrial DNA and their genographic and molecular genealogy (16–18). Some databases were focused on the nuclear DNA encoding the mitochondrial proteins such as MitoNuc (19) and MitoDat (20).

For mitochondrial protein databases, we have focused on databases that predict the subcellular location of mt-protein such as MitoP2 (21), HMPDb, MitoProteome (22), and MITOPRED (23) (see Table 2). In MitoP2, a support-vector machine (SVM) was trained with a reference set of mitochondrial proteins and a set of proteins belonging to other cellular compartments.

Table 1
Resources for mitochondrial- and nuclear-DNA encoding for mitochondrial proteins

Database name	Source	Web address	Basic characteristics
mtDNA-general			
MITOMAP (6)	mtDNA	http://www.mitomap.org/	A human mitochondrial genome database containing polymorphisms and mutations of the mtDNA. It also provides a global mtDNA mutational phylogeny.
mtDB (7)	mtDNA	http://www.genpat.uu.se/mtDB/	Human Mitochondrial Genome Database, a resource for population genetics and medical sciences. The tool enables searching for mitochondrial haplogroups and mtSNP information for many populations.
HmtDB (8)	mtDNA	http://www.hmdb.uniba.it/hmdb/index.jsp	A human mt genomic resource based on variability studies supporting population genetics and biomedical research.
MitoData	mtDNA	http://mitodata.org/DesktopDefault.aspx	A unique multinational database holding clinical, biochemical, and molecular genetic data on human mitochondrial diseases.
OMIM	mtDNA*/nuDNA	http://www.ncbi.nlm.nih.gov/sites/entrez	Online Mendelian Inheritance in Man (NCBI) * use "limit" item to find the mitochondrial OMIM.
GOBASE (9)	mtDNA	http://gobase.bcm.umontreal.ca/	A taxonomically broad organelle genome database that organizes and integrates diverse data on mitochondria and chloroplasts and representative bacteria.
The mtDNA Population Database (33)	mtDNA	http://www.fbi.gov/hq/lab/fsc/backissu/april2002/miller1.htm	Integrated software and db for forensic comparison.
mtDNA-SNP			
V-MitoSNP (10)	mtDNA	http://bio.kuas.edu.tw/v-mitosnp	mtSNP visualisation platform for primer design, PCR-RFLP, mtBLAST, and NCBI mtSNP ID (rs#) matching.
SNP-BLAST (12) in dbSNP of NCBI	mtDNA/nuDNA	http://www.ncbi.nlm.nih.gov/SNP/snp_blastByOrg.cgi	All SNP ID rs# are searchable. All SNPs are included in dbSNP and it is not necessary to select mt-specific function.
GiiB-JST mtSNP (11)	mtDNA	http://mtsnp.tmig.or.jp/mtsnp/index_e.shtml	A database of human mitochondrial genome polymorphisms.

(continued)

Table 1
(continued)

Database name	Source	Web address	Basic characteristics
SNP500Cancer (13)	mtDNA/ nuDNA	http://snp500cancer.nci.nih.gov	Database providing sequence and genotype assay information for candidate SNPs (includes mtSNP) useful in mapping diseases, such as cancer.
BLAT (14) in UCSC Genome Browser (15)	mtDNA*/ nuDNA	http://genome.ucsc.edu/cgi-bin/hgBlat?command=start	A sequence searching and alignment tool. BLAT is more accurate and faster than popular existing tools for sequence alignments. NCBI mtSNP ID (rs#) matching is also provided. * Chromosome MT included.
mtDNA-sample collection and others			
The Genographic Project public participation mtDNA database (16)	mtDNA	https://www3.nationalgeographic.com/genographic/	This database provides the genetic signatures of ancient human migrations, creating an open-source research database. It allows submission of personal samples for analysis.
EMPOP CR Database (17)	mtDNA	http://www.empop.org/empop.php?page=home&PHPSESSID=0c3e7272d-9c094e954ed-605c3f4026f1	Mitochondrial DNA Control Region Database. It aims at the collection, quality control, and the searchable presentation of mtDNA control region haplotypes from all over the world.
SMGF-mtDB (18)	mtDNA	http://www.smgf.org/pages/mtdatabase.jspx	The Sorenson Molecular Genealogy Foundation (SMGF) has built the world's foremost collection of mitochondrial DNA data and corresponding genealogies.
nuDNAs encoding the mitochondrial proteins			
MitoNuc (19)	nuDNA	http://bighost.area.ba.cnr.it/mitochondriome	Integrating mitochondrial data and databases, links to other mitochondrial sites and relevant information.
MitoDat (20)	nuDNA	http://www-lmmb.ncifcrf.gov/mitoDat/	Mendelian Inheritance and the Mitochondrion db (mitochondrial nuclear gene).

mtDNA mitochondrial genomic DNA, *nuDNA* nuclear genomic DNA encoding mitochondrial proteins

MitoP2 provides three tools, PSORT II (24), Predotar (25), and Subloc (26), to predict the subcellular translocation of proteins. MitoP2 is able to search with selection parameters and keywords but not sequence input. It also identifies the putative orthologous proteins between species to study evolutionarily conserved functions and pathways. In MITOPRED (23), prediction is based on

Table 2
Resources for mitochondrial- and nuclear-encoded mitochondrial proteins

Database name	Source	Web address	Basic characteristics
MitoP2 (21)	mt-protein/ nu-mt-protein	http://www.mitop.de:8080/mitop2	MitoP2 integrates information on mitochondrial proteins, their molecular functions, and associated diseases. It also provides prediction of subcellular location.
MitoProteome (22)	mt-protein/ nu-mt-protein	http://www.mitoproteome.org/	The database is generated from experimental evidence and public databases, and contains both mitochondrial- and nuclear-encoded entries. It also provides prediction of subcellular location.
HMPDb	mt-protein/ nu-mt-protein	http://bioinfo.nist.gov/hmpd/	HMPDb provides comprehensive data on human nuclear- and mt-encoded mitochondrial proteins involved in mitochondrial biogenesis and function. 2D-PAGE images, searching, and 3D structure views were included. It also provides prediction of subcellular location.
MITOPRED (23)	mt-protein/ nu-mt-protein	http://bioapps.rit.albany.edu/MITOPRED/	This predicts nuclear- and mitochondrial-encoded mitochondrial proteins from all eukaryotic species. It also provides prediction of subcellular location.

the occurrence patterns of Pfam domains (version 16.0) (27). In addition, MITOPRED links to many external resources, providing the necessary information for the study of mitochondrial genes, protein, and the associated diseases.

To date, a range of single nucleotide polymorphism (SNP) genotyping methods have been applied to mitochondrial genome studies, including TaqMan probe genotyping, PCR-restriction fragment length polymorphism (RFLP) analysis (28), and resequencing approaches (16). Among these methods, PCR-RFLP analysis is a commonly used laboratory method for the genotyping of mtSNPs; however, the need for restriction enzyme site mining can make this method inconvenient for some researchers. Moreover, the existing tools described above for mitochondrial analysis do not contain substantial information on mitochondrial RFLP restriction enzyme sites for mtSNP genotyping. In this review, we will introduce our developed software V-MitoSNP (10) for restriction enzyme mining of mtSNPs later to address this problem.

We will also introduce some fundamental mitochondrial informatics protocols, such as gene searching and mtSNP identification, by inputting mitochondrial DNA sequence (V-MitoSNP and BLAT), mtSNP RFLP restriction enzyme mining, and RFLP primer design (V-MitoSNP). We will also discuss methods for BLAST searching for mitochondrial genome and mtSNP sequences.

2. Materials

1. *Hardware*

   A standard personal computer platform with an Internet connection.

2. *Software*

   A regular Internet browser, such as Internet Explorer, is required. It should support JavaScript 1.1.

3. Methods

3.1. Implementation of V-MitoSNP

The database structure for the revised Cambridge Reference Sequence (rCRS) of the Human Mitochondrial DNA and mtSNPs in V-MitoSNP (10) is downloaded from MITOMAP (29) with permission, and the mtSNP rs# ID is downloaded from chromosome MT data in NCBI dbSNP version b123 (30). The database for RFLP restriction enzyme mining is downloaded from REBASE version 601 (31).

3.2. Program Workflow

The workflow of V-MitoSNP (10) consists of six modules as follows:

3.2.1. Input Module

V-MitoSNP uses two different input formats, namely a graphic input format and a text search input format. The graphic input format provides the selection of gene function. In the search input format, the gene locus, disease, NCBI SNP ID rs#, nucleotide range, and mtDNA sequence in IUPAC format are acceptable.

3.2.2. Display Module

The results of the input module are processed in the display module, which provides SNP, cancer, and disease information for the mtDNA as well as their RFLP availability for all mtSNPs.

3.2.3. Position Alignment Module

The input sequence is matched to the human mtDNA rCRS sequence (2). The mtBLAST for mtDNA sequence in IUPAC format is a gene-targeted search.

3.2.4. RFLP Analysis Module

V-MitoSNP provides a complete list of available restriction enzymes for each mtSNP. It provides the RFLP availability for both sense and antisense strands.

3.2.5. Primer Design Module

V-MitoSNP provides complete primer sets for all SNPs in mtDNA, such as the primer sets, for natural and mismatched PCR-RFLP for mt-SNP, respectively.

3.2.6. Virtual Electrophoresis Module

The full length of the PCR product using natural and mismatched primer sets are estimated by this module. Subsequently, in silico digestion by RFLP enzymes and in silico electrophoresis are possible, providing genotype information and the corresponding PCR-RFLP length.

3.3. An Example of Several Input Protocols and Their Outputs

These examples are mainly demonstrated using the tools in V-MitoSNP. However, some common features are present in some of the previously described resources, such as the UCSC Genome Browser and the NCBI BLAST service, so these are also reviewed in the following example. Since V-MitoSNP does not include protein structure and related information, we also review the protocol for prediction of mitochondrial proteins using the MitoP2, HMPDb and MitoProteome databases (URLs listed in Tables 1 and 2).

3.3.1. V-MitoSNP: Browsing for Genes of Interest

V-MitoSNP has a user-friendly graphical interface for visualization of mtSNPs. Taking the gene ND4 as an example, the gene can be selected by clicking on the gene region (see Fig. 1). Three catagories of SNPs are provided if data are available, including SNPs with no known cancer or disease association, SNPs with known cancer association, and SNPs with known disease association (see Notes 1 and 2).

3.3.2. Protocol for Keyword Input

V-MitoSNP can also be queried by keyword input (locus, disease, and SNP ID rs#). Fig 2. displays query examples for MT-CO1, LHON, and rs2853516, respectively. All the mitochondria-related diseases are abbreviated and hyperlinked to the full disease name (see Note 3). Disease output is not provided.

3.3.3. Protocol for Mitochondrial Gene Sequence Searching Using V-MitoSNP, BLAT, and BLAST

It is possible to search for genes, mtSNPs, and other elements in V-MitoSNP using a simple sequence search (see Note 4). The results of such a search are displayed in Fig. 3. Gene or genes within the input sequence are all shown on the panel of the mtSNP list. All the information, including the mtSNP nucleotide position, SNP ID rs#, nucleotide change, amino acid change, RFLP restriction enzyme, and primer design, is provided for each mtSNP. Where available, the mtSNPs are hyperlinked to view detailed information at dbSNP. These mtSNPs are also grouped into SNPs with no known cancer or disease association, SNPs

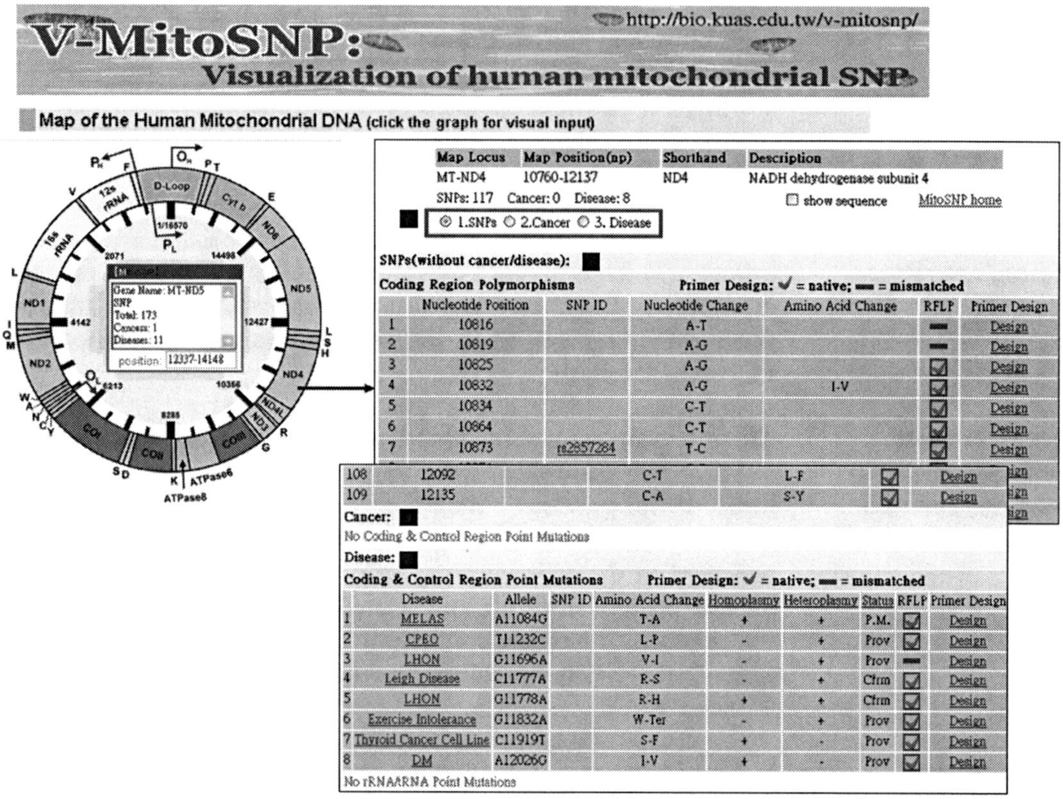

Fig. 1. The graphic input and output of V-MitoSNP (10) for selecting a gene of interest (such as ND4) from the mitochondrial genome. Genes can be selected using a point and click interface. The solid squares (■) indicate SNPs that have no known association with cancer or disease, SNPs with a known cancer association, and SNPs with a known disease association, respectively. Information for RFLP enzyme and primer (native and mismatched) design is provided by hyperlinking. Not all mtSNPs are listed

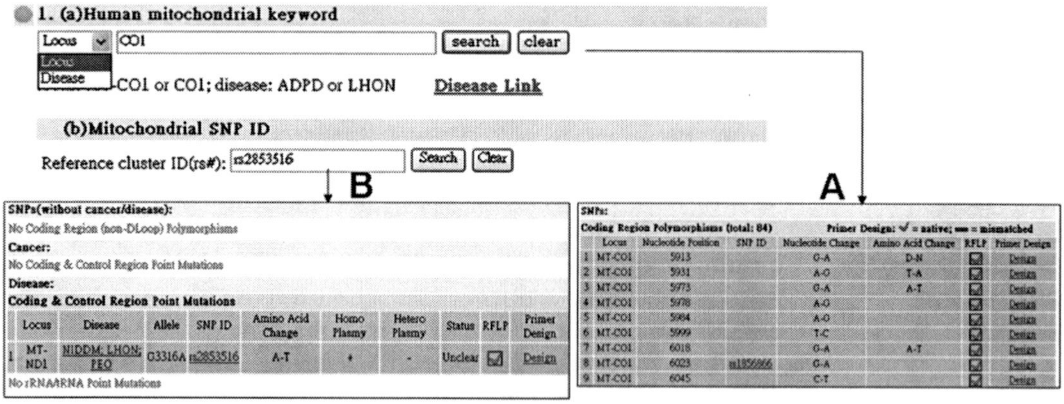

Fig. 2. The keyword input and output of V-MitoSNP. (**a**) Locus and disease input. MT-CO1 is an example of a locus. The representative output is provided. Output for disease input is not shown. Disease output provides a hyperlink to the full name for each mitochondrial-related disease. (**b**) The mtSNP ID rs# input such as rs2853516

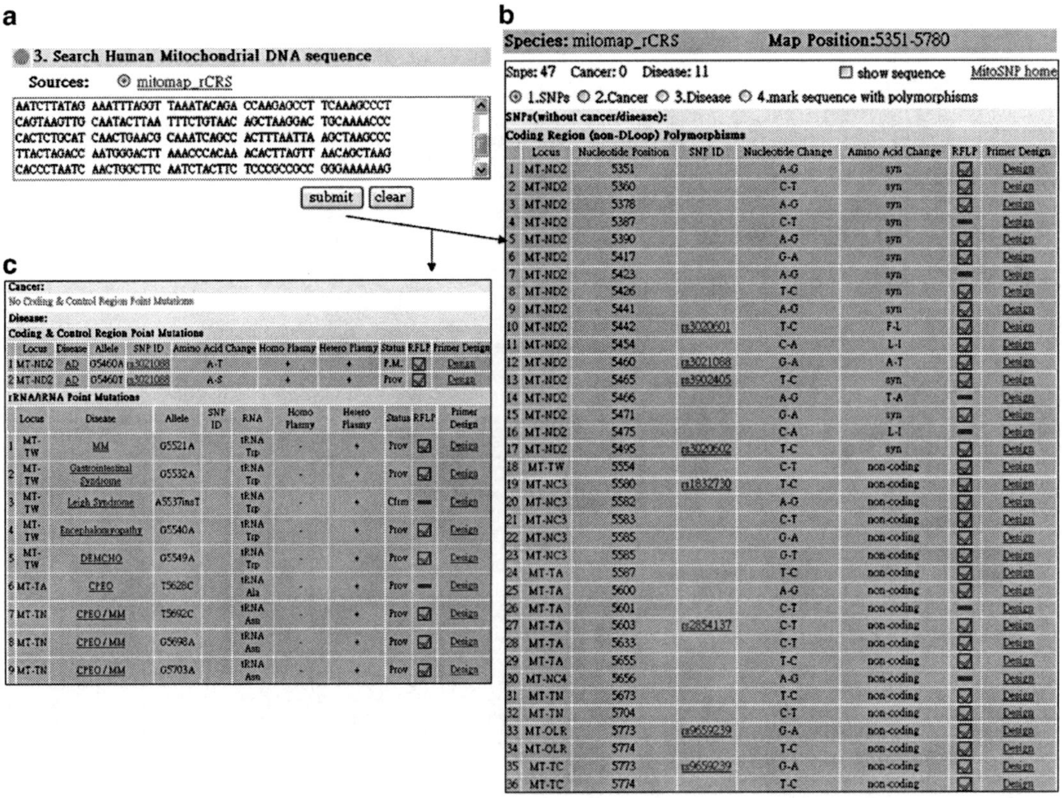

Fig. 3. Identifying genes and mtSNPs within the mitochondrial DNA sequence (see Note 1) input using V-MitoSNP. (**a**) Sequence input window. Sequence with IUPAC code nucleotides are acceptable. The external view for mitomap_rCRS is provided by hyperlink. Output results are shown in (**b**) and (**c**). The gene names (loci) are exactly provided within the input sequence, such as ND2, TW, NC3, TA, NC4, TN, OLR, and TC. (**b**) SNPs without cancer/disease association. (**c**) SNPs with known cancer and disease association. SNP version is dbSNP build 123

with known cancer association, and SNPs with known disease association (see Notes 1 and 2).

The BLAT tool hosted within the UCSC Genome Browser (14) is another easy method for searching mtDNA sequences (Fig. 4). The accession number for the gene or genes (marked with text – homolog sequence hits in Fig. 4c) within the input sequence are all shown in the UCSC genome visualization (see Note 5). The sequence hits that are homologous to other species are also provided. The mtSNPs are hyperlinked to view their detailed information.

Using NCBI BLAST to search mitochondrial genomic sequences is not as intuitive as BLAT; however, it does offer some additional functionality. First, using the query sequence in Note 4 to search the "Others (nr etc)" database, it is possible to search specific mtDNA haplotypes and view sequence level homology. Using this database has a problem, however, as the BLAST hit for

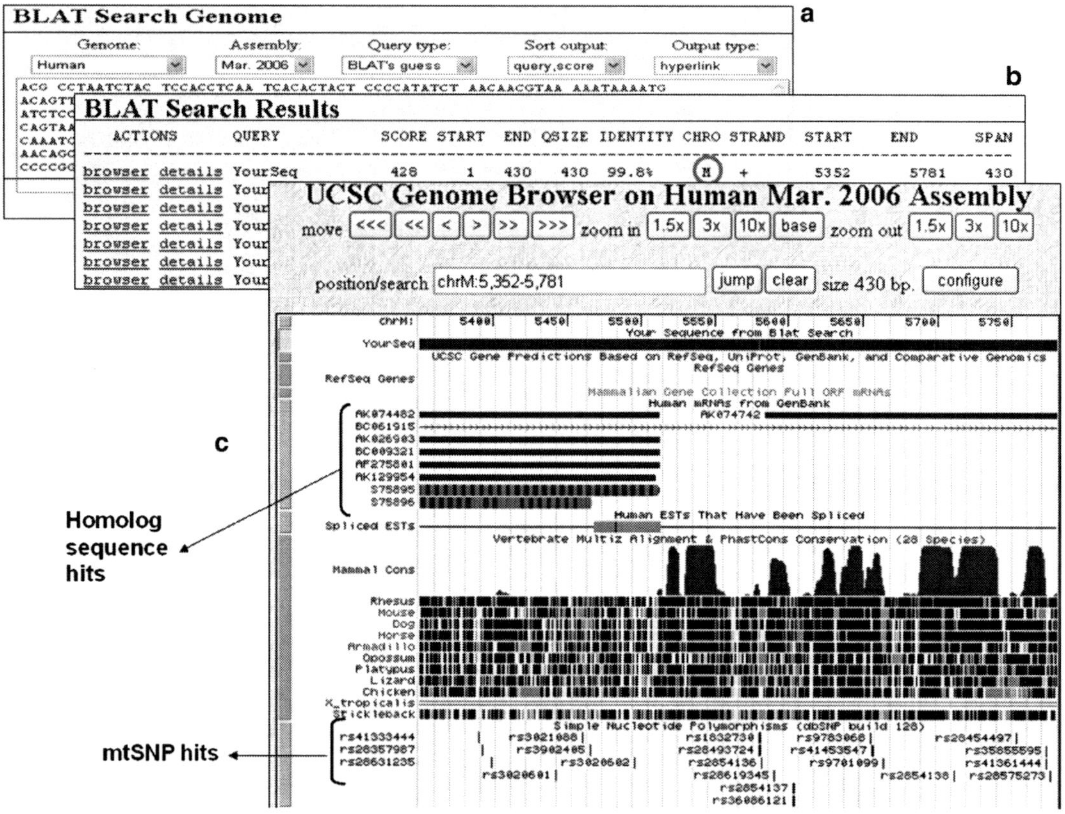

Fig. 4. Identifying genes and mtSNPs within the mitochondrial DNA sequence (see Note 1) using BLAT (14) from the UCSC Genome Browser (15). (**a**) Sequence input window. (**b**) BLAT search results. The top score is usually the best hit for homologous sequence indicated by circle (M is the mitochondrial; CHRO no. is the chromosome number). (**c**) Hits for homologous sequence and mtSNPs. The gene hits are provided in accession numbers belonging to the mt-genome. SNP version is dbSNP build 128

the mitochondrial query sequence is aligned against the whole mitochondrial genome, and no gene name is provided under this situation. However, if the mtDNA query is searched against the "Human genomic + transcript" database, then the rCRS sequence is returned as the top hit, and this can be viewed in the NCBI Map Viewer interface, which allows navigation to SNP and gene level (Fig. 5).

3.3.4. Sequence Searching for mtSNPs Using V-MitoSNP, BLAT, and SNP-BLAST

In the previous sequence searching example (see Subheading 3.3.3 and Note 4), V-MitoSNP, BLAT, and BLAST all displayed mtSNPs mapped across the query sequence (see Figs. 3–5, respectively). V-MitoSNP provided the mtSNP information in a tabular list, while BLAT and NCBI Map Viewer provided mtSNP information embedded in a graphical genome view. Each tool provides a hyperlink to dbSNP if a SNP ID rs# is available. The numbers of SNP IDs with rs# are different between V-MitoSNP and BLAT

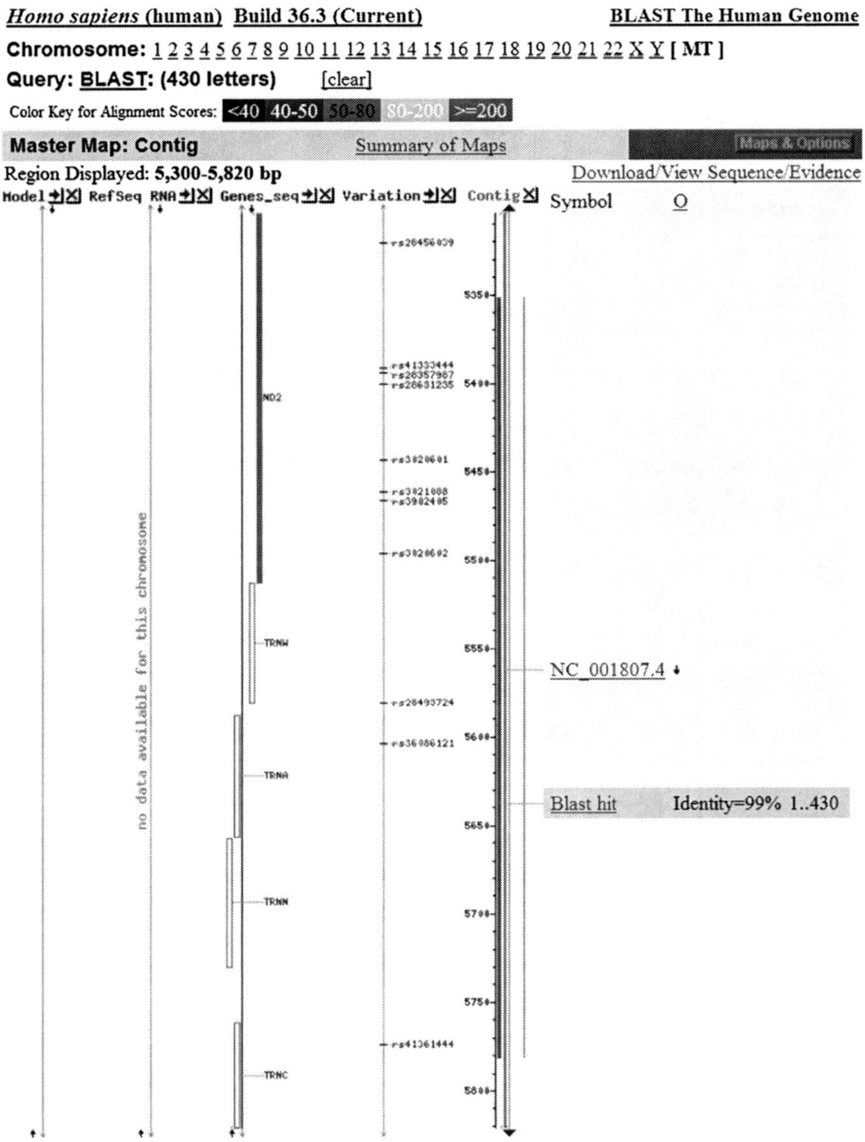

Fig. 5. NCBI Map Viewer results linked from a mtDNA query using the NCBI BLAST interface against the "Human genomic + transcript" database. The rCRS sequence is returned as the top hit and this can be viewed in the NCBI Map Viewer interface, which allows navigation to SNP and gene level

because different versions of dbSNP are utilized, i.e., dbSNP build 123 and 128, respectively. There are plans to update the dbSNP build for mtSNP in V-MitoSNP in the near future.

Using a short sequence as a query (see Note 6), the mtSNP searching output for V-MitoSNP and BLAT is the same as the longer sequence (not shown). In contrast, the NCBI SNP-BLAST algorithm has some problems for mtSNP searching using a short query sequence (see Note 7).

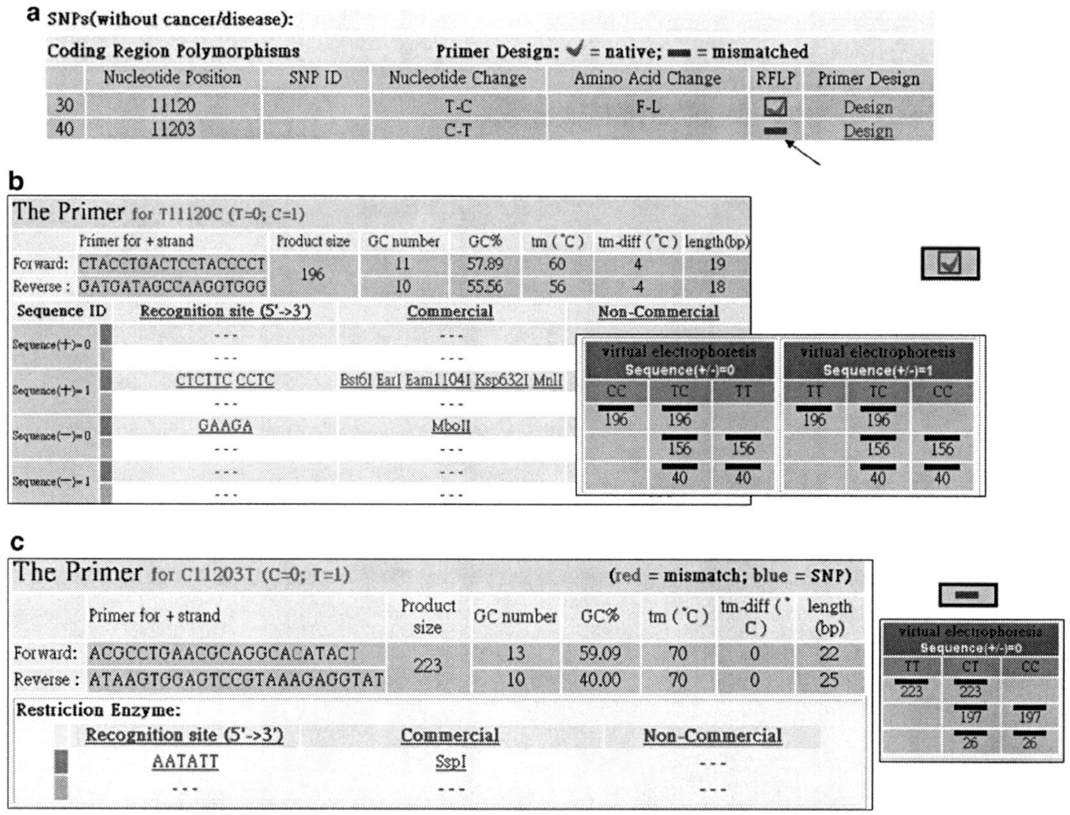

Fig. 6. A view of interfaces for restriction enzyme mining, primer design, and virtual electrophoresis for PCR-RFLP genotyping using V-MitoSNP. (**a**) Two typical RFLPs and their primer information. One has a restriction enzyme available for the mtSNP (RFLP), while the other has no restriction enzyme available for the mtSNP (*horizontal bar*). (**b**) Natural RFLP enzyme and its primer information. Enzymes for the sense and antisense strands are provided as well as their virtual electrophoresis information. All PCR-RFLP information is provided. (**c**) A mismatched primer was designed for creating the novel RFLP restriction enzyme site

3.3.5. Restriction Enzyme Mining, Primer Design, and Virtual Electrophoresis for PCR-RFLP Genotyping Using V-MitoSNP

The PCR-RFLP information for SNP genotyping using V-MitoSNP is displayed in Fig. 6. The restriction enzymes for the sense and antisense strand are not always the same as shown in Fig. 6b. Similarly, the restriction enzyme availability for an alternative SNP marked with "0" and "1" may be different. If no restriction enzyme site is available within the input sequence, V-MitoSNP automatically changes a nucleotide nearby to the SNP, which is assigned to a mismatch primer, thus creating a novel restriction enzyme site as shown in Fig. 6c. When a novel mutation is not available in the V-MitoSNP database, we recommend the SNP-RFLPing (32) tool for generic SNP RFLP assay design.

3.3.6. Prediction of Subcellular Location for Mitochondrial Proteins Using MitoP2, MitoProteome, and HMPDb

Although many databases related to prediction of the subcellular location of mitochondrial proteins have been developed, we recommend MitoP2 (21), HMPDb, and MitoProteome (21). Figure 7 compares the output of these three tools for prediction of the subcellular location. In MitoP2, the protein with the

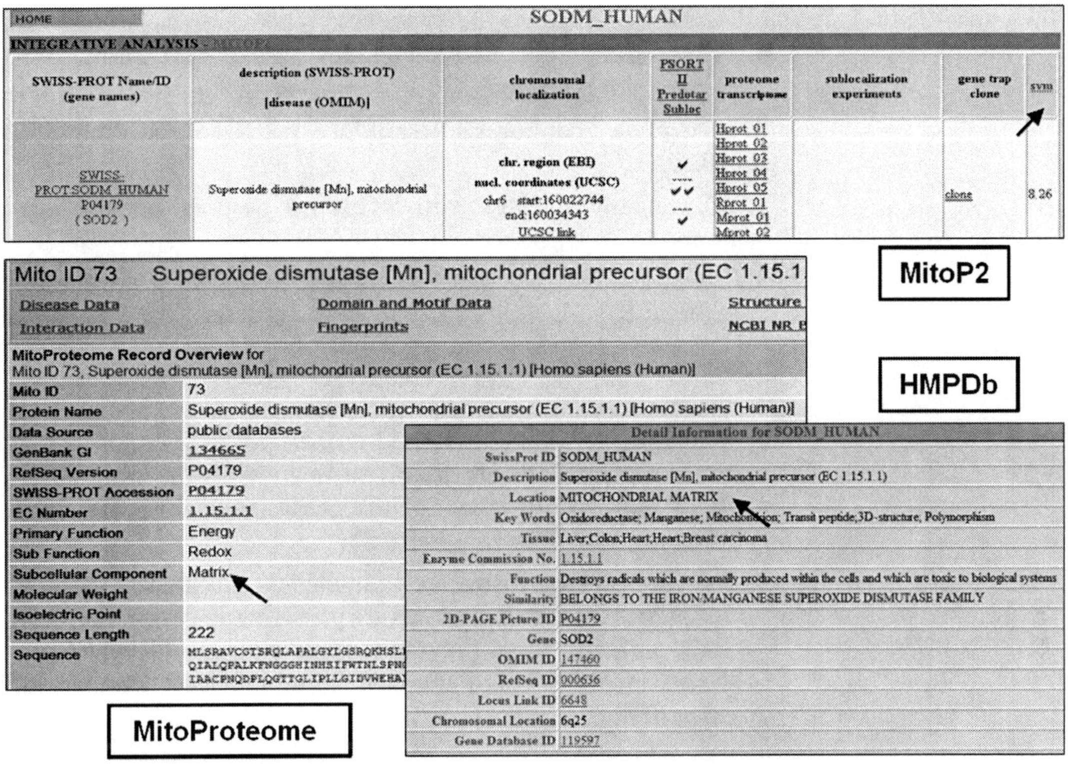

Fig. 7. An overview of the prediction for subcellular location of mt-protein in three common mitochondrial protein databases – MitoP2, HMPDb, and MitoProteome

highest support-vector machine (SVM) score has the strongest probability to be mitochondrial. In HMPDb and MitoProteome, the subcellular location is provided in text. There are many external links provided in these databases providing an array of useful information on the query mitochondrial protein.

In conclusion, there are many tools available for the analysis of mitochondrial proteins. We have also developed the V-MitoSNP software to provide a complete package for mtSNP association studies by RFLP genotype technology. Several advantageous characteristics distinguish V-MitoSNP from comparable software platforms. These improvements are listed below:

1. Interactive platform and graphic visualization for mtSNP related searches.
2. Long flanking sequences, up to 500bp, are provided for each mtSNP.
3. Sequence range input by clicking on a mitochondria map.
4. Keyword input for SNP retrieval related to cancers and/or diseases.
5. mtBLAST is provided to resolve mtDNA searching issues using NCBI BLAST.

6. Primer design for RFLP includes natural and mismatched primers for every mtSNP.
7. Virtual electrophoresis results are provided for each SNP RFLP.
8. All available mtSNPs with NCBI rs# number are also integrated in V-MitoSNP.

4. Notes

1. Not all mtSNPs are updated immediately to the database. If any novel mtSNPs are found to relate to some diseases, an RFLP genotyping assay can still be designed using the SNP-RFLPing tool (32) (see Note 8).
2. Not all polymorphisms are immediately designated with the dbSNP rs#; however, updates are performed on a regular basis. Some mitochondrial polymorphisms are not officially represented in dbSNP and therefore they may not have an rs# currently.
3. Full name of mitochondria-related diseases is provided by hyperlink to http://bio.kuas.edu.tw/v-mitosnp/Disease.jsp.
4. A mitochondrial gene input example for testing mtBLAST in V-MitoSNP, BLAT (UCSC), and BLAST (NCBI) is provided below in fasta format. The full length revised Cambridge Reference Sequence (rCRS) of the Human Mitochondrial DNA is available from the following URL (http://www.mitomap.org/mitoseq.html). In our example, we used a selection of sequence ranging from 5,351 to 5,780 bp is followed.

   >rCRS_5351_5780

   acgcctaatctactccacctcaatcacactactccccatatctaacaacgtaaaaataa
   aatgacagtttgaacatacaaaacccaccccattcctccccacactcatcgcccttacca
   cgctactcctacctatctccccttttatactaataatcttatagaaatttaggttaaataca-
   gaccaagagccttcaaagccctcagtaagttgcaatacttaatttctgtaacagct
   aaggactgcaaaaccccactctgcatcaactgaacgcaaatcagccactttaattaag
   ctaagcccttactagaccaatgggacttaaacccacaaacacttagttaacagct
   aagcaccctaatcaactggcttcaatctacttctcccgccgccgggaaaaaaggcgg-
   gagaagccccggcaggtttgaag
5. The nucleotide position between rCRS and BLAT of UCSC Genome Browser and NCBI BLAST has one nucleotide shift. Therefore, the sequence range shown in BLAT and BLAST is one more nucleotide than the rCRS.
6. FASTA sequence information for rs2857284 (http://www.ncbi.nlm.nih.gov/SNP/snp_ref.cgi?rs=2857284) for testing short sequence searches using the NCBI SNP-BLAST service

and V-MitoSNP. Y indicates the C/T polymorphism. The output is shown in Note 7.

>rs2857284

TTATTATTCTCATCCC**Y**CTACTATTTTTAACCAAAT

7. Results from a short sequence query using the NCBI SNP-BLAST service (http://www.ncbi.nlm.nih.gov/SNP/snp_blastByOrg.cgi) vary depending on the options specified in the search (results not shown). Using the megablast option, which is recommended only for sequences >28 bp, no correct mtSNP hit is returned. Without the megablast option, many mtSNPs are returned, including the correct one – rs2857284. The other SNP hits are returned because their flanking sequences overlapped the flanking sequence of rs2957284. This situation may confuse users, and it serves to emphasize the desirability of using larger flanking sequences for searching when using the BLAST Algorithm.

Acknowledgments

This work was partly supported by the National Science Council in Taiwan under grant NSC97-2311-B-037-003-MY3, NSC96-2221-E-214-050-MY3, NSC96-2622-E214-004-CC3, and the grant KMU-EM-98-1.4.

References

1. Anderson, S., Bankier, A.T., Barrell, B.G., de Bruijn, M.H., Coulson, A.R., Drouin, J., *et al.* (1981) Sequence and organization of the human mitochondrial genome. *Nature* **290**, 457–465.
2. Andrews, R.M., Kubacka, I., Chinnery, P.F., Lightowlers, R.N., Turnbull, D.M., and Howell, N. (1999) Reanalysis and revision of the Cambridge reference sequence for human mitochondrial DNA. *Nat Genet* **23**, 147.
3. Taylor, R.W., and Turnbull, D.M. (2005) Mitochondrial DNA mutations in human disease. *Nat Rev Genet* **6**, 389–402.
4. Zanssen, S., and Schon, E.A. (2005) Mitochondrial DNA mutations in cancer. *PLoS Med* **2**, e401.
5. Pakendorf, B., and Stoneking, M. (2005) Mitochondrial DNA and human evolution. *Annu Rev Genomics Hum Genet* **6**, 165–183.
6. Ruiz-Pesini, E., Lott, M.T., Procaccio, V., Poole, J.C., Brandon, M.C., Mishmar, D., *et al.* (2007) An enhanced MITOMAP with a global mtDNA mutational phylogeny. *Nucleic Acids Res* **35**, D823–D828.
7. Ingman, M., and Gyllensten, U. (2006) mtDB: Human Mitochondrial Genome Database, a resource for population genetics and medical sciences. *Nucleic Acids Res* **34**, D749–D751.
8. Attimonelli, M., Accetturo, M., Santamaria, M., Lascaro, D., Scioscia, G., Pappada, G., *et al.* (2005) HmtDB, a human mitochondrial genomic resource based on variability studies supporting population genetics and biomedical research. *BMC Bioinformatics* 6 Suppl. 4, S4.
9. O'Brien, E.A., Zhang, Y., Yang, L., Wang, E., Marie, V., Lang, B. F., and Burger, G. (2006) GOBASE – a database of organelle and bacterial genome information. *Nucleic Acids Res* **34**, D697–D699.
10. Chuang, L.Y., Yang, C.H., Cheng, Y.H., Gu, D.L., Chang, P.L., Tsui, K.H., and Chang,

H.W. (2006) V-MitoSNP: visualization of human mitochondrial SNPs. *BMC Bioinformatics* 7, 379.

11. Tanaka, M., Takeyasu, T., Fuku, N., Li-Jun, G., and Kurata, M. (2004) Mitochondrial genome single nucleotide polymorphisms and their phenotypes in the Japanese. *Ann N Y Acad Sci* **1011**, 7–20.
12. Johnson, M., Zaretskaya, I., Raytselis, Y., Merezhuk, Y., McGinnis, S., and Madden, T.L. (2008) NCBI BLAST: a better web interface. *Nucleic Acids Res* **36**, W5–W9
13. Packer, B.R., Yeager, M., Burdett, L., Welch, R., Beerman, M., Qi, L., *et al.* (2006) SNP500Cancer: a public resource for sequence validation, assay development, and frequency analysis for genetic variation in candidate genes. *Nucleic Acids Res* **34**, D617–D621.
14. Kent, W.J. (2002) BLAT – the BLAST-like alignment tool. *Genome Res* **12**, 656–664.
15. Hinrichs, A.S., Karolchik, D., Baertsch, R., Barber, G.P., Bejerano, G., Clawson, H., *et al.* (2006) The UCSC Genome Browser Database: update 2006. *Nucleic Acids Res* **34**, D590–D598.
16. Behar, D.M., Rosset, S., Blue-Smith, J., Balanovsky, O., Tzur, S., Comas, D., *et al.* (2007) The Genographic Project public participation mitochondrial DNA database. *PLoS Genet* **3**, e104.
17. Brandstatter, A., Niederstatter, H., Pavlic, M., Grubwieser, P., and Parson, W. (2007) Generating population data for the EMPOP database – an overview of the mtDNA sequencing and data evaluation processes considering 273 Austrian control region sequences as example. *Forensic Sci Int* **166**, 164–175.
18. Ritchie, K., Myres, N.M., Angerhofer, N., Hughes, R., Ekins, J., Perego, U.A., and Woodward, S.R. (2008) The Sorenson Molecular Genealogy Foundation mtDNA Database. *URL:* http://www.smgf.org/pages/mtdatabase.jspx
19. Attimonelli, M., Catalano, D., Gissi, C., Grillo, G., Licciulli, F., Liuni, S., Santamaria, M., Pesole, G., and Saccone, C. (2002) MitoNuc: a database of nuclear genes coding for mitochondrial proteins. Update 2002. *Nucleic Acids Res* **30**, 172–173.
20. Lemkin, P. F., Chipperfield, M., Merril, C., and Zullo, S. (1996) A World Wide Web (WWW) server database engine for an organelle database, MitoDat. *Electrophoresis* **17**, 566–572.
21. Prokisch, H., and Ahting, U. (2007) MitoP2, an integrated database for mitochondrial proteins. *Methods Mol Biol* **372**, 573–586.
22. Guda, P., Subramaniam, S., and Guda, C. (2007) Mitoproteome: human heart mitochondrial protein sequence database. *Methods Mol Biol* **357**, 375–383.
23. Guda, C., Guda, P., Fahy, E., and Subramaniam, S. (2004) MITOPRED: a web server for the prediction of mitochondrial proteins. *Nucleic Acids Res* **32**, W372–W374.
24. Horton, P., Park, K.J., Obayashi, T., Fujita, N., Harada, H., Adams-Collier, C.J., and Nakai, K. (2007) WoLF PSORT: protein localization predictor. *Nucleic Acids Res* **35**, W585–W587.
25. Small, I., Peeters, N., Legeai, F., and Lurin, C. (2004) Predotar: a tool for rapidly screening proteomes for N-terminal targeting sequences. *Proteomics* **4**, 1581–1590.
26. Chen, H., Huang, N., and Sun, Z. (2006) SubLoc: a server/client suite for protein subcellular location based on SOAP. *Bioinformatics* **22**, 376–377.
27. Finn, R.D., Tate, J., Mistry, J., Coggill, P.C., Sammut, S.J., Hotz, H.R., *et al.* (2008) The Pfam protein families database. *Nucleic Acids Res* **36**, D281–D288.
28. Yu, X., Koczan, D., Sulonen, A.M., Akkad, D.A., Kroner, A., Comabella, M., *et al.* (2008) mtDNA nt13708A variant increases the risk of multiple sclerosis. *PLoS One* **3**, e1530.
29. Brandon, M.C., Lott, M.T., Nguyen, K.C., Spolim, S., Navathe, S.B., Baldi, P., and Wallace, D.C. (2005) MITOMAP: a human mitochondrial genome database – 2004 update. *Nucleic Acids Res* **33**, D611–D613.
30. Sherry, S.T., Ward, M.H., Kholodov, M., Baker, J., Phan, L., Smigielski, E.M., and Sirotkin, K. (2001) dbSNP: the NCBI database of genetic variation. *Nucleic Acids Res* **29**, 308–311.
31. Roberts, R.J., Vincze, T., Posfai, J., and Macelis, D. (2005) REBASE – restriction enzymes and DNA methyltransferases. *Nucleic Acids Res* **33 Database Issue**, D230–D232.
32. Chang, H.W., Yang, C.H., Chang, P.L., Cheng, Y.H., and Chuang, L.Y. (2006) SNP-RFLPing: restriction enzyme mining for SNPs in genomes. *BMC Genomics* **7**, 30.
33. Monson, K.L., Miller, K.W.P., Wilson, M.R., DiZinno, J.A., and Budowle, B. (2002) The mtDNA population database: an integrated software and database resource for forensic comparison. *Forensic Sci Commun* **4**, April

Chapter 15

Web-Based Analysis of (Epi-) Genome Data Using EpiGRAPH and Galaxy

Christoph Bock, Greg Von Kuster, Konstantin Halachev, James Taylor, Anton Nekrutenko, and Thomas Lengauer

Abstract

Modern life sciences are becoming increasingly data intensive, posing a significant challenge for most researchers and shifting the bottleneck of scientific discovery from data generation to data analysis. As a result, progress in genome research is increasingly impeded by bioinformatic hurdles. A new generation of powerful and easy-to-use genome analysis tools has been developed to address this issue, enabling biologists to perform complex bioinformatic analyses online – without having to learn a programming language or downloading and manually processing large datasets. In this tutorial paper, we describe the use of EpiGRAPH (http://epigraph.mpi-inf.mpg.de/) and Galaxy (http://galaxyproject.org/) for genome and epigenome analysis, and we illustrate how these two web services work together to identify epigenetic modifications that are characteristics of highly polymorphic (SNP-rich) promoters. This paper is supplemented with video tutorials (http://tinyurl.com/yc5xkqq), which provide a step-by-step guide through each example analysis.

Key words: Bioinformatics, Genome analysis, Statistics, Machine learning, Computational epigenetics, Single nucleotide polymorphisms (SNPs), Evolutionary constraint

1. Introduction

Vertebrate gene expression is regulated at several levels of control, which are tightly interlinked with each other (1, 2). The key mechanism of DNA-based "genetic regulation" is transcription factor binding to sequence-specific recognition motifs, which are commonly located in promoter and enhancer regions (3). In contrast, chromatin-based "epigenetic regulation" comprises gene-regulatory mechanisms that are not directly controlled by the DNA sequence, such as chromatin condensation across an

Michael R. Barnes and Gerome Breen (eds.), *Genetic Variation: Methods and Protocols*, Methods in Molecular Biology, vol. 628, DOI 10.1007/978-1-60327-367-1_15, © Springer Science+Business Media, LLC 2010

entire gene cluster (4). Variation in genetic and epigenetic gene regulation plays a major role in common diseases (5) and contributes to inter-individual differences in gene expression observed among healthy individuals (6–8). With the recent development of high-throughput protocols such as ChIP-on-chip and ChIP-seq (9), it is now possible to analyze genome-epigenome interactions and their impact on gene expression at a truly genomic scale.

However, such genome-wide analyses pose significant bioinformatic challenges. The goal of this paper is to illustrate the use of a new generation of web toolkits that enable biologists to perform complex (epi-) genome analyses online, without having to learn a programming language or to download large datasets onto local computers. Specific focus will be put on a specialized tool for statistical analysis and prediction of (epi-) genome data, EpiGRAPH, and on a general-purpose platform for manipulating large sets of genomic regions, Galaxy. The EpiGRAPH web service (10) provides a standardized workflow for identifying characteristic DNA attributes that are enriched in a given set of genomic regions, and for predicting similar regions across mammalian genomes. It is particularly useful for explorative analysis and bioinformatic prediction, as is evident from applications to DNA methylation data (8, 11), DNA melting profiles (12), CpG island annotation (13), and SNP function inference (14). Compared to EpiGRAPH's focus on a specific task, the Galaxy web service (15, 16) is a general-purpose tool for processing any set of genomic regions. It provides simple and straightforward methods to join, merge, and intersect genomic regions, to map between formats, genome assemblies and species, and to perform basic statistical analyses. In addition, it provides a user interface for more specialized toolkits such as HyPhy (17), EMBOSS, (18) and EpiGRAPH. Here, we will illustrate the synergistic potential of EpiGRAPH and Galaxy for analyzing genome and epigenome datasets.

The remainder of the paper is structured as follows. First, the Materials section highlights the technical prerequisites for using EpiGRAPH and Galaxy; second, we give an overview of available software tools that facilitate the analysis of epigenome datasets; third, we introduce EpiGRAPH by a simple case study on DNA methylation analysis and prediction; fourth, we outline the use of Galaxy for performing calculations on sets of genomic regions; fifth, we describe an advanced case study that uses both EpiGRAPH and Galaxy in order to identify genomic and epigenomic characteristics that distinguish highly polymorphic promoter regions from their non-polymorphic counterparts. Finally, in the Notes section, we briefly comment on practical issues and highlight potential pitfalls of the methods that are outlined in this paper.

2. Materials

The user must have access to a computer with Internet access, which could for example be a PC running Microsoft Windows, an Apple computer running MacOS, or a UNIX workstation. Galaxy and EpiGRAPH are web toolkits and operated via a web browser, therefore it is important to have a sufficiently up-to-date web browser installed. Both toolkits have been tested with current versions of Mozilla Firefox (http://www.firefox.com/), Microsoft Internet Explorer (http://www.microsoft.com/ie/), Apple Safari (http://www.apple.com/de/safari/), KDE Konqueror (http://www.konqueror.org/) and Opera (http://www.opera.com/). Furthermore, the user should make sure that JavaScript and web browser cookies are enabled, since EpiGRAPH cannot be used without JavaScript, while Galaxy's user-friendliness will be reduced when JavaScript is switched off.

Beyond the essential web browser, it is recommended to install an advanced text editor, e.g., EMACS (http://www.gnu.org/software/emacs/) or Programmer's Notepad (http://www.pnotepad.org/), which can handle large files effectively and which simplifies any data formatting tasks that may be required for a given dataset. Similarly, it is often helpful to copy and paste datasets into a spreadsheet software such as Microsoft Excel (http://office.microsoft.com/en-us/excel/) or OpenOffice.org Calc (http://www.openoffice.org/product/calc.html), which facilitates adding, removing, or rearranging columns in table-style datasets. For advanced users who want to perform data preparation steps or follow-up analyses in a simple programming environment, it is also useful to install the R statistics software (http://www.r-project.org/).

Finally, to be able to view the tutorial videos that accompany this paper, the Macromedia Flash Player and Apple QuickTime browser plug-ins are required, which can be freely downloaded from http://www.adobe.com/products/flashplayer/ and http://www.apple.com/quicktime/download/, respectively.

3. Methods

3.1. A Workflow for Epigenome Data Analysis Using Web-Based Tools

The analysis of epigenome datasets is often performed in four subsequent steps, as outlined in Fig. 1. First, depending on the experimental method used to acquire the raw data, it is usually necessary to perform specific data normalization and quality control steps, before a reliable set of enriched genomic regions can be derived. Second, visual inspection of the processed dataset

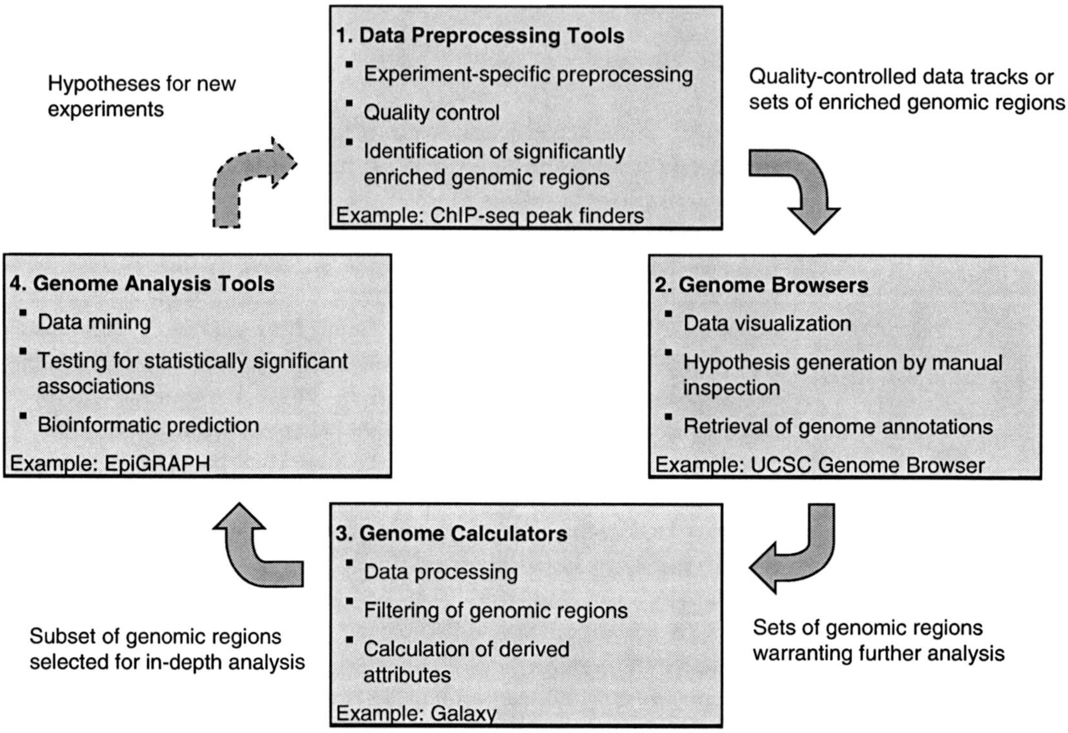

Fig. 1. Workflow for web-based analysis of epigenome datasets. This figure outlines a workflow for epigenome data analysis using publicly available tools and web services. After data preprocessing with software tools that address the specific properties of the experimental method used (*box 1*), the user uploads the newly generated dataset into a genome browser, in order to facilitate visualization and hypothesis generation by manual inspection (*box 2*). Next, he or she processes the data with a genome calculator such as Galaxy, in order to extract and prepare interesting regions for in-depth analysis (*box 3*). Finally, genome analysis tools such as EpiGRAPH can be used to test for significant associations with genome annotation data and to perform bioinformatic prediction (*box 4*), which might result in ideas for new experiments – driving the next iteration of the analytical circle

provides a starting point for data analysis and often gives rise to biological hypotheses that can subsequently be tested with more quantitative methods. Third, extensive data processing is often necessary in order to identify and extract a set of genomic regions that are relevant for a specific hypothesis. Fourth, statistical methods enable researchers to rigorously test the validity of a given hypothesis, and exploratory data mining can be used to identify as yet unknown associations of the input dataset with other genomic and epigenomic attributes. The data analysis step often gives rise to new hypotheses that can form the starting point for further experiments and the next iteration of the analytical circle.

The third and fourth steps of this analysis workflow are addressed by Galaxy and EpiGRAPH, respectively, and are discussed in more detail in subsequent sections of this paper. In the current section, we briefly highlight key software toolkits that contribute to the first two steps.

Experimental methods for epigenome mapping – including ChIP-on-chip (19), ChIP-seq (9), and DNA methylation analysis by bisulfite sequencing (1) – require a significant amount of data preprocessing and quality control (Step 1 in Fig. 1), which is addressed by specific software toolkits (reviewed in (20)). For ChIP-on-chip, data preprocessing starts with microarray data normalization, which is often performed with either the Bioconductor package (21) inside the R statistics software (http://www.r-project.org/) or a vendor-supplied tool. For ChIP-seq, the equivalent preprocessing step involves tag mapping to the genome assembly, which can be achieved using specialized BLAST-like tools such as Maq (http://maq.sourceforge.net/) or Bowtie (http://bowtie-bio.sourceforge.net/) For both ChIP-on-chip and ChIP-seq, data preprocessing results in genome-scale profiles of over-representation scores, which can be visualized as quantitative tracks inside genome browsers. However, because such profiles often carry significant levels of biological and technical noise, it is in most cases advisable to perform peak-detection on these profiles, i.e., to identify sets of genomic regions that are enriched with high confidence (22). A recent benchmarking study of several peak-detection methods suggests that vendor-supplied tools perform sufficiently well (23). Further tools that – in our opinion – provide a good balance between accuracy and user-friendliness are the web-based Splitter toolkit (http://zlab.bu.edu/splitter/) for NimbleGen and Agilent microarrays as well as the stand-alone MAT software (24) for Affymetrix microarrays. For DNA methylation analysis, two experimental strategies are widely used. Antibody-based methods such as MeDIP-chip and MeDIP-seq give rise to similar bioinformatic issues as ChIP-on-chip or ChIP-seq, which can be addressed with the same toolkits. In contrast, DNA methylation analysis by bisulfite sequencing requires dedicated software. The QUMA web service (25) provides a quick web-based solution for the analysis of clonal bisulfite sequencing data. In contrast, the BiQ Analyzer software (26) incorporates more extensive features for quality control and experiment documentation, but requires the user to download and install a small software tool.

Upon completion of data preprocessing, the logical next step is data visualization and initial manual inspection (step 2 in Fig. 1). This task is usually performed by uploading a preprocessed set of enriched genomic regions into a genome browser, from which it can be viewed and visually compared with other genome annotation data. To that end, a preprocessed set of enriched genomic regions is converted into the BED format (http://genome.ucsc.edu/FAQ/FAQformat.html#format1), which usually requires some reformatting that can be done by search and replace in a text editor, by grid-based processing in a spreadsheet software, by script-based processing with R (http://www.r-project.org/) or

Python (http://www.python.org/), or by a combination of these alternatives (see Note 1 for details). Next, the BED file has to be uploaded to a web server directory that is freely accessible from the internet and from which the genome browser can retrieve the dataset. This step requires write permission on a web server, or a public one-click web hosting service can be used (http://en.wikipedia.org/wiki/One-click_hosting). Alternatively, it is possible to upload the BED file directly to the UCSC Genome Browser, but this solution is less convenient and quickly reaches its limits when files become large. Finally, the URL(s) of the uploaded BED file(s) can be submitted to either the UCSC Genome Browser (27) or to Ensembl (28), which will then retrieve the dataset and visualize it alongside their default genome annotations. A more detailed description of the submission process and visualization options is available from the UCSC Genome Browser website (http://genome.ucsc.edu/goldenPath/help/hgTracksHelp.html#CustomTracks) and from the Ensembl website (http://www.ensembl.org/info/website/upload/index.html).

3.2. Predicting DNA Methylation: An Introduction Into EpiGRAPH (Supplemented by EpiGRAPH Video Tutorials 1 and 2)

DNA methylation is the only epigenetic modification that directly affects the DNA sequence, and it has been shown to correlate with specific aspects of the genomic DNA sequence, including DNA sequence patterns, structural properties of the DNA, and the distribution of repetitive DNA elements in the human genome (11, 13, 29, 30). For these reasons, DNA methylation is an interesting target for integrative genome and epigenome analysis using the EpiGRAPH web service. In the following case study, we demonstrate the use of EpiGRAPH for analyzing and predicting the DNA methylation status of CpG islands, essentially replicating the core bioinformatic analysis of a recent paper on DNA methylation prediction (11).

To make this case study as hassle-free as possible, all required data and settings are already preconfigured in the EpiGRAPH web service, and two video tutorials demonstrating the details of each step are available from EpiGRAPH's Background page (http://epigraph.mpi-inf.mpg.de/WebGRAPH/faces/Background.html#tutorial).

1. *Creating an account and logging into the EpiGRAPH web service.* EpiGRAPH's start page is available at http://epigraph.mpi-inf.mpg.de/, providing a brief summary of the web service and some suggestions for biologically relevant topics that can be addressed using EpiGRAPH. A click on the "Start EpiGRAPH" link brings us to the login page, which contains EpiGRAPH-related announcements as well as links to important background material (such as video tutorials and a documentation of EpiGRAPH's default attributes).

Clicking the "Register" button displays a standard registration page, and successful registration logs us into the EpiGRAPH web service. Alternatively, for getting a quick impression of EpiGRAPH, a guest account can be created by clicking the "Be a Guest" button.

2. *Specifying and launching an EpiGRAPH analysis on DNA methylation data.* Before starting an analysis, the first step on EpiGRAPH's overview page has to be the selection of the genome assembly to work on, using the choice box on the right of the page (underneath the EpiGRAPH logo). After selecting human genome assembly "hg18", we can click the "Define new analysis using this website" button, upon which EpiGRAPH will guide us through a three-step process specifying and launching a new EpiGRAPH analysis.

   On the first page, we upload a set of genomic regions to be used as input dataset for the EpiGRAPH analysis (Fig. 2). In this case study, a suitable dataset can be obtained simply by

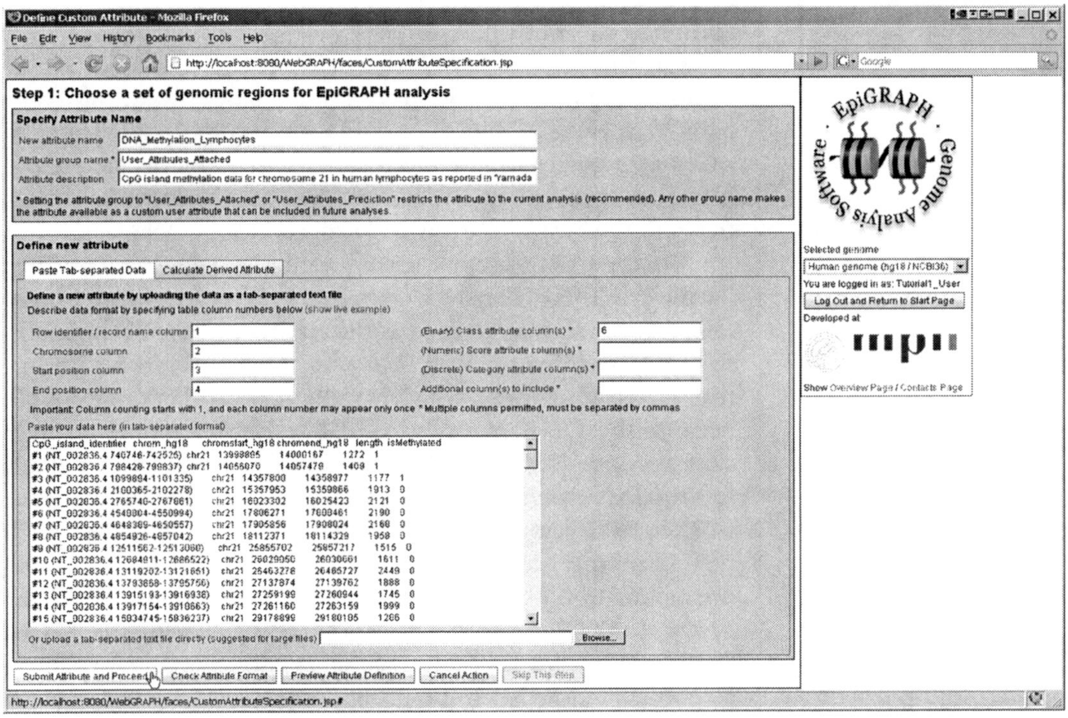

Fig. 2. Submitting a custom dataset for analysis with EpiGRAPH. This screenshot displays EpiGRAPH's attribute submission page, consisting of a brief attribute documentation (*top*), a set of text fields in which the column semantics are specified (e.g., which column contains the chromosome name and the start and end position for each genomic region) and a large text area into which a tab-separated table of genomic regions can be pasted. Due to different column widths, the columns of the table are not properly aligned, which is often the case and will not cause any problems. Importantly, each row in the table must correspond to exactly one genomic region, and its location in terms of chromosome name, start position and end position must be specified relative to the genome assembly selected in the choice box below the EpiGRAPH logo on the right of the screen ("hg18" in this case)

clicking the "Show live example" link. This dataset is in tab-separated format, containing one genomic region per row and mandatory columns for chromosome name (e.g., "chr21") as well as genomic start and end position (e.g., "13998895" and "14000167"). Two nonmandatory columns – a unique row identifier (first column) and a binary class attribute (last column) – are also included. The class attribute specifies whether or not the respective genomic region is methylated, based on an experimental analysis of DNA methylation on chromosome 21 (31). The input dataset is usually copied and pasted from a text editor or a spreadsheet software into the upload page's text area (see Note 1 for details on data preparation), and the content of each column (i.e., whether it contains the chromosome name, chromosome start or end position, or additional information) is specified by entering column names or column numbers into the corresponding text fields (as illustrated by the default entries made when clicking the "Show live example link"). In order to continue, we press the "Submit attribute and proceed" button.

On the second page, we could specify a control set of genomic regions to which our input dataset should be compared (see Note 2), but since the input dataset already contains two types of regions – methylated and unmethylated CpG islands as specified by the binary class column – we can press the "Skip this step" button and proceed to the next step.

On the third page, we specify a number of general settings for the EpiGRAPH analysis (Fig. 3): (1) We select which binary class column should be used as the target attribute of the analysis (i.e., for differentiation between positives/cases and negatives/control regions), which is straightforward in our example because the DNA methylation dataset includes only a single class column ("isMethylated"). (2) We confirm the default settings for down-sampling, a parameter that is important when working with large datasets (see Note 4). (3) We select which (epi-) genomic attributes to be included in the analysis. (4) For documentation purposes, we provide a title and a brief textual description of the analysis. In this case study, clicking the "Show live example" link will fill in all fields with appropriate values. In particular, four attribute groups are selected for inclusion in the analysis: all DNA sequence patterns of size two, several aspects of the predicted DNA structure, the overlap with repetitive DNA elements, and the overlap with annotated genes (better prediction accuracies at the expense of longer calculation time can be achieved by selecting all available attribute groups – see Note 6 for discussion). Finally, we click the "Start analysis" button and a confirmation page appears, indicating that the EpiGRAPH analysis has been started successfully.

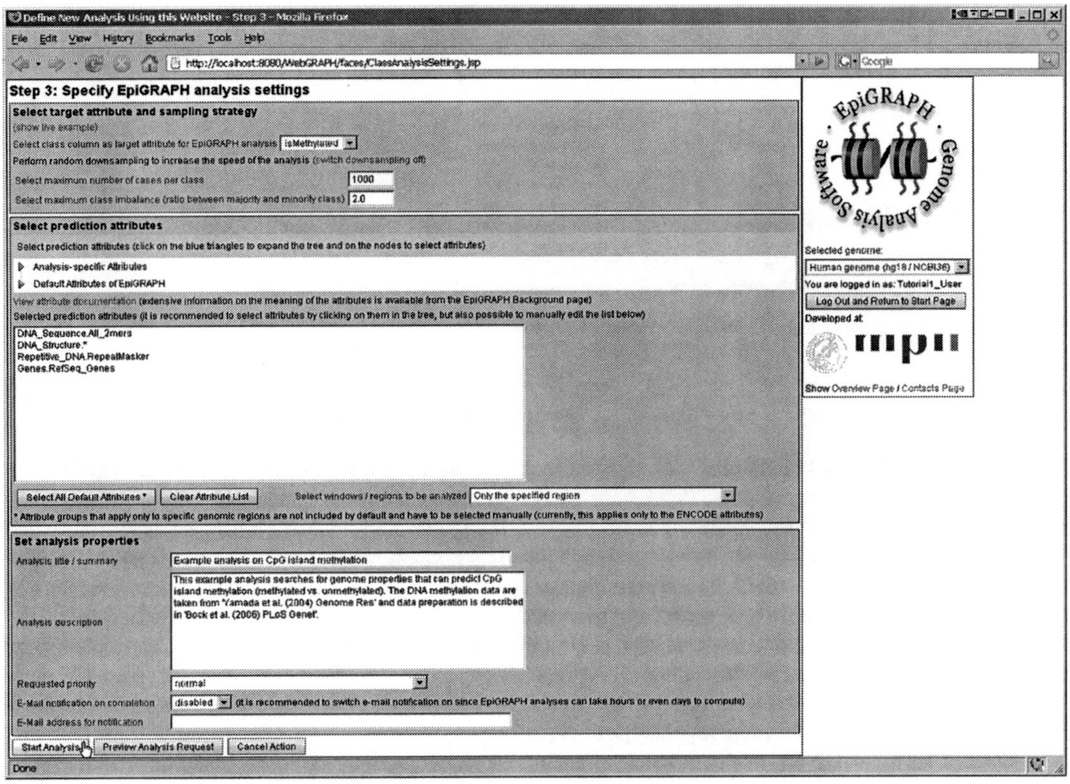

Fig. 3. Configuring and starting an EpiGRAPH analysis. This screenshot displays EpiGRAPH's analysis specification page. Here, the user can select which class attribute to use (if more than one class attribute was provided during the attribute submission steps), configure down-sampling, select prediction attributes, and enter a brief documentation of the analysis

3. *Interpreting the results of the EpiGRAPH analysis.* Returning to EpiGRAPH's overview page, the newly started analysis appears in the table of stored analyses at the bottom of the page, and its status is indicated as "queued" or "running". Clicking on the corresponding "Access" button opens the results overview page displaying the progress of the analysis. We wait a few minutes to give EpiGRAPH time to calculate the requested analysis and then press the "Refresh results" link at the top of the page, whereupon EpiGRAPH updates the results overview with a summary of all completed analyses.

   Interpreting these results, we first take a look at the outcome of the statistical analysis (Fig. 4). It highlights attributes that differ significantly between the sets of methylated CpG islands (class = 1) and unmethylated CpG islands (class = 0), according to pairwise statistical testing. Among the most significant genomic attributes are the frequencies of the DNA sequence patterns "CA" (over-represented in methylated CpG islands) and "CG" (over-represented in unmethylated CpG islands), a result that is consistent with current knowledge (11).

A. Statistical analysis comparing methylated and unmethylated CpG islands on chromosome 21

id	var name	att name	group name	P-val raw	sig bonf	sig fdr	mean class=0	mean class=1	method	select
1	Pat_CA_freq	All_2mers	DNA_Sequence	2.01e-11	Yes	Yes	0.113	0.156	wilcox	☐
2	rise	Predicted_Helix_Structure	DNA_Structure	5.32e-09	Yes	Yes	3.236	3.246	wilcox	☐
3	Pat_GC_std	All_2mers	DNA_Sequence	8.39e-08	Yes	Yes	0.056	0.042	wilcox	☐
4	Pat_AC_freq	All_2mers	DNA_Sequence	3.70e-07	Yes	Yes	0.086	0.116	wilcox	☐
5	Pat_CC_std	All_2mers	DNA_Sequence	3.81e-07	Yes	Yes	0.101	0.077	wilcox	☐
6	Pat_CG_freq	All_2mers	DNA_Sequence	5.23e-07	Yes	Yes	0.082	0.065	wilcox	☐
7	Pat_CG_std	All_2mers	DNA_Sequence	5.38e-07	Yes	Yes	0.055	0.043	wilcox	☐
8	repClass_Low_complexity_overlapTotalLength	RepeatMasker	Repetitive_DNA	9.80e-05	Yes	Yes	41.58	8.266	wilcox	☐
9	repFamily_Low_complexity_overlapTotalLength	RepeatMasker	Repetitive_DNA	9.80e-05	Yes	Yes	41.58	8.266	wilcox	☐
10	pk2_mean_std	Predicted_Solvent_Accessible_Surface	DNA_Structure	1.99e-04	Yes	Yes	0.312	0.303	wilcox	☐

B. Machine learning analysis predicting the DNA methylation status of CpG islands on chromosome 21

run	group name	#vars	prediction method	mean corr	corr sd	mean acc	acc sd	sens	spec	#cases
1	DNA_Sequence	30	svm_linear	0.484	0.056	0.777	0.025	0.621	0.852	43
2	DNA_Structure	24	svm_linear	0.418	0.095	0.756	0.040	0.514	0.872	43
3	Repetitive_DNA	81	svm_linear	0.350	0.058	0.740	0.018	0.343	0.931	43
4	Genes	10	svm_linear	0.029	0.114	0.653	0.032	0.093	0.924	43
5	DNA_Sequence+DNA_Structure+Repetitive_DNA+Genes	145	svm_linear	0.493	0.086	0.772	0.041	0.686	0.814	43

Fig. 4. Results of an EpiGRAPH analysis of DNA methylation at CpG islands. These screenshots display the results of an EpiGRAPH analysis comparing methylated CpG islands (class = 1) with unmethylated CpG islands (class = 0), based on a published dataset of DNA methylation on chromosome 21 (31). The results of the statistical analysis (*Panel A*) show that the "CG" sequence pattern is over-represented in unmethylated CpG islands, while the "CA" sequence pattern is over-represented in methylated CpG islands. Statistical testing was performed using the nonparametric Wilcoxon rank-sum test and *P*-values were adjusted for multiple testing using the highly conservative Bonferroni method (sig bonf) as well as the false discovery rate method (sig fdr). An explanation of the attribute names is available from http://epigraph.mpi-inf.mpg.de/WebGRAPH/faces/Background.html#attributes. The machine learning analysis (*Panel B*) confirms that these and other differences are sufficient to predict with relatively high accuracy whether or not a CpG island is methylated. The values in the bottom table correspond to the average performance of a linear support vector machine that was trained and evaluated in ten repetitions of a tenfold cross-validation, summarized by the mean correlation (mean corr), prediction accuracy (mean acc), sensitivity (sens), and specificity (spec). Additional columns display standard deviations observed among the repeated cross-validations with random partition assignment (corr sd and acc sd), the number of attribute variables in each attribute group (#vars), and the total number of genomic regions included in the analysis (#cases)

These differences can be visualized as boxplot diagrams by ticking the corresponding boxes in the "Select" column and pressing the "Calculate selected diagrams" button.

The boxplot diagrams – which appear on the results overview page after pressing the "Refresh results" link – provide an indication of the quantitative strength of association between these DNA sequence patterns and the DNA methylation status.

Further evidence that this association is not only significant but also relatively high in quantitative terms comes from the results of the machine learning analysis (see Note 5 for some background on machine learning). According to the performance evaluation table (Fig. 5), a support vector machine (32) is able to predict with an accuracy of 78% and a binary correlation coefficient of 0.5 whether or not a CpG island is methylated, based on the combination of all attribute groups that we selected when starting the analysis.

Note that this result provides important additional information beyond the *P*-values of the statistical analysis, for two reasons: First, correlation coefficients can be used as indicators of the quantitative strength of association, while *P*-value only assess the presence or absence of a statistically significant association

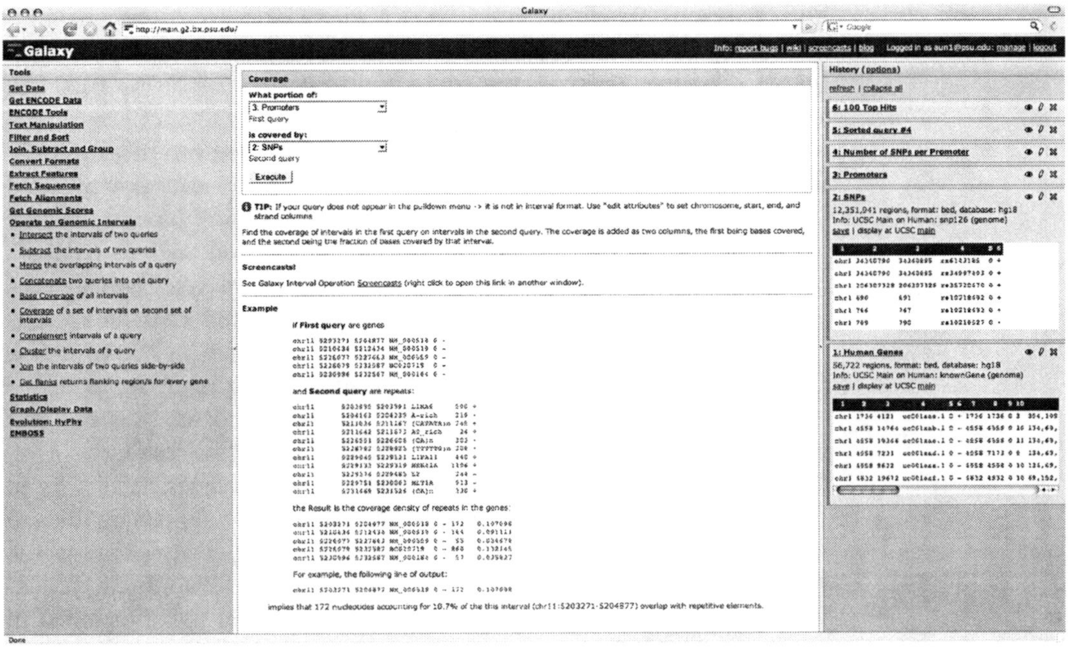

Fig. 5. Identification of highly polymorphic promoters using Galaxy. The Galaxy web interface consists of four areas: the upper bar, tool frame (*left column*), detail frame (*middle column*), and history frame (*right column*). The upper bar contains user account controls as well as help and contact links. The tool frame on the left lists the analysis tools and data sources available to the user. The middle frame displays the interface of the currently selected tool. The history frame on the right shows loaded datasets and results of analyses performed by the user. Pictured here are six history items representing two original datasets (1: Human Genes and 2: SNP) and results of their manipulations. Every action by the user generates a new history item, which can then be used in subsequent analyses, downloaded, or visualized

(*P*-values can be low even for small differences that hardly stand a chance of playing a biological role, under the condition that the differences are systematic and the sample size is large). Second, the machine learning analysis can quantify the collective predictiveness, or correlation, of an entire group of attribute (e.g., of all DNA sequence patterns of size two), while the statistical analysis treats all attributes separately.

After an initial inspection of the results, it is a good idea to save the completed analysis for documentation and further reference. To that end, we click the "Download XML documentation" button on the results overview page and save the XML documentation file to the local hard disk. This file constitutes a comprehensive account of the analysis settings and of all completed results, providing a suitable basis for sharing an EpiGRAPH analysis with colleagues (e.g., by including it in the supplementary material of a paper).

4. *Performing follow-up prediction based on a documented EpiGRAPH analysis.* For the sake of argument, let us assume that we obtained the XML documentation file saved at the end of step 3 from the supplementary material of a published paper on DNA methylation prediction and that we want to use its

results for predicting the DNA methylation status of a new list of CpG islands (see Note 6 for limitations of this approach). To that end, we return to EpiGRAPH's overview page (to make it more realistic, we could also log in as a different user) and click the button "Execute analysis based on existing XML file". On the next page, we select the previously downloaded XML documentation file using the "Browse" button, change the settings to "Retain previously calculated analysis results", and click the "Upload XML file and start analysis" button. As a result, the analysis documented in the uploaded XML file appears in the table of stored analyses at the bottom of the overview page. Note that the status of the analysis is already set to "completed", as we have uploaded a completed analysis and not requested EpiGRAPH to recalculate any of its results.

Clicking the corresponding "Access" button brings us to the results overview page, from where we could restart the statistical analysis and the machine learning analysis using the "Modify settings and recalculate" buttons, for example, reducing the number of (epi-) genomic attributes to be included in the analysis, setting a new *P*-value threshold, or selecting additional machine learning methods. However, we concentrate on the prediction analysis at the bottom of the page, clicking the "Start new prediction" button. On the next page, we upload a tab-separated table containing the genomic regions for which we want to predict the DNA methylation status (this table can be obtained from (http://epigraph.mpi-inf.mpg.de/WebGRAPH/faces/Background.html#tutorial). The table comprises the top-25% most methylated as well as the top-25% most unmethylated promoter regions from a recent study applying bisulfite sequencing to all promoter regions on chromosome 21 (33). The experimentally determined DNA methylation status of each region is provided in the table's "isMethylated" column.

Clicking the "Submit attribute and proceed" button brings us to a page on which we select all available attributes to be included in the prediction, and we specify that they should be used both separately and in combination (option five in the dropdown box). Next, we click the "Start prediction analysis with these settings" button, upon which EpiGRAPH will predict the DNA methylation status of all CpG islands in the new dataset, using a support vector machine trained on the input dataset originally uploaded in step 1. Furthermore, because we included a class column specifying an experimentally determined DNA methylation status, EpiGRAPH regards the new dataset as an independent test set and calculates several performance evaluation measures. We return to the results overview page and, after a few minutes, press the "Refresh results" link, prompting EpiGRAPH to update the results overview with a summary of the completed prediction analysis. The performance evaluation table indicates that the support vector machine accurately predicts DNA methylation status in a set of unseen genomic regions, for which the experimental DNA methylation status has been determined with a different experimental method and in a different lab. Finally, clicking the "Download cases list" button retrieves a tab-separated table containing individual DNA methylation predictions for each genomic region in the test set.

3.3. Genomics Analysis Using Galaxy (Supplemented by Galaxy Screencast "Promoters and SNPs")

Galaxy (http://galaxyproject.org/) provides a computational framework that addresses two key challenges of genome analysis, simplicity, and reproducibility. It enables bench researchers to rapidly access and analyze enormous datasets without installing or configuring any software. For software engineers and computational scientists, it provides a zero-configuration development framework that will immediately connect novel or existing analysis tool with their intended target audience – researchers.

(1) *A typical task in genomics: Identifying highly polymorphic promoters.* The utility of Galaxy is best illustrated by an example. A researcher wants to find human promoters showing evidence of adaptive evolution or relaxation of selective constraint. Such promoters are potentially interesting as they may point to genetic causes of human-specific gene expression. As single nucleotide polymorphisms (SNPs) are the most common source of genomic variation among the human population (34), a straightforward approach is to select the promoters that exhibit the highest density of SNPs. Such an analysis would involve the following steps:

   (a) Obtain gene and SNP annotations for the human genome from the UCSC Table Browser
   (b) Transform gene annotations into positions of potential promoters by selecting the region located 500 base pairs upstream of each gene's transcription start site
   (c) Calculate the intersection between each putative promoter and all SNPs
   (d) Compute the density of SNPs for each promoter region
   (e) Visualize the genomic vicinity of the ten promoters with highest SNP density.

Only the first and last steps can be performed using current genome browsers, while the researcher must find or build a custom solution to perform steps b through d. For most experimentalists, this presents a formidable barrier, preventing them from making effective use of existing datasets. Indeed, coordinates of SNPs are available from the UCSC Table Browser, but this dataset is enormous (millions of data points cannot be loaded into a desktop spreadsheet application) and effectively unusable by experimentalists who lack computational expertise or bioinformatic support. While designing

Galaxy we sought to enable experimentalists to perform such analysis without the need to install or configure anything.

(2) *Using Galaxy to identify highly polymorphic promoters.* Consider again the example of looking for human promoters showing evidence of adaptive evolution or relaxation of constraint. Usually, the initial step of such an analysis would involve downloading the coordinates of all genes and SNPs in the human genome onto one's personal computer. Next, the user would upload these data to an appropriate analysis tool (provided that it can handle this amount of data). Obviously, this procedure is inconvenient and often infeasible, once more highlighting the fundamental difficulty faced by experimental biologists every day: one first needs to download huge datasets (450 MB in the case of all human SNPs) and then re-uploads the same data to another Internet-based resource (if a suitable web service exists that can perform the analysis online) or install software that can perform the analysis on the local computer. It is much more efficient and practical to implement direct connections between analysis tools and data warehouses, which is what Galaxy does.

Here, we show how one can perform the search for rapidly evolving promoters using Galaxy (Fig. 5). First, we load coordinates of all human RefSeq genes (a conservative set of gene annotation) and SNPs (dbSNP release 126) into Galaxy using its direct connection to the UCSC Table Browser. Next, we transform coordinates of genes into coordinates of potential promoter regions by taking 500 base pairs immediate upstream on each gene's start. We use the coverage tool from Galaxy's "Operate on genomic intervals" tool category to compute the number of SNPs residing in each of the promoters we generated during the previous step. Finally, we use the sort tool and select 100 promoters with the highest number of SNPs. Figure 6 illustrates how all steps of this analysis are documented in Galaxy's history frame. The history starts from the datasets uploaded from UCSC Genome Browser, which are represented by the first two history items ("1: Human Genes" and "2: SNPs"). A detailed demonstration of this analysis is available as Galaxy screencast "Promoters and SNPs" on the following website: http://galaxyproject.org/screencasts.html.

3.4. An Advanced Case Study Combining the Use of Galaxy and EpiGRAPH (Supplemented by EpiGRAPH Video Tutorial 3)

In the following case study, we compare the genetic and epigenetic characteristics of highly polymorphic promoter regions with a control set of promoter regions that contain no more than a single SNP within the kilobase region upstream of the transcription start site of annotated genes. This case study is more complex than the previous two, making use of both Galaxy and EpiGRAPH to address a real-world biological question.

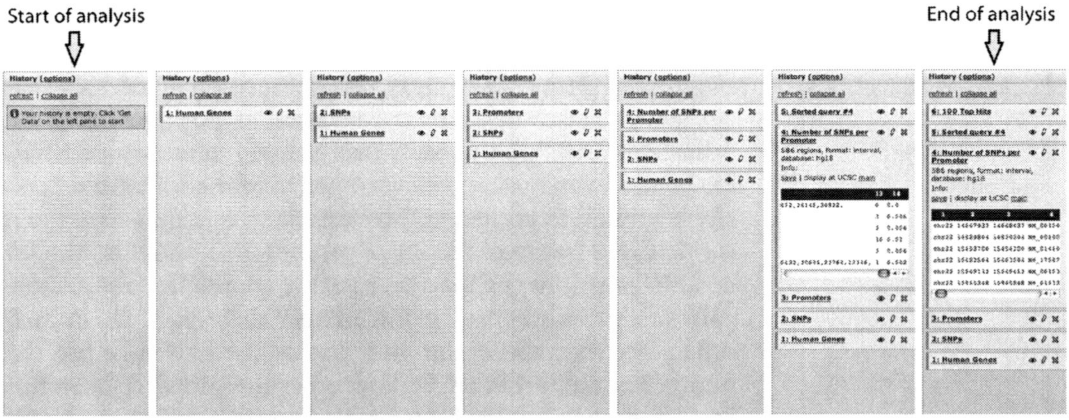

Fig. 6. Documentation of an analysis using Galaxy's history function. All actions performed within Galaxy are documented in the history frame, which contains uploaded data as well as calculated results. Original datasets are always preserved, and every subsequent analysis adds a new entry into the history frame. This screenshot illustrates how a user starts with an empty history, adds a dataset containing coordinates of human genes and SNPs, converts coordinates of genes into coordinates of promoter regions by selecting the region located 500 base pairs upstream of each gene, computes the number of SNPs per promoter, sorts the promoters by SNP density, and finally selects 100 top regions. In addition to documenting analyses, Galaxy's history frame allows the user to share a history with colleagues

Therefore, the following description has to focus on the main concepts, while we refer to video tutorial 3 from EpiGRAPH's Backgroundpage(http://epigraph.mpi-inf.mpg.de/WebGRAPH/faces/Background.html#tutorial) for a step-by-step guide.

(1) *Loading SNP and promoter region data into Galaxy.* Before we can use EpiGRAPH to identify genomic and epigenomic differences between polymorphic and nonpolymorphic promoter regions, it is necessary to derive suitable lists of positives/cases and negatives/control regions. As outlined in the previous section, the Galaxy web service provides us with a convenient solution for performing the necessary calculations online. In the first step, we load two sets of genomic regions from the UCSC Genome Browser into Galaxy, namely the genomic coordinates of all SNPs from the dbSNP database and the putative promoter regions of RefSeq-annotated genes (for practical reasons, the latter was defined as the kilobase region upstream of the annotated transcription start site). To increase the speed of calculation, we limit our analysis to a 1% subset of the human genome known as the ENCODE regions (35), although it would well be possible to perform the same analysis genome-wide. The video tutorial demonstrates two different ways in which data can be loaded into Galaxy, the first one being initiated from the UCSC Genome Browser and the second one initiated from within Galaxy (the effect of both methods is identical: the dataset becomes available for further processing in Galaxy).

(2) *Using Galaxy to derive sets of highly polymorphic and nonpolymorphic promoter regions.* Inside Galaxy, it is now possible to derive two sets of promoter regions, one comprising all regions that contain at least ten SNPs (to be used as positives) and the other comprising all regions that contain zero or one SNPs (to be used as negatives). Several generic functions are successively applied to complete this task. First, a region-based join is calculated between the set of promoter regions and the set of SNP positions, giving rise to a list containing all possible pairs of a promoter region and an SNP that overlap with each other. Second, the count function is used to quantify the number of times that a specific promoter region occurs in this list. Third, the resulting list is filtered according to the minimum or maximum SNP threshold (one and ten), respectively. Fourth, an identifier-based join with the original list of promoter regions is performed in order to recover the positional information that was lost during the counting step.

(3) *Specifying and launching an EpiGRAPH analysis on polymorphic promoter data.* Upon completion of the Galaxy analysis, we copy the resulting tab-separated tables of positives (highly polymorphic promoter regions) and negatives (nonpolymorphic promoter regions) into a spreadsheet and change a few column names for better readability, before pasting the data into a new EpiGRAPH analysis. The EpiGRAPH analysis is created as described in the first case study (see above), with two major differences. First, the sets of positives and negatives are uploaded as two different datasets on the first and second page of EpiGRAPH's analysis specification workflow, rather than being combined in a single input dataset and distinguished by a binary class column. Second, the current analysis includes all ten attribute groups that are available by default for the human genome assembly hg18, namely: (1) DNA sequence, (2) DNA structure, (3) repetitive DNA, (4) chromosome organization, (5) evolutionary history, (6) population variation, (7) genes, (8) regulatory regions, (9) transcriptome, and (10) epigenome and chromatin structure. As a result, calculation by EpiGRAPH takes substantially longer and it is highly recommended to switch on e-mail notification before starting the analysis.

(4) *Interpreting the results of the EpiGRAPH analysis.* After receiving an e-mail notification informing us about successful completion of the analysis, we click the direct link given in the e-mail, which logs us in automatically and opens the results overview page. Our inspection starts with a quick look at the results of the machine learning analysis. These results quantify how well EpiGRAPH could predict the target class from each of the ten attribute groups. In other words, they provide a measure for the combined predictiveness of each attribute group for

whether or not a specific promoter region is highly polymorphic. Reassuringly, the prediction performance is close to perfect for the "population variation" attribute group, which includes SNP data (96% prediction accuracy and a binary correlation of 0.92 between predictions and actual values). In addition to this expected result, a number of interesting attribute groups score highly, including "regulatory regions" and "epigenome and chromatin structure". An inspection of the results of the statistical analysis confirms this observation. While SNP-related attributes are again the most discriminatory features, there is also a clear tendency of nonpolymorphic promoter regions being associated with regulatory elements such as bona fide CpG islands (13) and conserved transcription factor binding sites. In contrast, they are depleted in terms of recombination hotspots.

As a follow-up analysis, it would be interesting to analyze whether the enrichment for bona fide CpG islands and transcription factor binding sites is a side effect of an elevated degree of evolutionary conservation among nonpolymorphic promoters or whether it constitutes a separate effect with an independent biological cause. Two options are available to address this question. First, we could restart the machine learning analysis with modified settings, assessing whether the combined predictiveness of the attribute groups "regulatory regions" and "evolutionary history" exceeds the predictiveness of "evolutionary history" alone (the latter group contains all conservation-related attributes). Second, we could click the "Download data table" button, download the table containing all calculated attribute values for all genomic regions included in the analysis, load this data table into a statistics software (such as R), and construct linear models in order to assess the significance of attributes from the "regulatory regions" group after statistically correcting for evolutionary conservation of the promoter region.

4. Notes

1. *Data preparation from diverse sources.* Epigenome analysis frequently incorporates genomic region data from a number of sources (collaboration partners, the supplementary material of published papers, output files of data preprocessing software, etc.), which come in a variety of formats (tab-separated or comma-separated tables, genome browser tracks, Excel sheets, etc.). Therefore, data preparation is an important step and requires caution and experience to prevent

errors that could invalidate all subsequent analyses. From our experience, the following tools can significantly facilitate data preparation: (1) The liftOver utility of the UCSC Genome Browser (http://genome.ucsc.edu/cgi-bin/hgLiftOver) maps genomic coordinates from one genome assembly to another, e.g., from human genome hg17 to hg18. (2) Advanced text editors provide various features for text file formatting, such as column-based editing and support for regular expressions when performing complex search-and-replace operations (a practical introduction into regular expressions is available from http://analyser.oli.tudelft.nl/regex/). (3) Adding, removing, and rearranging columns as well as cosmetic changes (e.g., renaming columns) is often done easiest within spreadsheet software, before saving the final table in tab-separated format or copying and pasting it directly into EpiGRAPH.

2. *Deriving an appropriate control set.* For the two EpiGRAPH case studies presented in this paper, the choice of an appropriate control set is obvious: unmethylated CpG islands complement methylated CpG islands and non-polymorphic promoter regions complement highly polymorphic promoter regions. However, for many applications, a control set must be derived by random sampling of genomic regions, which requires careful correction for potential confounding factors. Assume that we want to analyze the epigenetic characteristics of preferred retroviral integration sites (i.e., genomic regions at which viruses such as HIV are incorporated into the host DNA), based on sequenced integration sites (36). We will have to make sure that the control set does not contain more repetitive regions than the set of integration sites, because the latter dataset is artificially biased against repetitive regions, and this should be reflected in the control set. Furthermore, we may want to adjust the control set for the GC content of the genomic regions (which is a strong predictor for a wide range of genomic properties), in order to pick up more subtle differences. EpiGRAPH's attribute submission page provides functionality to derive "fair" random control sets. On the one hand, the chromosomal distribution and region-length distribution of the input dataset can be exactly matched by the control set; on the other hand, any deviation in terms of GC content, repeat content, and exon overlap can be limited to a pre-defined maximum.

3. *Working with custom attributes.* EpiGRAPH enables the user to define custom attributes that can be used in the same ways as the default attributes, i.e., not only as input datasets to be analyzed with EpiGRAPH, but also as prediction attributes for inclusion in the analysis of other datasets. The upload page for defining a new custom attribute can be accessed

using the "Upload custom attribute dataset" button on EpiGRAPH's overview page. It looks essentially identical to the attribute submission step when launching an EpiGRAPH analysis. Three ways of defining a custom attributes are available: (1) to upload the attribute data in tab-separated format, as we did in the two EpiGRAPH case studies above (e.g., useful for incorporating custom experimental data); (2) to derive the new attribute from a source attribute that is already contained in EpiGRAPH's database (either by default or as a custom attribute), specifying a filter criterion and a formula defining an additional column that is to be calculated by EpiGRAPH (e.g., useful for retrieving the DNA sequences of a set of genomic regions); (3) to request calculation of a matched control attribute for a given source attribute (e.g., useful when the same set of control regions are to be used in multiple EpiGRAPH analyses). All custom attributes are exclusively available to the user under whose account they were created. It is, however, possible to download a custom attribute in XML format and share this file with other researchers, who can then upload it into their own EpiGRAPH user accounts.

4. *Working with large datasets.* Genome analysis with Galaxy and EpiGRAPH can be performed on large datasets. However, analyses take longer and have to be planned more carefully when datasets are large. (1) It is usually advisable to perform a pilot study on a small subset of the dataset of interest before going large-scale. For this purpose, EpiGRAPH provides functionality to down-sample datasets to a given size. (2) The increase in prediction accuracy gained by including more than 1,000 genomic regions in the input dataset of an EpiGRAPH analysis is usually low and rarely worth the additional calculation time. (3) From our experience, Mozilla Firefox is the web browser that is most tolerant toward cutting and pasting large sets of genomic regions into text areas on a web page. (4) To process large tables with spreadsheet software, Microsoft Excel 2007 is often the best choice because it supports tables with up to 1,048,576 rows and up to 16,384 columns, while the limits of other spreadsheets are substantially lower and often insufficient. (5) It is rarely a good idea to submit more than five analyses in parallel to any given web service, and it is advisable to contact the scientists who operate the web service for advice before starting extremely large analyses.

5. *Understanding the basics of machine learning.* EpiGRAPH's machine learning analysis uses classification algorithms such as support vector machines and logistic regression models in order to assess the predictiveness of entire groups of attributes for a class value of interest. To that end, it tries to predict

whether a given genomic region is likely to belong to the set of positives or negatives, based on different combinations of prediction attributes. Technically, machine learning algorithms are methods for estimating or approximating a mathematical function that links the values of several (known) prediction attributes to a prediction of the (unknown) class value. The estimating function is learnt from a training dataset and its performance is evaluated on a test dataset. Because data is frequently scarce, EpiGRAPH applies a strategy called cross-validation to perform classifier training and testing on the same dataset – splitting it into ten partitions, training on nine partitions and testing on the tenth partition, and repeating this process ten times. An important concern when using machine learning methods is the risk of over-training, i.e., the danger that the classification algorithm "remembers" individual cases rather than learning generalizable concepts, which leads to over-optimistic prediction accuracies that are not sustainable on new datasets. While EpiGRAPH is implemented in a way that the risk of over-training is low, potential error sources remain (such as re-running the machine learning analysis based only on the top-scoring attributes from the statistical analysis) and it is recommended to consult further background texts on machine learning and / or discuss with an experienced bioinformatician before drawing far-reaching conclusions from the results of EpiGRAPH's machine learning analysis. A good practical introduction into machine learning is provided by Witten and Frank (37), while Hastie et al. (38) provide a more mathematical treatment. Further references are given by Tarca et al. in a recent primer on machine learning methods (39).

6. *Understanding DNA methylation prediction.* In this paper, we have used DNA methylation data to illustrate epigenome prediction with EpiGRAPH. While the use of support vector machines for predicting the DNA methylation status of CpG islands is well-established (11, 30), it is not recommended to use predictions calculated with the classifier from the first case study for any real applications, for two reasons: First, the dataset used for training the classifier is small and restricted to chromosome 21, while genome-scale datasets are now available as training data. Second, the prediction is based on only a small subset of relevant attributes, although it is known that additional attributes groups – such as more complex DNA sequence patterns – can increase prediction accuracy. To obtain more realistic DNA methylation predictions, EpiGRAPH should be applied to a larger and more representative DNA methylation dataset (e.g., (40)) and all EpiGRAPH's default attributes should be included in the prediction). Alternatively, a pre-calculated

genome-wide map of CpG island strength prediction can be used, which we derived previously (13) and which is available from http://neighborhood.bioinf.mpi-inf.mpg.de/CpG_islands_revisited (the higher a CpG island's predicted strength, the less likely it is methylated).

Acknowledgments

We would like to thank Joachim Büch for maintaining the IT infrastructure of EpiGRAPH, Yoichi Yamada and Sascha Tierling for providing DNA methylation data, and Martina Paulsen as well as Jörn Walter for helpful discussions. EpiGRAPH is partially funded by the European Union through the CANCERDIP project (HEALTH-F2-2007-200620; http://www.cancerdip.eu/). Galaxy is supported by NSF Grant DBI-0543285 and NIH Grant 5R01HG003646-02 as well as by funds from the Huck Institutes for Life Sciences at Penn State University and Pennsylvania Department of Health.

References

1. Bernstein, B.E., Meissner, A. and Lander, E.S. (2007) The mammalian epigenome. *Cell*, **128**, 669–681.
2. Chen, K. and Rajewsky, N. (2007) The evolution of gene regulation by transcription factors and microRNAs. *Nat. Rev. Genet.*, **8**, 93–103.
3. Zhang, M.Q. (2005) In: Pal, S. K. (ed.), *PReMI*. Springer-Verlag Berlin Heidelberg, Vol. 3776, pp. 31–38.
4. Frigola, J., Song, J., Stirzaker, C., Hinshelwood, R.A., Peinado, M.A. and Clark, S.J. (2006) Epigenetic remodeling in colorectal cancer results in coordinate gene suppression across an entire chromosome band. *Nat. Genet.*, **38**, 540–549.
5. Feinberg, A.P. (2007) Phenotypic plasticity and the epigenetics of human disease. *Nature*, **447**, 433–440.
6. Eckhardt, F., Lewin, J., Cortese, R., Rakyan, V.K., Attwood, J., Burger, M., *et al.* (2006) DNA methylation profiling of human chromosomes 6, 20 and 22. *Nat. Genet.*, **38**, 1378–1385.
7. Williams, R.B., Chan, E.K., Cowley, M.J. and Little, P.F. (2007) The influence of genetic variation on gene expression. *Genome Res.*, **17**, 1707–1716.
8. Bock, C., Walter, J., Paulsen, M. and Lengauer, T. (2008) Inter-individual variation of DNA methylation and its implications for large-scale epigenome mapping. *Nucleic Acids Res.*, **36**, e55.
9. Schones, D.E. and Zhao, K. (2008) Genome-wide approaches to studying chromatin modifications. *Nat. Rev. Genet.*, **9**, 179–191.
10. Bock, C., Halachev, K., Buch, J. and Lengauer, T. (2009) EpiGRAPH: User-friendly software for statistical analysis and prediction of (epi-) genomic data. *Genome Biol.*, **10**, R14.
11. Bock, C., Paulsen, M., Tierling, S., Mikeska, T., Lengauer, T. and Walter, J. (2006) CpG island methylation in human lymphocytes is highly correlated with DNA sequence, repeats, and predicted DNA structure. *PLoS Genet.*, **2**, e26.
12. Liu, F., Tostesen, E., Sundet, J.K., Jenssen, T.K., Bock, C., Jerstad, G.I., *et al.* (2007) The human genomic melting map. *PLoS Comput. Biol.*, **3**, e93.
13. Bock, C., Walter, J., Paulsen, M. and Lengauer, T. (2007) CpG island mapping by epigenome prediction. *PLoS Comput. Biol.*, **3**, e110.
14. Moser, D., Ekawardhani, S., Kumsta, R., Palmason, H., Bock, C., Athanassiadou, Z., *et al.* (2008) Functional analysis of a potassium-chloride co-transporter 3 (SLC12A6) promoter polymorphism leading to an additional DNA methylation site. *Neuropsychopharmacology*, **34**, 458–467.
15. Blankenberg, D., Taylor, J., Schenck, I., He, J., Zhang, Y., Ghent, M., *et al.* (2007)

A framework for collaborative analysis of ENCODE data: making large-scale analyses biologist-friendly. *Genome Res.*, **17**, 960–964.

16. Giardine, B., Riemer, C., Hardison, R.C., Burhans, R., Elnitski, L., Shah, P., *et al.* (2005) Galaxy: a platform for interactive large-scale genome analysis. *Genome Res.*, **15**, 1451–1455.
17. Pond, S.L., Frost, S.D. and Muse, S.V. (2005) HyPhy: hypothesis testing using phylogenies. *Bioinformatics*, **21**, 676–679.
18. Rice, P., Longden, I. and Bleasby, A. (2000) EMBOSS: the European Molecular Biology Open Software Suite. *Trends Genet.*, **16**, 276–277.
19. van Steensel, B. (2005) Mapping of genetic and epigenetic regulatory networks using microarrays. *Nat. Genet.*, **37 Suppl**, S18–24.
20. Bock, C. and Lengauer, T. (2008) Computational epigenetics. *Bioinformatics*, **24**, 1–10.
21. Gentleman, R.C., Carey, V.J., Bates, D.M., Bolstad, B., Dettling, M., Dudoit, S., *et al.* (2004) Bioconductor: open software development for computational biology and bioinformatics. *Genome Biol.*, **5**, R80.
22. Liu, X.S. (2007) Getting started in tiling microarray analysis. *PLoS Comput. Biol.*, **3**, 1842–1844.
23. Johnson, D.S., Li, W., Gordon, D.B., Bhattacharjee, A., Curry, B., Ghosh, J., *et al.* (2008) Systematic evaluation of variability in ChIP-chip experiments using predefined DNA targets. *Genome Res.*, **18**, 393–403.
24. Johnson, W.E., Li, W., Meyer, C.A., Gottardo, R., Carroll, J.S., Brown, M. and Liu, X.S. (2006) Model-based analysis of tiling-arrays for ChIP-chip. *Proc. Natl. Acad. Sci. USA.*, **103**, 12457–12462.
25. Kumaki, Y., M. Oda, and M. Okano. 2008. QUMA: quantification tool for methylation analysis. *Nucleic Acids Res* 36: W170–175.
26. Bock, C., Reither, S., Mikeska, T., Paulsen, M., Walter, J. and Lengauer, T. (2005) BiQ Analyzer: visualization and quality control for DNA methylation data from bisulfite sequencing. *Bioinformatics*, **21**, 4067–4068.
27. Karolchik, D., Kuhn, R.M., Baertsch, R., Barber, G.P., Clawson, H., Diekhans, M., *et al.* (2008) The UCSC Genome Browser Database: 2008 update. *Nucleic Acids Res.*, **36**, D773–779.
28. Flicek, P., Aken, B.L., Beal, K., Ballester, B., Caccamo, M., Chen, Y., *et al.* (2008) Ensembl 2008. *Nucleic Acids Res.*, **36**, D707–714.
29. Das, R., Dimitrova, N., Xuan, Z., Rollins, R.A., Haghighi, F., Edwards, J.R., *et al.* (2006) Computational prediction of methylation status in human genomic sequences. *Proc. Natl. Acad. Sci. U. S. A.*, **103**, 10713–10716.
30. Fang, F., Fan, S., Zhang, X. and Zhang, M.Q. (2006) Predicting methylation status of CpG islands in the human brain. *Bioinformatics*, **22**, 2204–2209.
31. Yamada, Y., Watanabe, H., Miura, F., Soejima, H., Uchiyama, M., Iwasaka, T., *et al.* (2004) A comprehensive analysis of allelic methylation status of CpG islands on human chromosome 21q. *Genome Res.*, **14**, 247–266.
32. Noble, W.S. (2006) What is a support vector machine? *Nat. Biotechnol.*, **24**, 1565–1567.
33. Zhang, Y., C. Rohde, S. Tierling, T.P. Jurkowski, C. Bock, D. Santacruz, S. Ragozin, R. Reinhardt, M. Groth, J. Walter, and A. Jeltsch. 2009. DNA methylation analysis of chromosome 21 gene promoters at single base pair and single allele resolution. PLoS Genet 5: e1000438.
34. Frazer, K.A., Ballinger, D.G., Cox, D.R., Hinds, D.A., Stuve, L.L., Gibbs, R.A., *et al.* (2007) A second generation human haplotype map of over 3.1 million SNPs. *Nature*, **449**, 851–861.
35. ENCODE Project Consortium. (2004) The ENCODE (ENCyclopedia Of DNA Elements) Project. *Science*, **306**, 636–640.
36. Wang, G.P., Ciuffi, A., Leipzig, J., Berry, C.C. and Bushman, F.D. (2007) HIV integration site selection: analysis by massively parallel pyrosequencing reveals association with epigenetic modifications. *Genome Res.*, **17**, 1186–1194.
37. Witten, I.H. and Frank, E. (2000) *Data mining : practical machine learning tools and techniques with Java implementations.* Morgan Kaufmann, San Francisco, Calif.
38. Hastie, T., Tibshirani, R. and Friedman, J.H. (2001) *The elements of statistical learning : data mining, inference, and prediction.* Springer, New York.
39. Tarca, A.L., Carey, V.J., Chen, X.W., Romero, R. and Draghici, S. (2007) Machine learning and its applications to biology. *PLoS Comput. Biol.*, **3**, e116.
40. Meissner, A., Mikkelsen, T.S., Gu, H., Wernig, M., Hanna, J., Sivachenko, A., *et al.* (2008) Genome-scale DNA methylation maps of pluripotent and differentiated cells. *Nature*, **454**, 766–770.

Chapter 16

Short Tandem Repeats and Genetic Variation

Bo Eskerod Madsen, Palle Villesen, and Carsten Wiuf

Abstract

Single nucleotide polymorphisms (SNPs) are widely distributed in the human genome and although most SNPs are the result of independent point-mutations, there are exceptions. When studying distances between SNPs, a periodic pattern in the distance between pairs of identical SNPs has been found to be heavily correlated with periodicity in short tandem repeats (STRs). STRs are short DNA segments, widely distributed in the human genome and mainly found outside known tandem repeats. Because of the biased occurrence of SNPs, special care has to be taken when analyzing SNP-variation in STRs.

We present a review of STRs in the human genome and discuss molecular mechanisms related to the biased occurrence of SNPs in STRs, and its implications for genome comparisons and genetic association studies.

Key words: SNPs, Short tandem repeat, Pattern, Variation, Mechanism, Mutation, Polymorphism

Abbreviations

SNP	single nucleotide polymorphism
bp	base pair
STR	short tandem repeat

1. Introduction

Single nucleotide polymorphisms (SNPs) are widely distributed in the human genome, and are not restricted to any type of genetic elements such as exons, transcripts, transposons or tandem repeats. There are 11.9 million reported SNPs in the human genome (dbSNP (1, 2), build 128) and panels of up to 650 k SNPs have been used as markers for genetic disease susceptibility variants in genome wide association studies (3–7). SNPs are generally thought to be the result of independent mutational events which subsequently have spread in the human population

Michael R. Barnes and Gerome Breen (eds.), *Genetic Variation: Methods and Protocols*, Methods in Molecular Biology, vol. 628, DOI 10.1007/978-1-60327-367-1_16, © Springer Science+Business Media, LLC 2010

and thereby lead to nucleotide diversity in the genome (8). Much effort has gone into identifying new SNPs in the human genome and studying the frequencies of SNPs in different human populations. For example in the HapMap project, 4.0 million non-redundant SNPs (release #23, January 2008) have been genotyped in 270 individuals from four different human populations from Africa, Asia and Central Europe (9). SNP information from the HapMap project has been used to select the SNP panels for genome wide association studies, and has contributed to the validation of some of the SNPs that have been reported to dbSNP.

By comparing genomes from different species or individuals, single nucleotide variation has been used to estimate how genomes evolve over time. Simplified models such as the Jukes-Cantor (10), Felsenstein (11) and HKY (12) models are typically applied to compare the evolution of different segments of the genome. Such comparisons can identify DNA segments that are highly conserved and/or under selection, and thereby identify functionally important elements in the genome. Knowledge about functional elements is then again used in studies of how genetic variation influences the resulting phenotype (e.g. disease).

In this review, we focus on how SNP occurrence may depend on periodicity in the nucleotide composition. Nucleotides occurring in a periodic manner are known as tandem repeats, microsatellites, simple repeats or simple sequence repeats (SSR). We especially focus on short (imperfect) tandem repeats (STRs) in the human genome, relate the findings to possible molecular mechanisms for generating STRs and discuss what implications the findings may have on genetic association studies and genome comparison studies.

2. Identification of Short Tandem Repeats

In this review, we use the definition of STRs given by Madsen et al. (13) (originally called periodic DNA). In brief, a DNA segment is defined as an STR if (1) it is at least 9 bp long, (2) the repeat-unit (e.g. AT in ATATATATAT) is repeated at least three times, (3) only a few base pairs in the segment do not match the repeat-unit. To allow for sequence ambiguity, all possible SNP alleles are used in the identification of an STR (see Fig. 1).

Several algorithms, such as Tandem Repeat Finder (14), mreps (15) and TROLL (16), have been developed for the identification of tandem repeats. These algorithms are designed for general identification of tandem repeats, but care should be taken since the algorithms differ significantly in what they detect as tandem repeats (17). None of the above mentioned algorithms incorporate

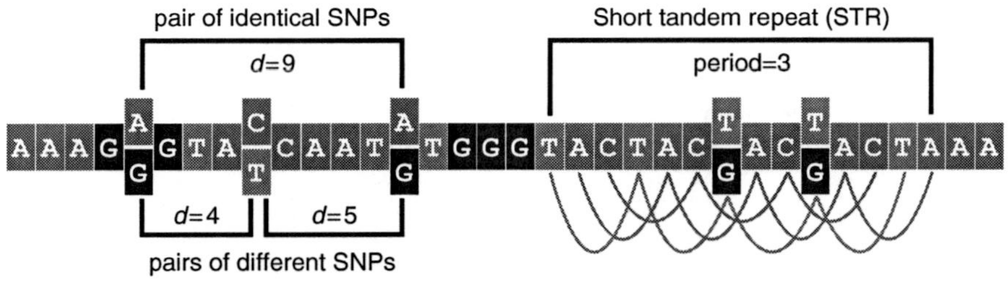

Fig. 1. Definitions of distances in pairs of SNP and an example of an STR. Distances are calculated between all pairs of SNPs, thus the figure shows three pairs with three distances. The distance (*d*) between any two SNPs is defined as the positive difference between the two genomic SNP positions, for example, $d=1$ corresponds to contiguous SNPs. A pair of identical SNPs is defined as two SNPs with identical alleles (here SNP1: A/G, SNP2: A/G, $d=9$). Pairs of different SNPs are defined as two SNPs with different alleles (here SNP1: A/G, SNP2: C/T, $d=4$; SNP1: C/T, SNP2: A/G, $d=5$). To the right, an example of an STR is shown. The period (*p*) is 3, and it is shown that SNPs are allowed in the pattern. Adapted in part from Madsen et al. (13), with permission from Genome Research

information about known SNP variations in the human genome, and we previously implemented a specialized algorithm for the identification of STRs (13).

STRs are widely distributed in the human genome; i.e. STRs make up 4.3% of the entire human genome, whereas 2.87% of exons and 4.3% of the entire transcribed regions are tagged as STRs (13) Furthermore, STRs are generally different from the "Simple Repeats" track from the UCSC Table Browser (18) (found using Tandem Repeat Finder (14)), as 97.17% of all STRs are found outside the track (13). The genomic content of tandem repeats in general has been investigated in several studies, and is described elsewhere (19–26).

3. A Periodic Pattern in SNP Distances

One feature of STRs is a periodic pattern in the distance between pairs of "identical SNPs" (SNPs with identical alleles). In contrast to non-STR, pairs of identical SNPs are common and clearly non-uniformly distributed in STRs (Fig. 2). If SNPs occur with the same probability independently at all sites in the genome, then the distance between two random SNPs is uniformly distributed. This does not hold true for immediately adjacent SNPs because of the high CpG mutation rate (27). Inside STR regions, pairs of identical SNPs positioned 2, 4, 6 or 8 bp apart are much more frequent than pairs of identical SNPs positioned 3, 5, 7 or 9 bp apart, whereas this pattern is completely absent for pairs of different SNPs (13). This 2, 4, 6, 8 pattern is most likely explained by biased introduction of SNPs in STRs (see Molecular mechanisms) and in concordance, there are found 1.8 times more SNPs in STRs than would have been expected by chance (13).

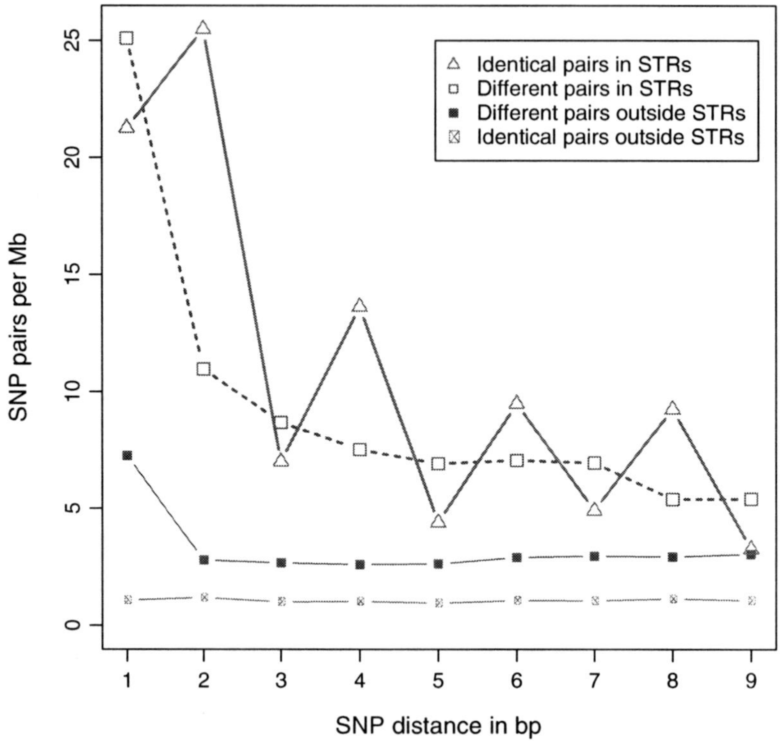

Fig. 2. Pairs of SNPs inside and outside STRs. Shown is the distance between pairs of SNPs inside and outside STRs. Both pairs of identical SNPs and pairs of different SNPs are overrepresented inside STRs when compared to outside STRs. Pairs of identical SNPs show the highest overrepresentation in STRs and identical SNPs 2, 4, 6 or 8 bp apart are much more common than identical SNPs 3, 5, 7 or 9 bp apart. Adapted in part from Madsen et al. (13), with permission from Genome Research

As for tandem repeats in general, the majority of STRs have periods of 1 or 2 bp (13, 28, 29). The 2, 4, 6, 8 pattern in SNP distances are therefore likely to be due to SNPs emerging according to the periods of STRs; i.e. if an A/G SNP is present in an STR segment ATATATATAT, then another A/G SNP in the same segment occurs more often than is expected by chance, generating pairs of identical SNPs 2, 4, 6 or 8 bp apart. If this biased emergence of SNPs is equally probable for all periods of STRs, then the 2, 4, 6, 8 pattern would be generated simply because STRs with period $p=2$ are common.

4. Molecular Mechanisms

Length variations in tandem repeats are generally thought to be generated by polymerase slippage and uneven cross over (30–35). Polymerase slippage is a mechanism, whereby the DNA polymerase skips one (or more) repeat-unit(s) in a tandem repeat, or copies a repeat-unit more than once from the template strand (36, 37). Uneven cross over is a mechanism whereby the two homologous DNA strands do not break in the same position before recombination, leading to a strand with a deletion of a segment and a strand with an insertion of the same segment (38). If these irregularities are not caught by the repair mechanisms, they lead to length variations in tandem repeats.

The observed 2, 4, 6, 8 pattern in STRs cannot be explained by misalignments of sequences due to length variations in STR segments, since only SNPs which are mapped to an exact location in the reference genome are used (13). However, this does not rule out that length variation mediates the bias towards an excess of pairs of identical SNPs in STRs. E.g. if a repeat-unit is inserted at the left side of the C in ATCTATATAT, generating the "temporary" sequence ATATCTATATAT, and a repeat-unit subsequently is removed on the right side of C, we get the two sequences ATCTATATAT and ATATCTATAT in the population, which will be interpreted as two A/C SNPs in distance $d=2$ bp (see Fig. 3). Repair mechanisms may tend to correct for insertions in the same meiotic cycle as they are introduced and thereby generate pairs of identical SNPs in STRs, as just explained. Alternatively, an inversion of 3 bp (e.g CTA) yields a pair of identical SNPs too. A second independent length-mutation in a STR can result in the

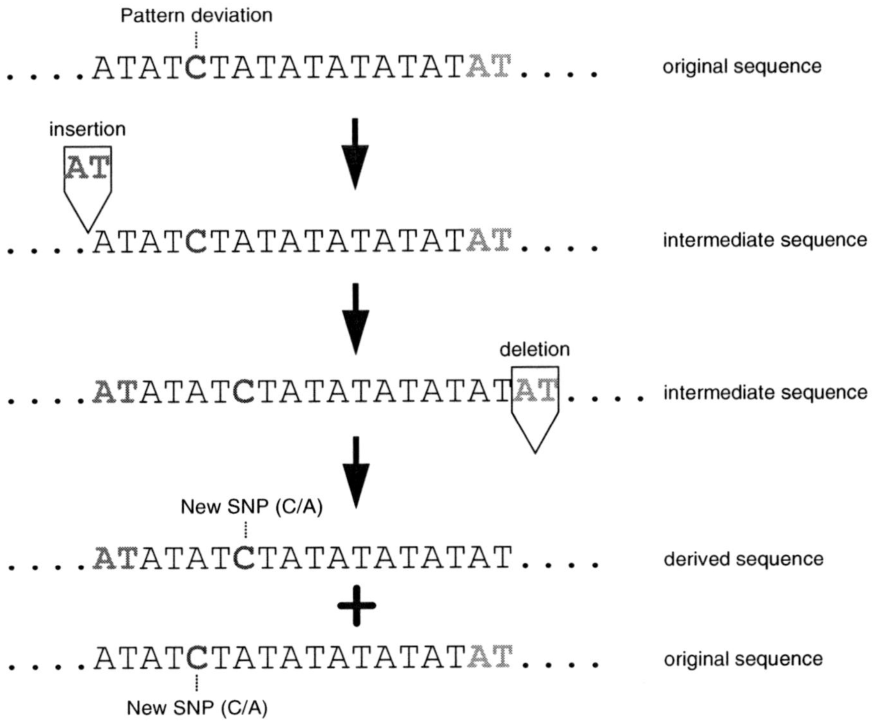

Fig. 3. A molecular mechanism for generating a pair of identical SNPs

same, but this scenario is less probable since two independent mutations are needed. Another possibility is gene conversion, where a DNA segment is copied to a new position without creating a length polymorphism (39, 40). Complex mechanisms of context dependent generation of point mutations could explain the observed pattern as well, but no such mechanism are known. It is worth noting, however, that the elevated mutation rate in CpG islands (27) is context dependent, and the importance of such a mechanism can not be ruled out per se.

5. Genetic Association Studies

Like other forms of genetic variation, insertion deletion polymorphisms (indels) are of great interest because they may influence gene function and cause disease. An example is Fragile X Syndrome that is caused by expansion of a three-nucleotide tandem repeat in the FMR-1 gene (41–44). Likewise, cystic fibrosis is frequently caused by a three bp deletion that eliminates a single amino acid from the protein encoded by the CFTR gene (45–48). Next-generation sequencing technologies may enable identification of new disease susceptibility variants by resequencing a large number of disease cases and controls. However, sequencing the entire genome of a large group of affected individuals may still be prohibitively expensive for years to come and identification of probable targets for disease causing variants may be useful. Hypermutable segments of functional genomic elements (exons) are probable targets for disease related mutations and may therefore be good candidates for resequencing studies. Tandem repeats are well known to be hypermutable and to have an excess of indels compared to the rest of the genome, but tandem repeats are rare in functional elements such as exons (20, 28, 35, 49–51). In contrast, STRs are widely distributed in the human genome (13, 28) and since they share the hypermutability of longer tandem repeats (unpublished results), they may be targets for disease causing mutations. If hypermutable segments are located in "junk" (uninformative) DNA, mutations are not disease causing. Tandem repeats are mainly thought to be "junk" DNA, but several studies have shown that tandem repeats can have a functional role. Examples of tandem repeat related functions are differentiated transcription activity of human genes (52), and the ability of pathogens to adapt to their host (26). Additional examples of functional tandem repeats are reviewed elsewhere (24, 35, 53–56).

The call-rate for genotyping SNPs in the HapMap (9) study has been shown to be significantly lower for SNPs located inside STRs (13). This supports that STRs are hypermutable and emphasizes that care should be taken when SNP studies are designed and analyzed. Besides affecting the call-rate, structural variants may lead to genotyping errors, if the DNA sequence is altered close to a SNP position and a wrong genomic position is read for the SNP. Such a bias may be difficult to identify and precautionary steps should be taken in the study design. A strategy to minimize the impact of STRs in genotyping studies is simply to avoid SNPs inside or near STRs. The downside of this strategy is that variants in some parts of the genome are poorly covered in the study and hence disease associated variants may be missed. Resequencing STR segments would solve the problem, but that approach may be too expensive in many studies.

As it has been debated for tandem repeats (24, 52, 54), STRs may serve functional roles in the genome. One possibility is that DNA and/or RNA fold according to the repeated sequence of STRs and thereby influence gene function (35). A mutation in such an STR may alter the folding and thus the function. Furthermore, hypermutable regions (e.g. in exons) may introduce a high level of phenotypic variation and thereby allow for fast adaptation to a changing environment. Although hypermutability in functional elements may have been beneficial throughout evolution, disease related variants may also be introduced in an elevated rate in such regions. Hypermutable segments with functional roles may be obvious candidates for resequencing studies, since a high density of rare disease susceptibility variants are expected.

6. Genome Comparison Studies

Models for genome comparison usually assume that mutations occur independently and a violation on this assumption may bias findings. The excess of pairs of identical SNPs in STRs clearly show that the assumption of independent mutations is not always valid, and hence care must be taken. Since it is not known whether the underlying molecular mechanism(s) is (are) restricted to STRs or just visible in these segments, excluding STRs from genome comparison studies may not guarantee that the analyzed variation have occurred independently.

7. Concluding Remarks

The presence of a periodic pattern of SNPs in STRs emphasizes that care should be taken when using SNPs in disease association studies and genome comparisons. Further studies are needed to clarify what mechanisms underlie the excess of pairs of identical

SNPs in STRs. Investigations of how common insertion or deletion of repeat-units is in STR regions may help to distinguish between some of the possible mechanisms, whereas identifying the exact mechanism(s) may be difficult.

Whether STRs are associated with gene function, or are a probable target for disease-causing mutations, remains an open question, but it is worth giving a second thought.

References

1. Sherry, S.T., Ward, M. and Sirotkin, K. (1999) dbSNP – database for single nucleotide polymorphisms and other classes of minor genetic variation. *Genome Res.*, **9**, 677–679.
2. Sherry, S.T., Ward, M.H., Kholodov, M., Baker, J., Phan, L., Smigielski, E.M. and Sirotkin, K. (2001) dbSNP: the NCBI database of genetic variation. *Nucleic Acids Res.*, **29**, 308–311.
3. Eberle, M.A., Ng, P.C., Kuhn, K., Zhou, L., Peiffer, D.A., Galver, L., *et al.* (2007) Power to detect risk alleles using genome-wide tag SNP panels. *PLoS Genet.*, **3**, e170.
4. Fan, J.-B., Chee, M.S. and Gunderson, K.L. (2006) Highly parallel genomic assays. *Nat. Rev. Genet.*, **7**, 632–644.
5. Easton, D.F., Pooley, K.A., Dunning, A.M., Pharoah, P.D.P., Thompson, D., Ballinger, D.G., *et al.* (2007) Genome-wide association study identifies novel breast cancer susceptibility loci. *Nature*, **447**, 1087–1093.
6. Sladek, R., Rocheleau, G., Rung, J., Dina, C., Shen, L., Serre, D., *et al.* (2007) A genome-wide association study identifies novel risk loci for type 2 diabetes. *Nature*, **445**, 881–885.
7. The Wellcome Trust Case Control Consortium. (2007) Genome-wide association study of 14,000 cases of seven common diseases and 3,000 shared controls. *Nature*, **447**, 661–678.
8. Stoneking, M. (2001) Single nucleotide polymorphisms. From the evolutionary past. *Nature*, **409**, 821–822.
9. The International HapMap Consortium. (2003) The International HapMap Project. *Nature*, **426**, 789–796.
10. Jukes, T.H. and Cantor, C.R. (1969) Evolution of protein molecules. In Munro, H.N. (ed.), *Mammalian Protein Metabolism*. Academic Press, New York.
11. Felsenstein, J. (1981) Evolutionary trees from DNA sequences: a maximum likelihood approach. *J. Mol. Evol.*, **17**, 368–376.
12. Hasegawa, M., Kishino, H. and Yano, T. (1985) Dating of the human-ape splitting by a molecular clock of mitochondrial DNA. *J. Mol. Evol.*, **22**, 160–174.
13. Madsen, B.E., Villesen, P. and Wiuf, C. (2007) A periodic pattern of SNPs in the human genome. *Genome Res.*, **17**, 1414–1419.
14. Benson, G. (1999) Tandem repeats finder: a program to analyze DNA sequences. *Nucleic Acids Res.*, **27**, 573–580.
15. Kolpakov, R., Bana, G. and Kucherov, G. (2003) mreps: efficient and flexible detection of tandem repeats in DNA. *Nucleic Acids Res.*, **31**, 3672–3678.
16. Castelo, A.T., Martins, W. and Gao, G.R. (2002) TROLL – tandem repeat occurrence locator. *Bioinformatics*, **18**, 634–636.
17. Leclercq, S., Rivals, E. and Jarne, P. (2007) Detecting microsatellites within genomes: significant variation among algorithms. *BMC Bioinformatics*, **8**, 125.
18. Karolchik, D., Hinrichs, A.S., Furey, T.S., Roskin, K.M., Sugnet, C.W., Haussler, D. and Kent, W.J. (2004) The UCSC Table Browser data retrieval tool. *Nucleic Acids Res.*, **32**, D493–D496.
19. Boby, T., Patch, A.M. and Aves, S.J. (2005) TRbase: a database relating tandem repeats to disease genes for the human genome. *Bioinformatics*, **21**, 811–816.
20. Borstnik, B. and Pumpernik, D. (2002) Tandem repeats in protein coding regions of primate genes. *Genome Res.*, **12**, 909–915.
21. O'Dushlaine, C., Edwards, R., Park, S. and Shields, D. (2005) Tandem repeat copy-number variation in protein-coding regions of human genes. *Genome Biol.*, **6**, R69.
22. Hancock, J.M. and Simon, M. (2005) Simple sequence repeats in proteins and their significance for network evolution. *Gene*, **345**, 113–118.
23. Alba, M.M. and Guigo, R. (2004) Comparative analysis of amino acid repeats in rodents and humans. *Genome Res.*, **14**, 549–554.

24. Kashi, Y. and King, D.G. (2006) Simple sequence repeats as advantageous mutators in evolution. *Trends Genet.*, **22**, 253–259.
25. Kelkar, Y.D., Tyekucheva, S., Chiaromonte, F. and Makova, K.D. (2008) The genome-wide determinants of human and chimpanzee microsatellite evolution. *Genome Res.*, 18, 30–38.
26. Mrazek, J., Guo, X. and Shah, A. (2007) Simple sequence repeats in prokaryotic genomes. *Proc. Natl. Acad. Sci. U.S.A.*, **104**, 8472–8477.
27. Hwang, D.G. and Green, P. (2004) Inaugural article: Bayesian Markov chain Monte Carlo sequence analysis reveals varying neutral substitution patterns in mammalian evolution. *Proc. Natl. Acad. Sci. U.S.A.*, **101**, 13994–14001.
28. Lai, Y. and Sun, F. (2003) The Relationship Between Microsatellite Slippage Mutation Rate and the Number of Repeat Units. *Mol. Biol. Evol.*, **20**, 2123–2131.
29. Almeida, P. and Penha-Goncalves, C. (2004) Long perfect dinucleotide repeats are typical of vertebrates, show motif preferences and size convergence. *Mol. Biol. Evol.*, **21**, 1226–1233.
30. Levinson, G. and Gutman, G.A. (1987) Slipped-strand mispairing: a major mechanism for DNA sequence evolution. *Mol. Biol. Evol.*, **4**, 203–221.
31. Pearson, C.E., Edamura, K.N. and Cleary, J.D. (2005) Repeat instability: mechanisms of dynamic mutations. *Nat. Rev. Genet.*, **6**, 729–742.
32. Ellegren, H. (2004) Microsatellites: simple sequences with complex evolution. *Nat. Rev. Genet.*, **5**, 435–445.
33. Chambers, G.K. and MacAvoy, E.S. (2000) Microsatellites: consensus and controversy. *Comp. Biochem. Physiol. B Biochem. Mol. Biol.*, **126**, 455–476.
34. Kruglyak, S., Durrett, R.T., Schug, M.D. and Aquadro, C.F. (1998) Equilibrium distributions of microsatellite repeat length resulting from a balance between slippage events and point mutations. *Proc. Natl. Acad. Sci. U.S.A.*, **95**, 10774–10778.
35. Mirkin, S.M. (2007) Expandable DNA repeats and human disease. *Nature*, **447**, 932–940.
36. Weber, J.L. and Wong, C. (1993) Mutation of human short tandem repeats. *Hum. Mol. Genet.*, **2**, 1123–1128.
37. Walsh, P.S., Fildes, N.J. and Reynolds, R. (1996) Sequence analysis and characterization of stutter products at the tetranucleotide repeat locus vWA. *Nucleic Acids Res.*, **24**, 2807–2812.
38. Jeffreys, A.J., Barber, R., Bois, P., Buard, J., Dubrova, Y.E., Grant, G., *et al.* (1999) Human minisatellites, repeat DNA instability and meiotic recombination. *Electrophoresis*, **20**, 1665–1675.
39. Holliday, R. (1964) A mechanism for gene conversion in fungi. *Genet. Res.*, **5**, 282–304.
40. Lewin, B. (2004) *Genes VIII.* Prentice Hall, New Jersey.
41. Warren, S.T., Zhang, F., Licameli, G.R. and Peters, J.F. (1987) The fragile X site in somatic cell hybrids: an approach for molecular cloning of fragile sites. *Science*, **237**, 420–423.
42. Kremer, E.J., Pritchard, M., Lynch, M., Yu, S., Holman, K., Baker, E., *et al.* (1991) Mapping of DNA instability at the fragile X to a trinucleotide repeat sequence p(CCG)n. *Science*, **252**, 1711–1714.
43. Verkerk, A.J.M.H., Pieretti, M., Sutcliffe, J.S., Fu, Y.-H., Kuhl, D.P.A., Pizzuti, A., *et al.* (1991) Identification of a gene (FMR-1) containing a CGG repeat coincident with a breakpoint cluster region exhibiting length variation in fragile X syndrome. *Cell*, **65**, 905–914.
44. Yu, S., Pritchard, M., Kremer, E., Lynch, M., Nancarrow, J., Baker, E., *et al.* (1991) Fragile X genotype characterized by an unstable region of DNA. *Science*, **252**, 1179–1181.
45. Collins, F.S., Drumm, M.L., Cole, J.L., Lockwood, W.K., Vande Woude, G.F. and Iannuzzi, M.C. (1987) Construction of a general human chromosome jumping library, with application to cystic fibrosis. *Science*, **235**, 1046–1049.
46. Kerem, B., Rommens, J.M., Buchanan, J.A., Markiewicz, D., Cox, T.K., Chakravarti, A., Buchwald, M., Tsui, L.C. (1989) Identification of the cystic fibrosis gene: genetic analysis. *Science*, **245(4922)**, 1073–1080.
47. Riordan, J.R., Rommens, J.M., Kerem, B., Alon, N., Rozmahel, R., Grzelczak, Z., Zielenski, J., *et al.* (1989) Identification of the cystic fibrosis gene: cloning and characterization of complementary DNA. *Science*, **245(4922)**, 1066–1073.
48. Rommens, J.M., Iannuzzi, M.C., Kerem, B., Drumm, M.L., Melmer, G., Dean, M., Rozmahel, R., *et al.* (1989) Identification of the cystic fibrosis gene: chromosome walking and jumping. *Science*, **245(4922)**, 1059–1065.
49. Ellegren, H. (2000) Microsatellite mutations in the germline: implications for evolutionary inference. *Trends Genet.*, **16**, 551–558.
50. Toth, G., Gaspari, Z. and Jurka, J. (2000) Microsatellites in different eukaryotic genomes: survey and analysis. *Genome Res.*, **10**, 967–981.

51. International Human Genome Sequencing Consortium. (2001) Initial sequencing and analysis of the human genome. *Nature*, **409**, 860–921.
52. Lawson, M.J. and Zhang, L. Housekeeping and tissue-specific genes differ in simple sequence repeats in the 5′-UTR region. *Gene*, 407, 54–62.
53. Thomas, E.E. (2005) Short, local duplications in eukaryotic genomes. *Curr. Opin. Genet. Dev.*, **15**, 640–644.
54. Li, Y.-C., Korol, A.B., Fahima, T. and Nevo, E. (2004) Microsatellites within genes: structure, function, and evolution. *Mol. Biol. Evol.*, **21**, 991–1007.
55. Sutherland, G.R. and Richards, R.I. (1995) Simple tandem DNA repeats and human genetic disease. *Proc. Natl. Acad. Sci. U.S.A.*, **92**, 3636–3641.
56. Zuckerkandl, E. (2002) Why so many noncoding nucleotides? The eukaryote genome as an epigenetic machine. *Genetica*, **115**, 105–129.

Chapter 17

Bioinformatic Tools for Identifying Disease Gene and SNP Candidates

Sean D. Mooney, Vidhya G. Krishnan, and Uday S. Evani

Abstract

As databases of genome data continue to grow, our understanding of the functional elements of the genome grows as well. Many genetic changes in the genome have now been discovered and characterized, including both disease-causing mutations and neutral polymorphisms. In addition to experimental approaches to characterize specific variants, over the past decade, there has been intense bioinformatic research to understand the molecular effects of these genetic changes. In addition to genomic experimental assays, the bioinformatic efforts have focused on two general areas. First, researchers have annotated genetic variation data with molecular features that are likely to affect function. Second, statistical methods have been developed to predict mutations that are likely to have a molecular effect. In this protocol manuscript, methods for understanding the molecular functions of single nucleotide polymorphisms (SNPs) and mutations are reviewed and described. The intent of this chapter is to provide an introduction to the online tools that are both easy to use and useful.

Key words: Single nucleotide polymorphism, SNP, Genetic disease, Candidate gene, Genome, Bioinformatics, Machine learning

1. Introduction

Over the past decade, considerable effort has been placed on understanding how genetic changes give rise to the molecular effects that cause diseases and phenotypes (1–3). These efforts have given rise to many databases, web resources, and tools for prioritizing candidate single nucleotide polymorphisms (SNPs) or hypothesizing the molecular causes of genetic disease. In this review, these resources and online tools are described within the genomic context of the annotations they provide. Most of the focus is on human annotations, although some resources provide insight into SNP data from model organisms such as mouse, fruit fly, or chimpanzee.

Michael R. Barnes and Gerome Breen (eds.), *Genetic Variation: Methods and Protocols*, Methods in Molecular Biology, vol. 628, DOI 10.1007/978-1-60327-367-1_17,

There are now many databases that provide access to SNP or disease mutation data. Most SNP data is eventually deposited in the de facto central SNP database, The Single Nucleotide Polymorphism database (dbSNP, http://www.ncbi.nlm.nih.gov/SNP/). There are also now many genotype-phenotype databases available as well including the Human Gene Mutation Database (HGMD, http://www.hgmd.cf.ac.uk/ac/index.php) (4), Online Mendelian Inheritance in Man (OMIM, http://www.ncbi.nlm.nih.gov/sites/entrez?db=omim) (5), the Pharmacogenetics Knowledge Base (PharmGKB, http://www.pharmgkb.org/) (6), and database of Genotype and Phenotype (dbGAP, http://www.ncbi.nlm.nih.gov/sites/entrez?db=gap) (7). There are also a growing number of databases of resequencing polymorphism data including the SeattleSNPs project (http://pga.mbt.washington.edu/) and sequencing of somatic mutations in cancer (8, 9). This has led to a wealth of genetic variation data.

Typically, SNP data is used as a marker in the context of a linkage or population-based association study. Here, we are focusing on SNPs as the elements that cause disease and alter phenotypes through alteration of some molecular function. There are a number of challenges to identifying these so-called functional variants. First, the marker variants themselves are likely in linkage disequilibrium (or linkage, depending on the study) with the causal variant. Second, identification of candidate disease genes may be the first challenge to narrowing a region for SNP prioritization. Finally, our understanding of how SNPs disrupt molecular function is poorly understood. Here, we focus on two important areas, identification of candidate genes that may have causal variants and identification of candidate causal SNPs.

2. Materials

In general, most of the tools here are deployed as a website or a web resource, requiring only a computer with an internet connection. Occasionally, other software may be required. For visualization of protein structure, UCSF Chimera (10) or Delano Scientific PyMOL (http://delanoscientific.com/) maybe useful. Some tools require Flash or Scalable Vector Graphics (SVG).

3. Methods

3.1. Prediction of Genes Likely to Cause or be Associated with Disease

A recent disease gene prioritization tool is FitSNPs (Functionally interpolated SNPs; http://fitsnps.stanford.edu/) (11). The tool is claimed to provide a new way to distinguish disease-associated genes from false positives in genome-wide association studies.

The feature is based on human microarray data, and it reveals the association between gene expression and disease-associated variants.

Another relatively recent addition to the library of tools that use biochemical information to aid in genetic studies are algorithms for identification of candidate disease genes or genes likely to have associated disease causing genetic changes. These approaches are generally supervised, that is they require knowledge of genes that cause a disease. Most tools use several points of data to infer candidates, and here we discuss web-based tools available for prioritization. One comprehensive example is the Endeavouralgorithm(http://homes.esat.kuleuven.be/~bioiuser/endeavour/), first published in 2006 in *Nature Biotechnology* (12). Endeavour uses a variety of publicly "–omic" features to predict candidate genes including protein interaction, gene expression, function, sequence, and literature. The tool consists of either Java or web-based clients and is easy to use. It requires a list of training set genes and a list of test set genes.

GeneSeeker (http://www.cmbi.ru.nl/geneseeker/) (13) produces a list of candidate disease genes based on cytogenetic localization and expression/phenotypic data from various human and mouse databases. GeneSeeker connects to these databases directly online to guarantee the user to be able to access the most recent data instead of having to download the updated repositories periodically. Although this tool is best for Mendelian diseases that show difference in gene expression patterns in affected tissues, it can also be used to predict candidate genes in other complex diseases.

Gene2Disease (G2D, http://www.ogic.ca/projects/g2d_2/) (14), a system that identifies the candidate disease gene by doing a homology search on Gene Ontology (GO, http://www.geneontology.org/) (15) annotated disease-associated genes. G2D uses biomedical literature searches and associated disease conditions with GO terms.

The automated server, SUSPECTS (http://www.genetics.med.ed.ac.uk/suspects/) (16) combines the scores from PROSPECTR (http://www.genetics.med.ed.ac.uk/prospectr/, based on sequence features) (17), Gene Ontology (GO), InterPro (18) and expression libraries to rank candidate genes in large regions of interest. This tool assumes that the candidate genes in general, share similar domains, annotation, and expression pattern. It provides a 3-D graphical output of the region of interest and hyperlinks to enhance the depth of information about the gene.

Transcriptomics of OMIM (TOM, http://www-micrel.deis.unibo.it/~tom/) (19) identifies candidate genes involved in inherited diseases. The algorithm uses mapping, expression, and functional online data repositories. This tool, in general, can be used to predict gene-locus and locus-locus query. It offers flexibility to the user to be able to make a choice between expression data alone or functional analysis using GO terms or both to filter candidate disease genes.

PRIORITIZER ((http://pcdoeglas.med.rug.nl/prioritizer/) (20) uses a Bayesian approach to classify genes that are associated in diseases. This tool uses data from GO, the Kyoto Encyclopedia of Genes and Genomes (KEGG, http://www.genome.ad.jp/kegg/) (21), Biomolecular Interaction Network Database (BIND, http://binddb.org/) (22), the Human Protein Reference Database (HPRD, http://www.hprd.org/) (23, 24), and Reactome, predicted PPI and expression data. Disease genes are identified through common interactions of proteins in multiple disease intervals that have common phenotypes. This method is based on the assumption that candidate genes are functionally closely related.

Gentrepid (https://www.gentrepid.org/) (25) aims to improve some of the existing methods for candidate gene prediction by using structural bioinformatics and system biology approaches such as domain comparison, pathways, and protein–protein interaction data. This tool is based on two assumptions. First, newly identified disease genes and the known disease genes participate in the same complex or pathway. Second, candidate genes that have same phenotype as known disease genes have similar functions. Gentrepid is reported to have better performance than the updated version of the G2D tool which outperformed earlier tools.

PhenoPred (http://www.phenopred.org/) (26) utilizes publically available protein interaction, gene function, sequence features, and disease information to prioritize genes associated with disease. The authors have automatically mapped protein-disease annotations to the Disease Ontology (SVM) hierarchy. Then, for each disease, a support vector machine is trained using random genes as negative examples. Then, each of the SVMs is applied back to genes not used in training, and the prediction scores are ranked. A web service for all of the annotations is available on the website, and either genes or diseases can be queried. Several of these tools have been compared and used in concert to identify genes in complex diseases including type 2 diabetes and obesity (27). It should be noted that how each of these methods compared to each other is unclear. Each method is listed in Table 1.

It is worth being aware of the drawbacks of using the various features in the described tools. The disadvantage of the tools that rely mainly on GO terms is that GO annotation is not complete due to the ongoing process of annotation and also includes a bias to well characterized or studied diseases. Earlier tools such as SUSPECTS, POCUS (28), and G2D are based on descriptive keyword search to identify candidate disease gene. In the case of prediction tools based on structural characteristics of

Table 1
Bioinformatic tools for prioritization of genetic disease candidate genes

Name	URL	Features
FitSNPs	http://fitsnps.stanford.edu/	Human Gene expression
Endeavour	http://homes.esat.kuleuven.be/~bioiuser/endeavour/	Gene expression, protein interaction, protein sequence and domain, Kyoto Encyclopedia of Genes and Genomes (KEGG), literature, others
Gene2Disease (G2D)	http://www.ogic.ca/projects/g2d_2/	Gene Ontology (GO) and biomedical literature searches using MEDLINE
GeneSeeker	http://www.cmbi.ru.nl/geneseeker/	Cytogenetic localization and gene expression patterns from mouse and human databases
Gentrepid	https://www.gentrepid.org/	Domain comparison, pathways and protein–protein interaction.
PhenoPred	http://www.phenopred.org/	Protein interaction, gene function, sequence features and disease information
PRIORITIZER	http://pcdoeglas.med.rug.nl/prioritizer/	GO, KEGG, Biomolecular Interaction Network Database (BIND), Human Protein Reference Database (HPRD), Reactome, predicted PPI and expression data
SUSPECTS	http://www.genetics.med.ed.ac.uk/suspects/	Sequence, GO, InterPro and expression libraries
Transcriptomics of OMIM (TOM)	http://www-micrel.deis.unibo.it/~tom/	GO, genomic location and expression data

gene products, one can leave out the specificity of the gene-by-gene insight that is available in the case of ontology based tools. TOM tries to merge both the methods.

3.2. Prioritization of Functional SNPs and Mutations

The first useful approach a researcher should undertake for identification of functional sites near genetic variation data is to identify functional features that reside on or near the site of variability. This will enable hypothesis generation and guide the researcher toward the first experiments to assay a potential functional effect. The first approach is almost always visualization upon a genome browser, such as UCSC Genome Database (http://genome.ucsc.edu/) (29) or Ensembl (http://ensembl.org/) (30). However, in addition to these resources, several SNP or mutation specific databases have been developed that provide a variety of genomic annotations. Below they are described, separated by the types of genomic features they can provide, such as at the protein level, at the mRNA/transcript level, and at the genomic level. Each of the following resources is generally freely available and can be accessed on the Internet.

3.2.1. Protein Level

3.2.1.1. Protein Structure Annotation

One of the most common annotations of a SNP is identification of its location on a known or predicted protein structure (see review (31) for understanding the importance of protein structure in genetics). Several web-based databases annotate protein structure and provide a variety of services to query, and these include Large Scale human SNP annotation (LS-SNP, http://modbase.compbio.ucsf.edu/LS-SNP//) (32), SNPs3D (http://snps3d.org/) (33), MutDB (http://www.mutdb.org/) (34), and PolyDoms (http://polydoms.cchmc.org/polydoms/) (35). LS-SNP stands out as a useful and unique resource because it provides annotations of nsSNPs that have been mapped to homology models from the MODBASE (http://modbase.compbio.ucsf.edu/modbase-cgi/index.cgi) (36) dataset.

While visualizing protein structure is useful to an expert in the biochemistry of that protein, it may or may not be useful for hypothesizing the effects that an amino acid substitution will have on that site. This is because effects on protein structure can be very subtle and may be visually nonobvious.

3.2.1.2. Annotation of Known Functional Sites

Many bioinformatic tools are available to predict functional sites upon protein sequences and structures. These tools generally are developed in laboratories of individual researchers and are widely distributed. Examples include prediction of catalytic residues in enzymes (37), protein and DNA binding residues (38), and post-translational modifications (39). Several papers have discussed the importance of stability (40), protein interaction (41), and other functions, such as posttranslational modifications, on disease proteins (42). Reviewing all of them is beyond the scope of this chapter. However, there are some resources that integrate several annotations together for a more comprehensive analysis. First, the Universal Protein Resource (Uniprot, http://www.pir.uniprot.org/) database (43) contains annotations of both variation (VARIANT features) and sites of interest, such as posttranslational modification sites. Second, several datasets directly integrate genetic variation data and known protein functional sites such as the SNP Function Portal (44) and SNPeffect (http://snpeffect.vib.be/) (45). The SNPeffect and PupaSuite (http://pupasuite.bioinfo.cipf.es/) (46) tools have been updated to combine annotations and provide predictions of functional site disruption on protein sequences and structures (47).

If any of these predictive tools are used, however, the accuracies of the methods should be scrutinized by referring back to the paper that originally described the method. Again, these methods should be used to hypothesize the effect, and should not be

Table 2
Useful annotation resources for characterization and hypothesizing of SNP function. The following resources aggregate many annotations from other resources

		Genome	Transcript	Protein
LS-SNP	http://modbase.compbio.ucsf.edu/LS-SNP/			X
MutDB	http://mutdb.org/			X
PolyDoms	http://polydoms.cchmc.org/			X
PolyMAPr	http://pharmacogenomics.wustl.edu/	X	X	X
PromoLign	http://polly.wustl.edu/promolign/main.html	X		
PupaSuite	http://pupasuite.bioinfo.cipf.es/	X	X	X
SNP function portal	http://brainarray.mbni.med.umich.edu/Brainarray/Database/SearchSNP/snpfunc.aspx	X	X	X
SNP@Promoter	http://variome.kobic.re.kr/SNPatPromoter/	X		
SNPPer	http://snpper.chip.org/	X		
SNPs3D	http://www.snps3d.org/			X
SNPSeek	http://snp.wustl.edu/cgi-bin/SNPseek/index.cgi	X	X	X

considered definitive or causative, because they generally have high false discovery rates and sensitivity may be low (1, 2). Further biochemical experiments are almost always required for confirmation. These methods are summarized in Table 2.

3.2.1.3. Prediction of Whether an Amino Acid Substitution Will Affect Protein Function or Phenotype

Many tools have been developed to prioritize a given amino acid substitution and many analyses have been applied to understanding the effects of nsSNPs and mutations that are not included in the tools below (48–52). These tools are all supervised, that is, they use a training set of positive and negative examples to "learn" sites. They usually use features based on sequence, structure, or known function. Some of these tools use experimental amino acid substitutions as training (49, 50, 53), others use substitutions based on disease-associated human alleles (32, 33, 54, 55). Two of the first published methods were Sorting Intolerant from Tolerant (SIFT, http://blocks.fhcrc.org/sift/SIFT.html) (56) and Polymorphism Phenotyping (PolyPhen, http://genetics.bwh.harvard.edu/pph/) (55), and both are widely accepted and easy to use. SIFT uses conservation in a multiple sequence alignment as its sole feature, and experimental mutations as its training data. PolyPhen includes protein structure data and other features, while its training is based on human allele data. More recently, other methods have been developed and deployed online, including SNPs3D (33), LS-SNP (32), PMut (http://mmb2.pcb.ub.es:8080/PMut/) (54), the SAP prediction method (http://sapred.cbi.pku.edu.cn/) (57), Screening for Nonacceptable Polymorphisms (SNAP, http://cubic.bioc.columbia.edu/services/SNAP/) (58), Predicting the Amino Acid Replacement Probability (Parepro, http://www.mobioinfor.cn/parepro/) (59) and Protein Analysis Through Evolutionary Relationships (PANTHER, http://www.pantherdb.org/) (60). For a recent comparison of most of these methods, see the review of Ng and Henikoff (2). The SVM utilized by LS-SNP (32) and the method SNAP (58) are two more recent additions to this library of tools that have web sites available for prediction.

Two considerations should be made when choosing a tool to use. First, training sets used for prediction are an important issue to consider when choosing a method; recently, an overview of this issue was published (53). Second, the approach for classification should also be considered, although in general, more recent machine learning approaches appear to be more accurate. Overall, characterizing protein amino acid substitutions remains the most well studied area of predicting the effects of genetic variation. Current research efforts are focusing on improving accuracy through better features, training sets, and classification approaches. The methods described here are summarized in Table 3.

3.2.2. Transcript Level

3.2.2.1. Annotation of Sites that May Affect Splicing

Several resources annotate SNPs with transcript level features. It is well understood that pathogenic mutations can occur in splicing factor binding sites such as intron–exon splice sites, exonic splicing enhancers (ESE) and silencers (ESS). A recent review highlights the importance of splicing function on genetic disease (61). There are now several tools available for annotation of splicing effects including Polymorphism Mining and Annotation Programs (PolyMAPr, http://pharmacogenomics.wustl.edu/) (62), SNPSeek (http://snp.wustl.edu/cgi-bin/SNPseek/index.cgi), PupaSuite (http://pupasuite.bioinfo.cipf.es/) (46) and the SNP Function Portal (44). These resources generally use motif or position specific scoring matrix (PSSM) based prediction of splicing signals or known sites in humans or comparative sites in model organisms such as ESEFinder (63).

3.2.3. Genome Level

3.2.3.1. Identification of Genomic Features near a Candidate SNP

It is now clear that genetic variation affects gene expression and can affect phenotype (see introduction of (64) for brief review). The molecular mechanisms underlying changes in gene expression continue to be unclear, although there are now insights. One challenge in identification of human functional SNPs is that many

Table 3
Tools for predicting functional nonsynonymous single nucleotide polymorphisms (nsSNP)

Name	URL	Training set
LS-SNP	http://modbase.compbio.ucsf.edu/LS-SNP/	Human allele
PANTHER	http://www.pantherdb.org/tools/csnpScoreForm.jsp	Evolution/human allele
Parepro	http://www.mobioinfor.cn/parepro/	Human allele
PMut	http://mmb2.pcb.ub.es:8080/PMut/	Human allele
PolyPhen	http://genetics.bwh.harvard.edu/pph/	Human allele
SAPRED	http://sapred.cbi.pku.edu.cn/	Human allele
SIFT	http://blocks.fhcrc.org/sift/SIFT.html	Experimental[a]
SNAP	http://cubic.bioc.columbia.edu/servers/SNAP/	Experimental[b]
SNPs3D	http://www.snps3d.org/	Human allele

[a]Training set consists of saturation mutagenesis experimental data in *LacI*, HIV-1 protease, T4 lysozyme
[b]Training set consists of amino acid substitutions in the Protein Mutant Database (73) and Swiss-Prot

SNPs may be in linkage disequilibrium (LD) with each other. That is, pairs, or groups, of SNPs may be highly correlated within a population, preventing accurate statistical identification of the causal element. This challenge has kept experimentally validated functional SNPs for use as bioinformatic training data for predicting expression altering SNPs elusive (65).

There are several SNP browsing tools that can identify features in the promoter region and relate that information to SNPs that are present upon them. These include the Ensembl, NCBI, and UCSC genome databases (29, 30, 66), SNPper (67), SNPSeek, SNP@Promoter (http://variome.kobic.re.kr/SNPatPromoter/) (68), the SNP Function Portal (44), PupaSuite (46), and PolyMAPr (62). Generally, these tools can provide annotations of sequence conservation from genome alignments, transcription factor binding sites using databases such as the Transcription Factor Database (TRANSFAC, http://www.biobase-international.com/pages/index.php?id=transfac) (69), CpG islands, and other genomic features. Other features shown to be of interest, such as microRNA binding sites, are currently not available outside of the genome browsers (70).

3.2.3.2. Identification of SNPs that May Affect Expression of Genes

Although this is still an ongoing area of research, there are now insights into the mechanisms of cis-acting alleles. A recent survey of features for prediction of regulatory SNPs found that distance to the transcription start site, local repetitive content, sequence conservation, allele frequency, and CpG islands were the most important features for discrimination of regulatory SNPs (71). Transacting regulation appears to be more complicated (64).

Accurate prediction of genetic regulatory networks appears to be in its infancy. Recently, sequence based prediction of expression was shown to be feasible in Drosophila using the sequences of transcription factor binding sites (72). However, this approach has not been shown to work for changes as small as a SNP.

4. Conclusions

In summary, there are now many resources for prediction of candidate genes (Table 1) and functional SNPs (Tables 2 and 3). Much research has been performed in predicting the effects of protein amino acid substitutions. Many functional SNPs are synonymous or fall outside of coding regions. This has led to more research focus on predicting the effects of these variants, and we are now beginning to understand the features that are important for determining molecular disruption.

Acknowledgments

We are graciously supported by K22LM009135 (PI: Mooney), R01LM009722 (PI: Mooney), P01AG018397 (PI: Econs), U01GM061373 (PI: Flockhart), and the Indiana Genomics Initiative. The Indiana Genomics Initiative (INGEN) is supported in part by the Lilly Endowment.

References

1. Mooney, S. (2005) Bioinformatics approaches and resources for single nucleotide polymorphism functional analysis. *Brief Bioinform*, **6**, 44–56.
2. Ng, P.C. and Henikoff, S. (2006) Predicting the effects of amino Acid substitutions on protein function. *Annu Rev Genomics Hum Genet*, **7**, 61–80.
3. Steward, R.E., MacArthur, M.W., Laskowski, R.A. and Thornton, J.M. (2003) Molecular basis of inherited diseases: a structural perspective. *Trends Genet*, **19**, 505–513.
4. Cooper, D.N., Stenson, P.D. and Chuzhanova, N.A. (2006) The Human Gene Mutation Database (HGMD) and its exploitation in the study of mutational mechanisms. *Curr Protoc Bioinformatics*, **Chapter 1**, Unit 1.13.
5. Hamosh, A., Scott, A.F., Amberger, J., Valle, D. and McKusick, V.A. (2000) Online Mendelian Inheritance in Man (OMIM). *Hum Mutat*, **15**, 57–61.
6. Altman, R.B. (2007) PharmGKB: a logical home for knowledge relating genotype to drug response phenotype. *Nat Genet*, **39**, 426.
7. Mailman, M.D., Feolo, M., Jin, Y., Kimura, M., Tryka, K., Bagoutdinov, R., *et al.* (2007)

The NCBI dbGaP database of genotypes and phenotypes. *Nat Genet*, **39**, 1181–1186.

8. Sjoblom, T., Jones, S., Wood, L.D., Parsons, D.W., Lin, J., Barber, T.D., *et al.* (2006) The consensus coding sequences of human breast and colorectal cancers. *Science*, **314**, 268–274.
9. Greenman, C., Stephens, P., Smith, R., Dalgliesh, G.L., Hunter, C., Bignell, G., *et al.* (2007) Patterns of somatic mutation in human cancer genomes. *Nature*, **446**, 153–158.
10. Pettersen, E.F., Goddard, T.D., Huang, C.C., Couch, G.S., Greenblatt, D.M., Meng, E.C. and Ferrin, T.E. (2004) UCSF Chimera – a visualization system for exploratory research and analysis. *J Comput Chem*, **25**, 1605–1612.
11. Chen, R., Morgan, A.A., Dudley, J., Deshpande, T., Li, L., Kodama, K., Chiang, A.P. and Butte, A.J. (2008) FitSNPs: highly differentially expressed genes are more likely to have variants associated with disease. *Genome Biol*, **9**, R170.
12. Aerts, S., Lambrechts, D., Maity, S., Van Loo, P., Coessens, B., De Smet, F., *et al.* (2006) Gene prioritization through genomic data fusion. *Nat Biotechnol*, **24**, 537–544.
13. van Driel, M.A., Cuelenaere, K., Kemmeren, P.P., Leunissen, J.A., Brunner, H.G. and Vriend, G. (2005) GeneSeeker: extraction and integration of human disease-related information from web-based genetic databases. *Nucleic Acids Res*, **33**, W758–W761.
14. Perez-Iratxeta, C., Wjst, M., Bork, P. and Andrade, M.A. (2005) G2D: a tool for mining genes associated with disease. *BMC Genet*, **6**, 45.
15. Ashburner, M., Ball, C.A., Blake, J.A., Botstein, D., Butler, H., Cherry, J.M., *et al.* (2000) Gene ontology: tool for the unification of biology. The Gene Ontology Consortium. *Nat Genet*, **25**, 25–29.
16. Adie, E.A., Adams, R.R., Evans, K.L., Porteous, D.J. and Pickard, B.S. (2006) SUSPECTS: enabling fast and effective prioritization of positional candidates. *Bioinformatics*, **22**, 773–774.
17. Adie, E.A., Adams, R.R., Evans, K.L., Porteous, D.J. and Pickard, B.S. (2005) Speeding disease gene discovery by sequence based candidate prioritization. *BMC Bioinformatics*, **6**, 55.
18. Mulder, N.J., Apweiler, R., Attwood, T.K., Bairoch, A., Bateman, A., Binns, D., *et al.* (2007) New developments in the InterPro database. *Nucleic Acids Res*, **35**, D224–D228.
19. Rossi, S., Masotti, D., Nardini, C., Bonora, E., Romeo, G., Macii, E., *et al.* (2006) TOM: a web-based integrated approach for identification of candidate disease genes. *Nucleic Acids Res*, **34**, W285–W292.
20. Franke, L., van Bakel, H., Fokkens, L., de Jong, E.D., Egmont-Petersen, M. and Wijmenga, C. (2006) Reconstruction of a functional human gene network, with an application for prioritizing positional candidate genes. *Am J Hum Genet*, **78**, 1011–1025.
21. Ogata, H., Goto, S., Sato, K., Fujibuchi, W., Bono, H. and Kanehisa, M. (1999) KEGG: Kyoto Encyclopedia of Genes and Genomes. *Nucleic Acids Res*, **27**, 29–34.
22. Bader, G.D., Betel, D. and Hogue, C.W. (2003) BIND: the Biomolecular Interaction Network Database. *Nucleic Acids Res*, **31**, 248–250.
23. Peri, S., Navarro, J.D., Kristiansen, T.Z., Amanchy, R., Surendranath, V., Muthusamy, B., *et al.* (2004) Human protein reference database as a discovery resource for proteomics. *Nucleic Acids Res*, **32**, D497–D501.
24. Mishra, G.R., Suresh, M., Kumaran, K., Kannabiran, N., Suresh, S., Bala, P., *et al.* (2006) Human protein reference database – 2006 update. *Nucleic Acids Res*, **34**, D411–D414.
25. George, R.A., Liu, J.Y., Feng, L.L., Bryson-Richardson, R.J., Fatkin, D. and Wouters, M.A. (2006) Analysis of protein sequence and interaction data for candidate disease gene prediction. *Nucleic Acids Res*, **34**, e130.
26. Radivojac, P., Peng, K., Clark, W.T., Peters, B.J., Mohan, A., Boyle, S.M. and Mooney, S.D. (2008) An integrated approach to inferring gene-disease associations in humans. *Proteins*, **72**, 1030–1037.
27. Tiffin, N., Adie, E., Turner, F., Brunner, H.G., van Driel, M.A., Oti, M., *et al.* (2006) Computational disease gene identification: a concert of methods prioritizes type 2 diabetes and obesity candidate genes. *Nucleic Acids Res*, **34**, 3067–3081.
28. Turner, F.S., Clutterbuck, D.R. and Semple, C.A. (2003) POCUS: mining genomic sequence annotation to predict disease genes. *Genome Biol*, **4**, R75.
29. Karolchik, D., Baertsch, R., Diekhans, M., Furey, T.S., Hinrichs, A., Lu, Y.T., *et al.* (2003) The UCSC Genome Browser Database. *Nucleic Acids Res*, **31**, 51–54.
30. Birney, E., Andrews, D., Bevan, P., Caccamo, M., Cameron, G., Chen, Y., *et al.* (2004) Ensembl 2004. *Nucleic Acids Res*, **32 Database issue**, D468–D470.
31. Laskowski, R.A. and Thornton, J.M. (2008) Understanding the molecular machinery of

genetics through 3D structures. *Nat Rev Genet*, **9**, 141–151.

32. Karchin, R., Diekhans, M., Kelly, L., Thomas, D.J., Pieper, U., Eswar, N., *et al.* (2005) LS-SNP: large-scale annotation of coding non-synonymous SNPs based on multiple information sources. *Bioinformatics*, **21**, 2814–2820.
33. Yue, P., Melamud, E. and Moult, J. (2006) SNPs3D: candidate gene and SNP selection for association studies. *BMC Bioinformatics*, 7, 166.
34. Singh, A., Olowoyeye, A., Baenziger, P.H., Dantzer, J., Kann, M.G., Radivojac, P., *et al.* (2007) MutDB: update on development of tools for the biochemical analysis of genetic variation. *Nucleic Acids Res*, **36 (Database issue)**, D815–D819.
35. Jegga, A.G., Gowrisankar, S., Chen, J. and Aronow, B.J. (2007) PolyDoms: a whole genome database for the identification of non-synonymous coding SNPs with the potential to impact disease. *Nucleic Acids Res*, **35**, D700–D706.
36. Pieper, U., Eswar, N., Braberg, H., Madhusudhan, M.S., Davis, F.P., Stuart, A.C., *et al.* (2004) MODBASE, a database of annotated comparative protein structure models, and associated resources. *Nucleic Acids Res*, **32 Database issue**, D217–D222.
37. Youn, E., Peters, B., Radivojac, P. and Mooney, S.D. (2006) Evaluation of features for catalytic residue prediction in novel folds. *Protein Sci*, **16**, 216–226.
38. Ofran, Y. and Rost, B. (2003) Predicted protein–protein interaction sites from local sequence information. *FEBS Lett*, **544**, 236–239.
39. Iakoucheva, L.M., Radivojac, P., Brown, C.J., O'Connor, T.R., Sikes, J.G., Obradovic, Z. and Dunker, A.K. (2004) The importance of intrinsic disorder for protein phosphorylation. *Nucleic Acids Res*, **32**, 1037–1049.
40. Wang, Z. and Moult, J. (2001) SNPs, protein structure, and disease. *Hum Mutat*, **17**, 263–270.
41. Ye, Y., Li, Z. and Godzik, A. (2006) Modeling and analyzing three-dimensional structures of human disease proteins. *Pac Symp Biocomput*, **11**, 439–446.
42. Radivojac, P., Baenziger, P.H., Kann, M.G., Mort, M.E., Hahn, M.W. and Mooney, S.D. (2008) Gain and loss of phosphorylation sites in human cancer. *Bioinformatics*, **24**, i241–i247.
43. UniProt Consortium (2008) The universal protein resource (UniProt). *Nucleic Acids Res*, **36**, D190–D195.
44. Wang, P., Dai, M., Xuan, W., McEachin, R.C., Jackson, A.U., Scott, L.J., *et al.* (2006) SNP Function Portal: a web database for exploring the function implication of SNP alleles. *Bioinformatics*, **22**, e523–e529.
45. Reumers, J., Maurer-Stroh, S., Schymkowitz, J. and Rousseau, F. (2006) SNPeffect v2.0: a new step in investigating the molecular phenotypic effects of human non-synonymous SNPs. *Bioinformatics*, **22**, 2183–2185.
46. Conde, L., Vaquerizas, J.M., Santoyo, J., Al-Shahrour, F., Ruiz-Llorente, S., Robledo, M. and Dopazo, J. (2004) PupaSNP Finder: a web tool for finding SNPs with putative effect at transcriptional level. *Nucleic Acids Res*, **32**, W242–W248.
47. Reumers, J., Conde, L., Medina, I., Maurer-Stroh, S., Van Durme, J., Dopazo, J., *et al.* (2008) Joint annotation of coding and non-coding single nucleotide polymorphisms and mutations in the SNPeffect and Pupa Suite databases. *Nucleic Acids Res*, **36**, D825–D829.
48. Cai, Z., Tsung, E.F., Marinescu, V.D., Ramoni, M.F., Riva, A. and Kohane, I.S. (2004) Bayesian approach to discovering pathogenic SNPs in conserved protein domains. *Hum Mutat*, **24**, 178–184.
49. Chasman, D. and Adams, R.M. (2001) Predicting the functional consequences of non-synonymous single nucleotide polymorphisms: structure-based assessment of amino acid variation. *J Mol Biol*, **307**, 683–706.
50. Krishnan, V.G. and Westhead, D.R. (2003) A comparative study of machine-learning methods to predict the effects of single nucleotide polymorphisms on protein function. *Bioinformatics*, **19**, 2199–2209.
51. Saunders, C.T. and Baker, D. (2002) Evaluation of structural and evolutionary contributions to deleterious mutation prediction. *J Mol Biol*, **322**, 891–901.
52. Vitkup, D., Sander, C. and Church, G.M. (2003) The amino-acid mutational spectrum of human genetic disease. *Genome Biol*, **4**, R72.
53. Care, M.A., Needham, C.J., Bulpitt, A.J. and Westhead, D.R. (2007) Deleterious SNP prediction: be mindful of your training data! *Bioinformatics*, **23**, 664–672.
54. Ferrer-Costa, C., Gelpi, J.L., Zamakola, L., Parraga, I., de la Cruz, X. and Orozco, M. (2005) PMUT: a web-based tool for the annotation of pathological mutations on proteins. *Bioinformatics*, **21**, 3176–3178.
55. Ramensky, V., Bork, P. and Sunyaev, S. (2002) Human non-synonymous SNPs: server and survey. *Nucleic Acids Res*, **30**, 3894–3900.

56. Ng, P.C. and Henikoff, S. (2003) SIFT: predicting amino acid changes that affect protein function. *Nucleic Acids Res*, **31**, 3812–3814.
57. Ye, Z.Q., Zhao, S.Q., Gao, G., Liu, X.Q., Langlois, R.E., Lu, H. and Wei, L. (2007) Finding new structural and sequence attributes to predict possible disease association of single amino acid polymorphism (SAP). *Bioinformatics*, **23**, 1444–1450.
58. Bromberg, Y. and Rost, B. (2007) SNAP: predict effect of non-synonymous polymorphisms on function. *Nucleic Acids Res*, **35**, 3823–3835.
59. Tian, J., Wu, N., Guo, X., Guo, J., Zhang, J. and Fan, Y. (2007) Predicting the phenotypic effects of non-synonymous single nucleotide polymorphisms based on support vector machines. *BMC Bioinformatics*, **8**, 450.
60. Mi, H., Lazareva-Ulitsky, B., Loo, R., Kejariwal, A., Vandergriff, J., Rabkin, S., *et al.* (2005) The PANTHER database of protein families, subfamilies, functions and pathways. *Nucleic Acids Res*, **33**, D284–D288.
61. Wang, G.S. and Cooper, T.A. (2007) Splicing in disease: disruption of the splicing code and the decoding machinery. *Nat Rev Genet*, **8**, 749–761.
62. Freimuth, R.R., Stormo, G.D. and McLeod, H.L. (2005) PolyMAPr: programs for polymorphism database mining, annotation, and functional analysis. *Hum Mutat*, **25**, 110–117.
63. Smith, P.J., Zhang, C., Wang, J., Chew, S.L., Zhang, M.Q. and Krainer, A.R. (2006) An increased specificity score matrix for the prediction of SF2/ASF-specific exonic splicing enhancers. *Hum Mol Genet*, **15**, 2490–2508.
64. Yvert, G., Brem, R.B., Whittle, J., Akey, J.M., Foss, E., Smith, E.N., *et al.* (2003) Trans-acting regulatory variation in *Saccharomyces cerevisiae* and the role of transcription factors. *Nat Genet*, **35**, 57–64.
65. Hudson, T.J. (2003) Wanted: regulatory SNPs. *Nat Genet*, **33**, 439–440.
66. Pruitt, K.D. and Maglott, D.R. (2001) RefSeq and LocusLink: NCBI gene-centered resources. *Nucleic Acids Res*, **29**, 137–140.
67. Riva, A. and Kohane, I.S. (2002) SNPper: retrieval and analysis of human SNPs. *Bioinformatics*, **18**, 1681–1685.
68. Kim, B.C., Kim, W.Y., Park, D., Chung, W.H., Shin, K.S. and Bhak, J. (2008) SNP@Promoter: a database of human SNPs (single nucleotide polymorphisms) within the putative promoter regions. *BMC Bioinformatics*, **9 Suppl 1**, S2.
69. Matys, V., Fricke, E., Geffers, R., Gossling, E., Haubrock, M., Hehl, R., *et al.* (2003) TRANSFAC: transcriptional regulation, from patterns to profiles. *Nucleic Acids Res*, **31**, 374–378.
70. Chen, K. and Rajewsky, N. (2006) Natural selection on human microRNA binding sites inferred from SNP data. *Nat Genet*, **38**, 1452–1456.
71. Montgomery, S.B., Griffith, O.L., Schuetz, J.M., Brooks-Wilson, A. and Jones, S.J. (2007) A survey of genomic properties for the detection of regulatory polymorphisms. *PLoS Comput Biol*, **3**, e106.
72. Segal, E., Raveh-Sadka, T., Schroeder, M., Unnerstall, U. and Gaul, U. (2008) Predicting expression patterns from regulatory sequence in *Drosophila* segmentation. *Nature*, **451**, 535–540.
73. Kawabata, T., Ota, M. and Nishikawa, K. (1999) The Protein Mutant Database. *Nucleic Acids Res*, **27**, 355–357.

Chapter 18

Analysis of the Impact of Genetic Variation on Human Gene Expression

Elin Grundberg, Tony Kwan, and Tomi M. Pastinen

Abstract

Interindividual variation in gene expression has been convincingly shown to be controlled, in part, by genetic differences. Determining the architecture of genetic variation, the underlying gene expression may allow deeper insight into complex phenotypes, such as differences in disease susceptibility. Mapping genetic variants accounting for expression phenotypes in human cell and tissue panels has rapidly progressed from proof-of-principle experiments to general tools in biomedical discovery. We discuss the general approach and critical considerations for carrying out expression quantitative trait mapping in human tissues.

Key words: SNP, eQTL, Regulatory variation, DNA microarrays

1. Introduction

Technical and conceptual breakthroughs in human genomics during the past 15 years include DNA microarrays to assess transcriptome or genetic variation on a genome-wide scale, delineation of common patterns of genetic variation across human populations (HapMap Consortium 2003), and the development of statistical frameworks to work with large-scale association data. The correlation of human phenotypic variation (e.g., disease status or intermediate phenotypes such as plasma lipid levels) with dense genotyping data across the genome has proven to be very successful in finding common genetic variants linked to common phenotypic traits such as height (1, 2) and hair color (3), or diseases/conditions such as type I and type II diabetes, bipolar disorder, and Crohn's disease (4). These genome-wide association

Michael R. Barnes and Gerome Breen (eds.), *Genetic Variation: Methods and Protocols*, Methods in Molecular Biology, vol. 628, DOI 10.1007/978-1-60327-367-1_18,

studies (GWAS) typically yield genomic regions that span tens to hundreds of kilobases, yet contain a sparse number of genes and very few clear functional candidate variants that could explain the observed phenotypic differences. Functional variants can be subdivided into two main categories: (1) coding polymorphisms altering protein structure, which has now been shown to account for the minority of variants identified to date, and (2) noncoding polymorphisms in inter- and intragenic regions that affect some key regulatory pathway associated with the phenotype being studied. Thus, advances and refinements in GWAS need to be made to address two related questions: how to identify the gene(s) that alter disease risk, and how to pinpoint the causal coding/noncoding variant(s)? Tackling these key issues will aid in elucidating true disease mechanisms and novel therapeutic targets.

The identification of a strong genetic contribution to regulation of gene expression in yeast (5), mouse (6, 7), and human (8), as well as demonstration that these heritable changes can be mapped to genetic loci, commonly referred to as expression quantitative trait loci, (eQTLs) is providing a new avenue for the identification of functional variation in complex genomes. Consequently, there is a strong drive to comprehensively study genetic variation underlying human gene expression differences at the population level in tissues or cells. Human eQTL studies can link noncoding variants to expression differences at the cellular level and indirectly implicate specific genes and mechanisms underlying differential disease susceptibility (9). Since eQTLs typically have size effects (level of population variance explained by genetic markers) an order of magnitude higher than clinical phenotypes, eQTLs are more amenable for fine-mapping causal variants. Finally, eQTL data can assist in developing insight into gene networks underlying complex cellular processes (10).

This chapter describes and summarizes different large-scale technologies and approaches on how to analyze the impact of genetic variation on human gene expression (Fig. 1). Our emphasis is on association-based methods in unrelated individuals, since sample ascertainment for such studies is considerably more straightforward, particularly when complex tissues or primary cells from diverse tissues are considered. Key results from linkage-based studies are nevertheless included in our discussion, since it

are prepared according to manufacturer's protocols and then hybridized onto microarray chips from one of the two main suppliers, Affymetrix and Illumina. Unbound, excess material is then washed off to reduce background noise, followed by scanning in a chip reader. This yields gene (or exon) expression data along with genotyping data, for the RNA and DNA extracts, respectively. A *cis*- or *trans*-association analysis is typically performed, using a linear regression model where the SNP genotypes are coded as factors (0, 1, or 2) and regressed against expression scores for the samples. Multiple testing correction is applied in order to determine a significant *p*-value cutoff, which defines an initial list of significant eQTLs, followed by validation and further examination of the eQTL and eSNP hits

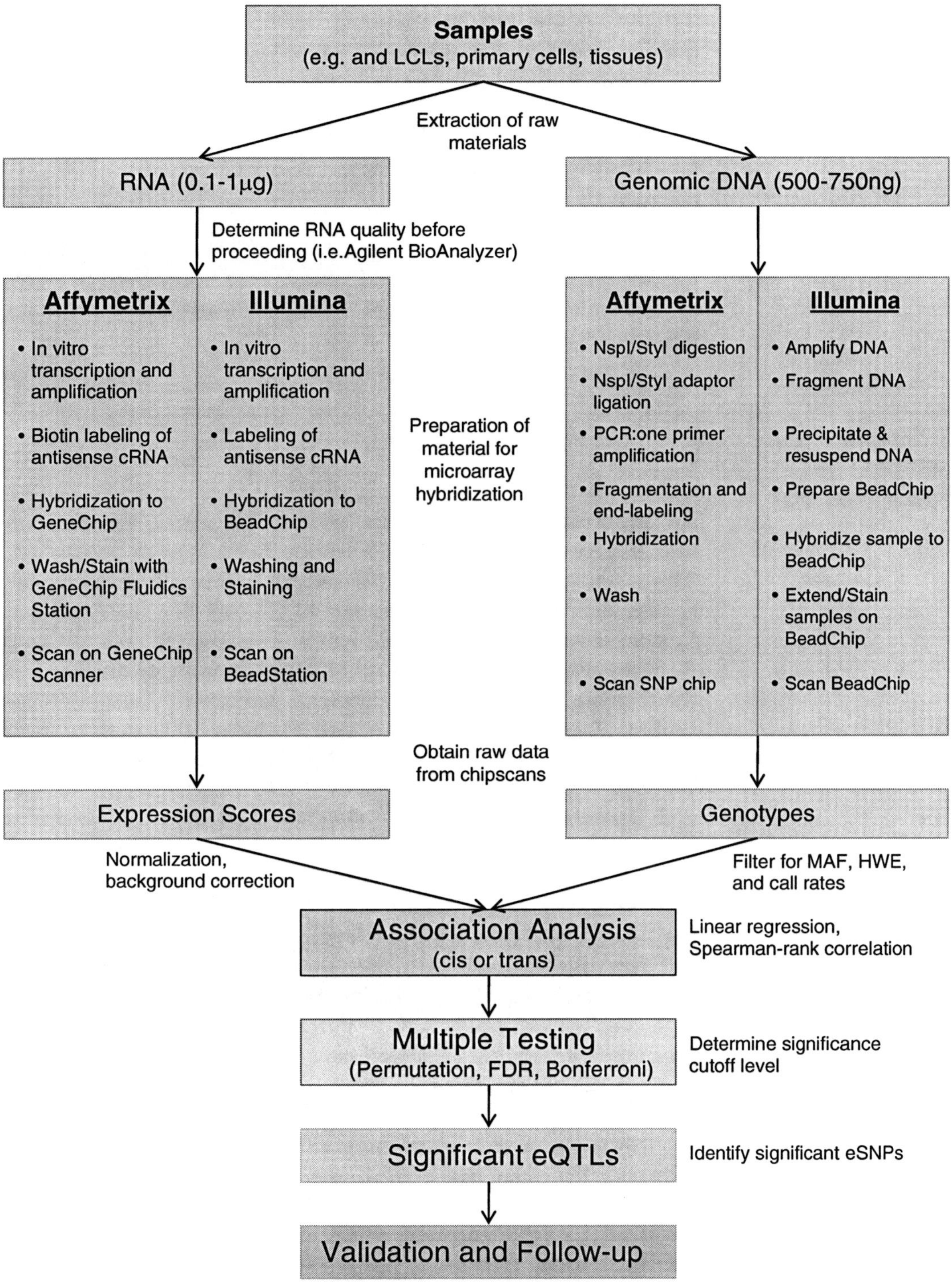

Fig. 1. General flow chart for examining genetic variants affecting eQTLs. Flow chart indicating the general steps from sample selection to analysis of the final results. From the panel of samples, RNA and DNA are extracted using the appropriate kits recommended by the microarray chip manufacturer. Genomic DNA needs to be extracted if genotyping information is unavailable. High RNA quality is critical to ensure optimal hybridization to the chip and reduction of possible background noise and this can be ascertained using appropriate equipment (i.e., Agilent BioAnalyzer). Both RNA and DNA

remains a useful method especially when dealing with cohorts of families. Some of the topics that we will focus on include:

- Sample selection and power analysis
- Whole-genome expression profiling
- Whole-genome genotyping
- Statistical approaches

In the last section, we have summarized the results from some of the key publications in the field, and discuss how different facets of human eQTL analysis were addressed using the approaches described herein.

2. Sample Selection and Power Analysis

An important step prior to the analysis of genetic variation on human gene expression is the selection of samples to be studied. The access to diverse human tissues and cells poses obvious limitations and many of the human eQTL studies have utilized Epstein–Barr virus transformed lymphoblastoid cell lines (LCLs). Publicly available collections of LCLs such as those analyzed in the Human Haplotype Map (HapMap) project (11) facilitate these studies. The HapMap project selected human LCLs derived from four major world populations of Northern European (30 trios), Yoruban African (30 trios), Chinese (45 unrelated individuals) and Japanese (45 unrelated individuals) ancestry and provides high-resolution genotyping information for these populations (Fig. 2). Genome-wide genotyping has been carried out on all these samples for nearly four million SNPs and the data is publicly available (http://www.hapmap.org).

The disadvantages of eQTL studies carried out using these immortalized HapMap LCLs include potential artificial phenotypic and epigenetic alterations induced by immortalization and prolonged cell culture, as well as lack of associated phenotypic (such as disease state) information. Even when LCLs are derived from a disease cohort of interest (12), they offer limited access to the landscape of regulatory variation as only the portion of the transcriptome regulated in LCLs can be analyzed. Consequently, the emphasis is now on measuring gene expression in diverse human cell types or tissues, which in many cases are relevant to disease models. However, utilization of tissues or primary cells poses additional challenges in terms of tissue heterogeneity, limited sample quantity (and quality) as well as finite sample sizes.

Choosing the proper sample size for these genetic association studies is challenging and requires power calculations taking allele frequencies as well as estimated effect sizes into account.

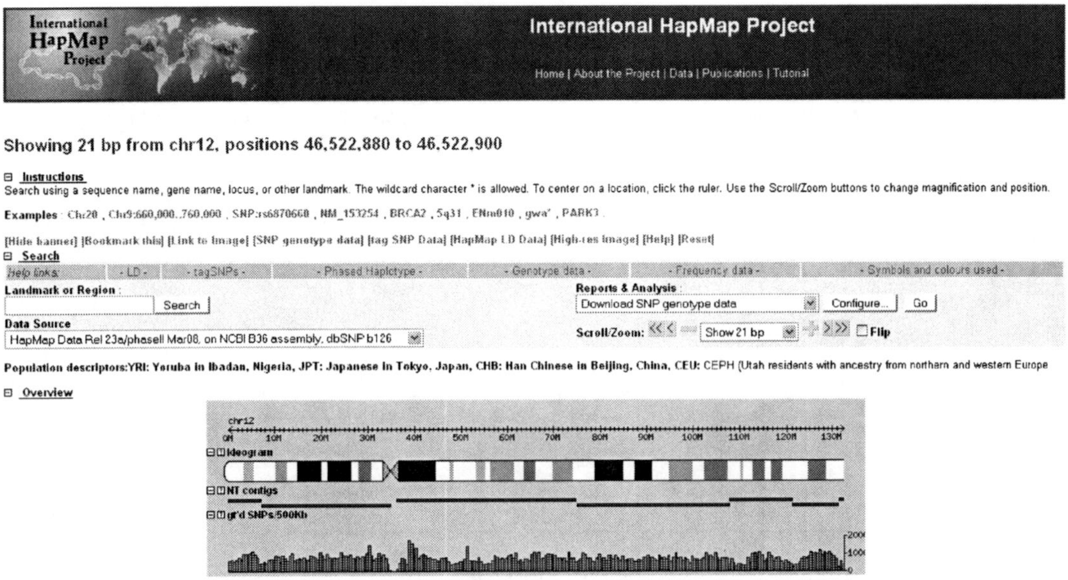

Fig. 2. International HapMap Project. Example screen capture of the International HapMap Project website. Shown is an ideogram of chromosome 12, and underneath are all the HapMap SNPs (release 23a, phase II) for which the HapMap populations (CEU, CHB, JPT, YRI) have been genotyped

In GWAS of clinical phenotypes, large samples with thousands of subjects are needed due to the low penetrance of clinical traits as well as the large numbers of SNPs usually tested. However, sample size estimates used to evaluate power in phenotype-driven GWAS are not appropriate for cellular phenotypes as gene expression. Gene expression phenotypes measured at the molecular level show considerably lower impact from nongenetic confounders (or polygenic inheritance), and work to date has demonstrated the occurrence of hundreds of expression traits where well over 10% (i.e., r^2 in a linear regression model correlating genotype with expression is above 0.1) of the population variance in gene expression is explained by a single genetic marker. Therefore, sample sizes in the low hundreds can be quite well powered to uncover eQTLs (13). However, in order to detect smaller (not necessarily less important) regulatory effects as well as regulatory variation in more heterogeneous human tissues, larger sample sizes are needed (Fig. 3).

3. Whole-Genome Expression Profiling

Several commercially available microarray platforms are currently employed for the study of human whole-genome expression profiling. The three most commonly utilized technologies, are provided by Agilent (Santa Clara, CA), Affymetrix (Santa Clara, CA)

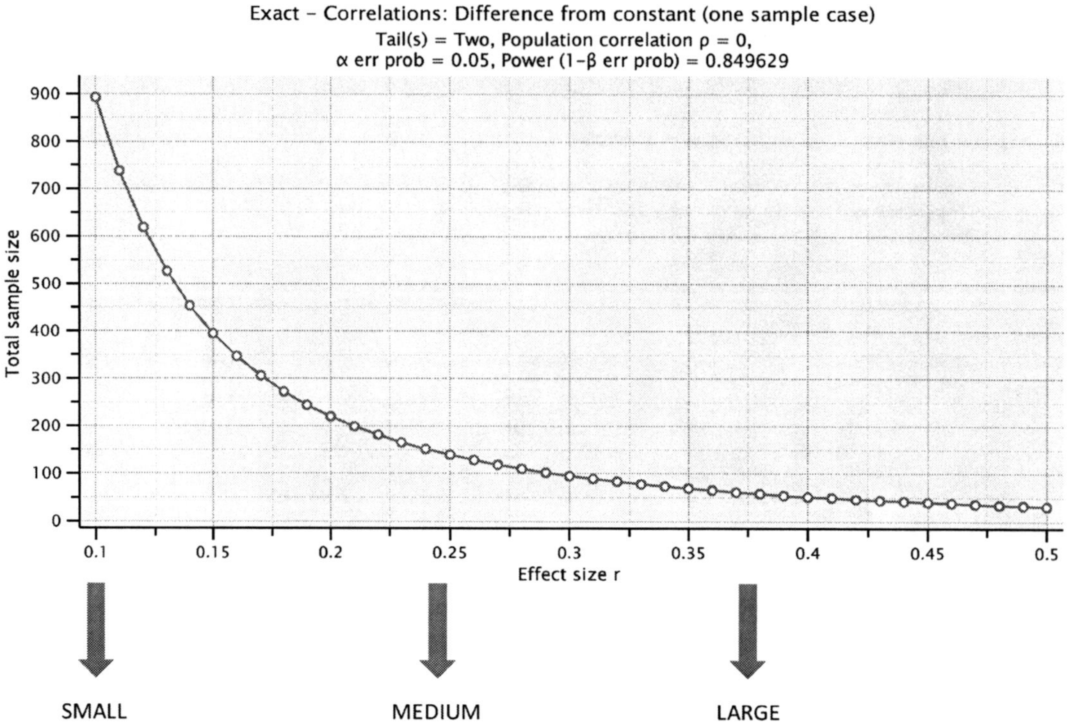

Fig. 3. Power Analysis. Graph showing the power analysis of sample size versus the effect size (r). Using small sample sizes (<100 individuals) allows one to detect large effects (higher r). As the number of samples increases, the analysis will be able to detect smaller and smaller regulatory effects (lower r) in more heterogenous human tissues

Table 1
Key features of the Affymetrix GeneChip and Illumina BeadArray Technologies

Platform	Affymetrix GeneChip	Illumina BeadArrays
Chip format	Single array/chip	6–8 arrays/chip
Probe length	25 mer	50 mer
Probe design	Multiple probes/gene	Multiple copies of the same probe/array
Probe attachment	Microbeads	In situ synthesis

and Illumina (San Diego, CA). In this chapter we focus on the latter two technologies: Affymetrix GeneChips and Illumina BeadArrays (Table 1). These systems differ by methods of generating the oligonucleotide arrays, redundancy of sampling transcripts by independent probes, and the lengths of the probes as well as labeling of target nucleic acids. However, our aim is not to compare these technologies in depth but rather to give case examples on how eQTL analysis is carried out by different approaches.

On the Affymetrix GeneChips, each transcript is targeted by multiple oligonucleotide probes (usually 25-mers), which are synthesized in a light-directed manner on the surface of a glass slide. The hybridized probe intensities are summarized across probes interrogating the same transcript to obtain an overall expression score for the gene that is being targeted. Current GeneChips available for genome-wide expression analysis include the HG-U133 arrays targeting predominantly the 3′ ends of over 47,000 transcripts using more than 1,300,000 distinct oligonucleotide features. A recently released product (GeneChip® Human Gene 1.0 ST Array) targets transcripts along the length of the gene.

The Illumina BeadArray Technology is based on silica beads covered with hundreds of thousands of copies of specific oligonucleotides. The silica beads are randomly assembled in microwells on either of two substrates: fiber optic bundles or planar silica slides. The technology was first used for high-throughput genotyping (see Subheading 4 below) but has recently been applied for whole-genome expression profiling. Using the Illumina BeadArray, each probe is synthesized as a 50 bp long oligonucleotide and similar to the Affymetrix HG-U133 design, is also mapped to the 3′ end of the gene. Typically, a single probe sequence is used to target one transcript; however this probe is included on 20–30 independent beads on the same array. Today, Illumina offers two types of bead chips: the HumanWG-6 and HumanRef-8. The Illumina HumanWG-6 has six arrays per chip and 48,000 bead types (targeting 24,000 RefSeq + 24,000 supplemental genes) per array where the Illumina HumanRef-8 has eight arrays per chip and 24,000 bead types targeting only genes from the RefSeq database.

However, these arrays focus on a single isoform of annotated genes and do not take into account the true landscape of the human transcriptome. The best current estimate on the number of genes found in the human genome is 20,000–25,000; however, it is now known that alternative splicing occurs in approximately 60–70% of genes, generating proteomic diversity (14, 15). The modular nature of gene structure in humans and other mammalian species lends itself to the potential selective inclusion or exclusion of individual exons to generate new transcript isoforms with potentially different functions. On the most commonly used gene expression platforms, most of the known isoforms are not independently analyzed, nor is the complexity of unannotated transcripts and isoforms taken into account in the probe design. While unbiased expression analysis will require sequencing-based approaches (16), more comprehensive expression array platforms also exist such as the Affymetrix Human Exon 1.0 ST GeneChip Array. All known RefSeq exons are represented on this array by a probeset, which comprises up to four individual 25-mer probes.

This allows exons to be interrogated independently of one another and therefore one can detect the presence of isoforms where a single exon is differentially expressed. The Exon Array can also provide information on whole gene expression levels by summarizing the scores from all probesets comprising the gene, and may potentially be a more useful way of measuring gene and isoform level expression.

Along with the introduction of new expression technologies, concerns have been raised about the reliability of the data when comparing between different platforms. Using very high quality RNA sources, the Affymetrix and Illumina gene expression analysis platforms yield highly correlated results (17). An unsolved question is how this technical reproducibility relates to human eQTL studies where the RNA is likely of heterogeneous quality and small differences across hundreds of samples are measured.

4. Whole-Genome Genotyping

Genome-wide assessment of eQTLs using association analyses requires high-density genotyping in parallel with the transcriptome analysis on expression arrays. Solutions for genome-wide genotyping at sufficient density are offered by Affymetrix and Illumina. The Affymetrix genotyping arrays are based on allele-specific hybridization and include up to 900K polymorphic SNPs, as well as additional probes designed to interrogate copy number variations (CNVs) (Genome-Wide Human SNP Array 6.0). Illumina's genotyping system on its BeadChips platform is based on allele discrimination by enzymatic primer extension and can similarly be utilized to interrogate SNPs at very high density, up to 1.07 million SNPs genotyped on a single chip (Human1M BeadChip). Details of SNP selection and genomic coverage of the genome-wide genotyping products can be found at the manufacturers' respective websites (www.affymetrix.com and www.illumina.com). Based on analyses of earlier, lower density products for Caucasian populations, either of the latest genotyping arrays should achieve over ~85% coverage (18).

Genotyping on the Affymetrix arrays uses total genomic DNA (~500 ng) that is digested with *NspI* and *StyI* restriction enzymes and ligated to adaptors which recognize the overhangs. All fragments resulting from the restriction enzyme digestion, regardless of size, are substrates for adaptor ligation. A generic primer that recognizes the adaptor sequence is then used to amplify adaptor-ligated DNA fragments. PCR amplification products for all fragments from each restriction enzyme digest are then combined and purified using polystyrene beads. Finally, the amplified DNA is then fragmented, labeled, and hybridized to the chip.

The most recent releases of genome-wide genotyping arrays based on Illumina BeadChips technology are based on the Infinium® II Assay, which starts with amplification of the genomic DNA (~750 ng). The amplified DNA is then fragmented, precipitated, and hybridized to the BeadChips. Similar to the gene expression protocol, the BeadChips are covered by specific 50-mer oligonucleotides that anneal to the amplified and fragmented DNA. After hybridization, allelic specificity and labeling are achieved by enzymatic primer extension.

5. Statistical Approaches

The availability of both expression data and genotyping data allows us to perform genome-wide association studies linking the genetic variants, sometimes referred as "eSNPs" or "eQTNs" (expression quantitative trait nucleotides), to expression profiles (eQTLs). The association analysis can be performed using a number of different statistical methods. The most commonly used is a codominant model where the genotypes (assuming additive effects) for each individual are coded as a factor (0, 1, or 2) and regressed using a standard linear model against the expression scores for a particular gene. One can also use a nonparametric model to be less conservative, where a codominant or additive model is not assumed, and this can be estimated using the Spearman-rank correlation.

Association studies are generally subdivided into two basic categories: (1) *cis*-associations comprising SNPs within a limited and defined region from the gene being examined, and is typically 1 Mb or less from the 5′ and 3′ ends of genes, and (2) *trans*-associations comprising SNPs at distances greater than the limit defined for *cis*-associations, and includes SNPs that are found on other chromosomes. Some studies are restricted to *cis*-variants, since these are more amenable to detection due to typically larger size effects and smaller statistical penalties for multiple testing.

5.1. Cis-Associations

The range of SNPs to be tested in *cis*-association studies is relatively arbitrary but usually encompasses the entire gene and 1 Mb flanking both the 5′ and 3′ ends of genes. This is a fairly conservative region for testing and may allow for the detection of long-range *cis*-effects; however, it has been shown in numerous studies that the majority (>60%) of *cis*-effects are located within close proximity to the gene itself (<100 Kb) (Fig. 4) (9, 19, 20). This is due to the expectation that most noncoding polymorphisms that affect gene expression would map to regulatory elements, such as transcription factor binding sites, and enhancer elements that are typically located close to the transcript. Therefore, one

Fig. 4. Example of *cis*-associations for eQTL at SERPINB10 locus. (**a**). Vertical lines represents all SNPs (on Illumina Hapmap550 chip) located within 250 Kb flanking each side of the SERPINB10 gene. The height of each line indicates the significance of association from the linear regression analysis of SNP genotypes versus expression of SERPINB10 gene in the CEPH CEU population (parents of the 30 trios). (**b**) Genome map of the region surrounding SERPINB10 (highlighted in box). The majority of significant *cis*-associations (highest vertical bars) are all located with the first 100 Kb upstream of the SERPINB10 transcription start site, most likely due to linkage disequilibrium of a haplotype block

can limit the amount of SNPs to this smaller interval in order to decrease multiple testing problems inherent to large-scale association study approaches.

5.2. Trans-Associations

Trans-associations are typically defined as genetic variants at great distances (>1 Mb) and whose genotypes are correlated with expression of the gene. Despite the small number of validated *trans*-associations to date, carrying out genome-wide association studies may still identify SNPs in *trans* that directly or indirectly affect the expression of a given gene. For example, a SNP may be associated with the expression of a transcription factor that directly controls expression of another gene located on another chromosome. *Trans*-associations may provide valuable information for understanding gene networks and downstream effects of disease associated variants. Two problems make detection of true *trans*-variants challenging: since the search space is three orders of magnitude larger than the one typically involved in analysis of *cis*-variants, the number of false positives is also higher, and more stringent statistical cutoffs (and in some cases prohibitively so) need to be employed making detection of subtle variants impossible. Secondly, validation of *trans*-eQTLs requires perturbation of a network of genes believed to underlie the higher order effect, whereas *cis*-eQTLs can be validated by relatively straightforward reporter gene assays or other in vitro approaches targeting the regulatory sequences of a single gene.

5.3. Linear Regression Analysis

The association analysis can be performed using a number of freely available (R, PLINK) or commercially available (MatLab, SAS) software packages, or any type of software that can implement either a parametric (linear regression) or a more conservative nonparametric (Spearman-rank correlation) model.

The R software package is an overall excellent statistical program and for which a lot of packages have been developed for the analysis and comprehension of genomic data (i.e., BioConductor (21)). However, R is computationally slow and does not perform well for truly large, genome-wide scale association analyses with millions of individual tests. A better alternative for performing linear regression analysis is PLINK (22), which has been written specifically for association testing and requires less computation time than R. This makes a tremendous difference, particularly when carrying out genome-wide analysis for *trans*-eQTL identification as well as for generating permutation matrices for multiple testing corrections.

Whether *cis*- or *trans*-associations are being examined, the SNPs must first be filtered through certain criteria before testing. This is done to eliminate noise from putative genotyping errors as well as removing rare SNPs, which would not have enough power for detection of expression associations. SNPs are generally filtered for and kept if they have minor allele frequency (MAF) >1–5%, Hardy–Weinberg equilibrium (HWE) p-values >0.05, and call rates ≥95% across all samples. Using the PLINK software package, the user can easily specify these parameters in the analysis. For each SNP to be tested against a gene, each genotype is coded as a factor and a linear regression analysis is then performed against the expression scores for the gene across all samples. The analysis generates a number of statistics for the gene–SNP test, and all the relevant output should be retained, including the *p*-value, estimate (slope), and r^2 for the regression equation, the direction of the effect, i.e., which genotype is overexpressed and underexpressed, genotype counts and frequencies, and mean expression scores for each of the genotypes.

5.4. Multiple Testing Corrections

After performing millions of individual SNP–gene tests, the important question is which of these eQTLs are significantly associated with SNPs, and which are significant results obtained just by chance and thereby false positives. One approach for multiple testing simply adjusts for all tests performed (Bonferroni), which is computationally straightforward. But since tested SNPs are not independent, the Bonferroni correction is often considered overly conservative. Therefore, other methods, such as permutation (23) and False Discovery Rate (FDR) (24) based multiple testing corrections, are more commonly employed but come with added computational cost.

The first method and most time-consuming from a computational perspective is permutation testing. In this method, for each gene–SNP combination, typically 1,000–100,000 permutations are carried out wherein expression values are shuffled relative to the genotypes and a linear regression is performed, and the best *p*-value from this set of permutations is retained.

An empirical *p*-value is then obtained by comparing the observed (nonpermuted) *p*-value to the distribution of permuted *p*-values for the same gene. Typically, an empirical *p*-value threshold of 0.0001 is considered significant but the cut-offs are arbitrary.

The second method is the FDR, defined to be the expected proportion of false positive associations among all those called significant. The distribution of all *p*-values (all genes and all SNPs) for *cis*-associations are taken and used to calculate the FDR and to assess the significance of each *p*-value in the distribution. Similar to the *p*-value, a Q-value is calculated for a particular feature as the expected proportion of false positives incurred when calling the feature significant. Normally, this is not done for genome-wide association tests because the number of *p*-values makes it computationally prohibitive. Signals are considered significant if the corresponding Q-value is less than 0.05, if we assume that 5% of all significant hits are false positives.

5.5. Systematic Biases

The use of expression arrays and other arrays based on probe hybridization technology is not without its limitations. A truly reliable signal is based on a perfect 100% match between the probe sequence and the target DNA; a single mismatch or an indel will result in suboptimal binding conditions, consequently leading to a lower hybridization signal. This is a potential problem when examining samples from different individuals, where we know there is natural variation within every individual, and these SNPs are underlying probes on the array, potentially causing false eQTL associations (25). One aspect to keep in mind is that the probability for the presence of SNP effects also increases with larger sample sizes and allele frequencies. The Affymetrix platforms are more susceptible to SNP effects on probe hybridization than the Illumina or Agilent platforms due to the shorter probe lengths (25 bp for Affymetrix versus 50 bp for Illumina/ Agilent); however, one cannot discount the effects of mismatches on hybridization signals with the longer probe lengths. One solution is to cross-reference all the probes on the expression arrays with the location of all SNPs within the dbSNP and/or HapMap databases, and "mask out" or remove probes that overlap SNPs from the analysis. This has been performed in recent studies using the exon arrays to reduce the rate of false positive associations (26, 27). Although this may reduce coverage of the genome, masking out individual probes on the Exon Array that overlap HapMap SNPs result in the loss of <0.5% of probesets (exons), which is very minor and more than acceptable considering the trade-off of reduced false positives. There will still remain unannotated SNPs overlapping probes, but aside from the impractical solution of resequencing all probes on the array for all the samples, this remains the best current solution.

In line with the rapid introduction of large-scale genotyping approaches and GWAS, extensive debate has arisen about the risk of population stratification. Population stratification refers to a situation when individuals within a certain study population have different allele frequencies due to difference in ancestry, which can lead to biased or spurious results. Due to this, it is important to either: (1) demonstrate that population stratification is negligible in the data or (2) if population structure *is* detected, to adequately adjust for it. This can be done using freely available software packages (e.g., EIGENSTRAT (28)), which detects and corrects for population stratification. The approach behind EIGENSTRAT is powerful and based on the principal component analysis that models ancestry differences for thousands of genotype markers.

6. Genetics of Gene Expression: A Literature Review

The pioneering studies examining genetic variation in human gene expression on a genome-wide scale were published in 2004 and focused on linkage analysis in LCLs from the CEPH families (29, 30) (Table 2). Both studies used genotyping information from publicly available databases with genetic markers at a relatively low density, given the unavailability of genome-wide genotyping technology at that time. For measurements of gene expression, early versions of genome-wide oligonucleotide arrays by Affymetrix and Agilent were used. The heritability analyses were restricted to the most differentially expressed genes, and both studies defined a *cis*-eQTL as being located within 5 Mb of the target gene. Monks et al. reported that 31% of expressed genes were detected as heritable with a median heritability of 0.34 (29). Furthermore, they found that 30% of the QTNs were located in *cis*, whereas Morley et al. found that 19% of the eQTLs were linked with *cis*-acting transcriptional regulators (30). Morley et al. further used a population of 94 unrelated samples for the confirmation of the identified eQTLs, and 14/17 phenotypes were found to be significant ($p < 0.005$) using conventional linear regression of the log2-transformed expression scores versus the SNP genotypes. Although these studies were not classical association studies, the concordance between significance in linkage studies and association studies for the small set of tested genes indicated that large unrelated samples, which are much easier to obtain than families, could potentially be used for larger scale association studies. The estimation of heritability of gene expression was later confirmed by two independent studies published in 2007 that used larger sample sizes and whole-genome expression profiling, i.e., Affymetrix HG-U133 (12) and Illumina WG-6 BeadArrays (31).

Table 2
Summary of important papers in the analysis of genetic variation on eQTLs

	Sample size	Sample type	Study population	Genotyping platform	Expression platform	Statistical design
Monks et al. (2004) (29)	15 families (167 individuals)	LCL	European[a]	Marker genotypes[b]	Agilent custom made	Linkage (*cis/trans*)
Morley et al. (2004) (30)	14 families (139 individuals) 94 unrelated	LCL	European[a]	Marker genotypes[b]	Affymetrix Human Genome Focus Array	Linkage (*cis/trans*) Association (*cis/trans*)
Stranger et al. (2005) (32)	60 unrelated	LCL	European[a]	HapMap II release 16b	Illumina custom made	Association (*cis/trans*)
Dixon et al. (2007) (12)	206 families (830 individuals)	LCL	European (British descent)	Illumina Human-1/Illumina Human-Hap300	Affymetrix HG-U133P2	Linkage (*cis/trans*)
Stranger et al. (2007) (20)	30 trios 30 trios 45 unrelated 45 unrelated	LCL	European[a] Yoruban Chinese Japanese	HapMap II release 21	Illumina HumanWG-6	Association (*cis/trans*)
Göring et al. (2007) (31)	42 families (1240 individuals)	Lymphocytes	Mexican Americans	Marker genotypes[c]	Illumina HumanWG-6	Linkage (*cis/trans*)
Myers et al. (2007)	193 unrelated	Brain cortex	European	Affymetrix 500K Array	Illumina HumanRef-8	Association (*cis/trans*)
Spielman et al. (2007)	60 unrelated 41 unrelated 41 unrelated	LCL	European[a] Chinese Japanese	HapMap II release 19	Affymetrix Human Genome Focus Array	Association (*cis/trans*)

Kwan et al. (2008) (26)	60 unrelated	LCL	European[a]	HapMap II release 21	Affymetrix Human Exon 1.0 ST Array	Association (*cis*)
Schadt et al. (2008) (19)	427 unrelated	Liver	Caucasian	Affymetrix 500K	Agilent custom made	Association (*cis/trans*)
Emilsson et al. (2008) (9)	209 families (938 individuals) 124 families (570 individuals)	Blood Adipose tissue	Icelanders	Illumina Human-Hap300	Agilent custom made	Linkage (*cis/trans*) Association (*cis/trans*)

[a]Centre d'Etude du Polymorphisme Humain (CEPH) families of northern/western European origin
[b]Markers obtained from The SNP Consortium
[c]Marker genotypes using the Human MapPairs genome-wide screening set version 6 and 8 from Research Genetics

The first large-scale genetic analysis on population variation in eQTLs using dense genotyping information was published in 2005 and used the HapMap LCL panel from 60 unrelated Utah residents of Northern and Western European ancestry (labeled CEU) (32). At this point in time, the International HapMap project had generated genotyping information of more than one million SNPs for the LCL panels. The expression profiling was based on a custom-made Illumina BeadArray of 630 genes from the ENCODE (33) regions (except the HSA21 ENCODE region), all RefSeq genes on chromosome 21, and all RefSeq and curated genes in a 10 Mb interval between 20q12 and 20q13. From the 630 genes, they performed association studies on a subset of 374 genes that showed the most variable expression amongst all samples as well as having probe hybridization scores significantly above background. After filtering, 753K HapMap SNPs with a minor allele frequency > 5% were used in the regression analyses. *Cis*-associations were defined as <1 Mb distance between the SNP and the ends of the gene, and *trans*-associations were all other SNPs outside of this range. The issue of multiple-test correction was addressed by the three methods that have been mentioned in this review: permutation testing, FDR, and Bonferroni correction, with an overall high concordance between the three methods for assigning statistical significance to the regression analyses. The permutation tests yielded the least stringent cutoff of the three methods and identified 12% of genes to comprise a *cis*-genetic effect, and overall, the proximal *cis*-associations were more abundant than the *trans*-signals. Despite the limited number of genes interrogated in this study, it illustrated the power of using the HapMap populations for SNP–eQTL association studies. A subsequent and more extensive analysis was performed on all four HapMap populations using the Illumina WG-6 expression BeadArrays and the high density genotype information (HapMap Phase II), which include over two million common SNPs per population (20). Using this genome-wide data set, Stranger et al. confirmed their early results that approximately 10% of the genes have a cis-genetic effect and the detection of *trans*-effects is limited.

Perhaps, the most comprehensive study to date not involving the HapMap populations was a collaborative effort between deCODE genetics, Rosetta Inpharmatics, and Merck whose aim was to study clinical obesity and other diseases related to adipose tissue (9). They collected a population-based cohort of blood ($n = 673$) and adipose tissue ($n = 1,002$) from Icelandic subjects, and also measured a number of clinical traits such as body mass index (BMI) and percentage body fat (PBF). Expression levels for over 23,000 Ensembl transcripts were measured using a custom Agilent microarray. A multiple linear regression model was used to correlate gene expression with the clinical traits in both cohorts.

After fixing FDR at 5%, 3–9% of gene expression traits were correlated with the clinical traits in blood, and 63–72% of genes were correlated in adipose tissue. A subset of 150 unrelated subjects were genotyped using the Illumina Hap300, and this genotype data was used in a standard linear regression analysis with 21,000 genes, adjusting for sex, age, cell count (blood), and BMI (adipose tissue). At 5% FDR, 2,714 (11%) genes in blood and 3,364 (14%) genes in adipose tissue showed association with *cis*-eSNPs (within 2 Mb centered on the gene), and to a much lesser degree, *trans*-eSNPs. Overall, the results indicated a strong genetic effect of proximal signals, with significant overlap between the two tissues. A networks analysis of the highly associated genes identified networks showing enrichment for genes involved in the inflammatory and immune response system, as well as being associated to obesity-related traits. This systems biology approach in conjunction with eSNP–eQTL identification is beginning to provide tremendous insight not only into genes involved in diseases but entire pathways.

As mentioned previously, the majority of genome-wide QTL analyses use 3′ expression arrays, which is a gene-centric approach. A number of groups have taken an exon-centric approach using exon arrays for the analysis of genetic variants. A study by Kwan et al. used LCLs derived from the HapMap CEU population and looked for *cis*-effects (<50 Kb) on exon expression levels of all known RefSeq transcripts (26). They showed that natural variation is associated with eQTLs at the exon level in addition to the gene level, which, in a broader context, indicated that genetic variation is associated with expression of different transcript isoforms, i.e., alternatively spliced transcripts from the same gene locus that differ in the inclusion/exclusion of a cassette exon within the coding region, or changes in the 5′ and 3′ UTR that may have effects on gene regulation. This study also highlighted one of the design biases of 3′ expression arrays: since the probes are targeted towards the 3′ UTR, these arrays cannot detect a large number of alternatively spliced transcripts that represent a significant portion of the human transcriptome, including isoform changes differing at the 3′ UTR itself. This is an important consideration to remember when choosing the type of array for expression studies and also highlights the need for validation of these eQTL associations.

Progress in the field of genetic variation and its effects on gene and transcript isoform expression, and how this relates to disease associations or natural phenotypic traits has advanced significantly over the last few years, aided in part by the tremendous and rapid development of microarray technologies. There still remain some technical and statistical hurdles on how to handle such large data sets, especially with the trend towards larger and larger sample sizes, but these issues are constantly being addressed and refined as more and more studies are made available. Identifying associations is still a relatively straightforward task, but how to define what is considered significant will remain a much-debated issue. Although there is a trend towards specific tissue types that may be more biologically relevant to disease phenotypes, the use of LCLs from the HapMap populations is still providing a tremendous amount of information for the scientific community at large and remains an invaluable resource.

Acknowledgments

The authors thank Dominique Verlaan for critical reading of this review and Genome Quebec and Genome Canada for funding. TP holds a Canada Research Chair (Tier 2).

References

1. Lettre, G., Jackson, A. U., Gieger, C., Schumacher, F. R., Berndt, S. I., Sanna, S. et al (2008) Identification of ten loci associated with height highlights new biological pathways in human growth. *Nat Genet* **40**, 584–591
2. Weedon, M. N., Lango, H., Lindgren, C. M., Wallace, C., Evans, D. M., Mangino, M. et al (2008) Genome-wide association analysis identifies 20 loci that influence adult height. *Nat Genet* **40**, 575–583
3. Han, J., Kraft, P., Nan, H., Guo, Q., Chen, C., Qureshi, A. et al (2008) A genome-wide association study identifies novel alleles associated with hair color and skin pigmentation. *PLoS Genet* **4**, e1000074
4. Wellcome Trust Case Control Consortium. (2007) Genome-wide association study of 14,000 cases of seven common diseases and 3,000 shared controls. *Nature* **447**, 661–678
5. Brem, R. B., Yvert, G., Clinton, R., and Kruglyak, L. (2002) Genetic dissection of transcriptional regulation in budding yeast. *Science* **296**, 752–755
6. Bystrykh, L., Weersing, E., Dontje, B., Sutton, S., Pletcher, M. T., Wiltshire, T. et al (2005) Uncovering regulatory pathways that affect hematopoietic stem cell function using 'genetical genomics'. *Nat Genet* **37**, 225–232
7. Chesler, E. J., Lu, L., Shou, S., Qu, Y., Gu, J., Wang, J. et al (2005) Complex trait analysis of gene expression uncovers polygenic and pleiotropic networks that modulate nervous system function. *Nat Genet* **37**, 233–242
8. Cheung, V. G., Spielman, R. S., Ewens, K. G., Weber, T. M., Morley, M., and Burdick, J. T. (2005) Mapping determinants of human gene expression by regional and genome-wide association. *Nature* **437**, 1365–1369
9. Emilsson, V., Thorleifsson, G., Zhang, B., Leonardson, A. S., Zink, F., Zhu, J. et al (2008) Genetics of gene expression and its effect on disease. *Nature* **452**, 423–428
10. Schadt, E. E., Lamb, J., Yang, X., Zhu, J., Edwards, S., Guhathakurta, D. et al (2005) An integrative genomics approach to infer causal associations between gene expression and disease. *Nat Genet* **37**, 710–717
11. Altshuler, D., Brooks, L. D., Chakravarti, A., Collins, F. S., Daly, M. J., Donnelly, P. et al (2005) A haplotype map of the human genome. *Nature* **437**, 1299–1320
12. Dixon, A. L., Liang, L., Moffatt, M. F., Chen, W., Heath, S., Wong, K. C. et al (2007) A genome-wide association study of global gene expression. *Nat Genet* **39**, 1202–1207
13. Long, A. D., and Langley, C. H. (1999) The power of association studies to detect the contribution of candidate genetic loci to variation in complex traits. *Genome Res* **9**, 720–731
14. Lander, E. S., Linton, L. M., Birren, B., Nusbaum, C., Zody, M. C., Baldwin, J. et al (2001) Initial sequencing and analysis of the human genome. *Nature* **409**, 860–921

15. Modrek, B., Resch, A., Grasso, C., and Lee, C. (2001) Genome-wide detection of alternative splicing in expressed sequences of human genes. *Nucleic Acids Res* **29**, 2850–2859
16. Wilhelm, B. T., Marguerat, S., Watt, S., Schubert, F., Wood, V., Goodhead, I. et al (2008) Dynamic repertoire of a eukaryotic transcriptome surveyed at single-nucleotide resolution. *Nature* **453**, 1239–1243
17. Barnes, M., Freudenberg, J., Thompson, S., Aronow, B., and Pavlidis, P. (2005) Experimental comparison and cross-validation of the Affymetrix and Illumina gene expression analysis platforms. *Nucleic Acids Res* **33**, 5914–5923
18. Eberle, M. A., Ng, P. C., Kuhn, K., Zhou, L., Peiffer, D. A., Galver, L. et al (2007) Power to detect risk alleles using genome-wide tag SNP panels. *PLoS Genet* **3**, 1827–1837
19. Schadt, E. E., Molony, C., Chudin, E., Hao, K., Yang, X., Lum, P. Y. et al (2008) Mapping the genetic architecture of gene expression in human liver. *PLoS Biol* **6**, e107
20. Stranger, B. E., Nica, A. C., Forrest, M. S., Dimas, A., Bird, C. P., Beazley, C. et al (2007) Population genomics of human gene expression. *Nat Genet* **39**, 1217–1224
21. Gentleman, R. C., Carey, V. J., Bates, D. M., Bolstad, B., Dettling, M., Dudoit, S. et al (2004) Bioconductor: open software development for computational biology and bioinformatics. *Genome Biol* **5**, R80
22. Purcell, S., Neale, B., Todd-Brown, K., Thomas, L., Ferreira, M. A., Bender, D. et al (2007) PLINK: a tool set for whole-genome association and population-based linkage analyses. *Am J Hum Genet* **81**, 559–575
23. Churchill, G. A., and Doerge, R. W. (1994) Empirical threshold values for quantitative trait mapping. *Genetics* **138**, 963–971
24. Benjamini, Y., and Hochberg, Y. (1995) Controlling the False Discovery Rate: A Practical and Powerful Approach to Multiple Testing. *Journal of the Royal Statistical Society. Series B (Methodological)* **57**, 289–300
25. Alberts, R., Terpstra, P., Li, Y., Breitling, R., Nap, J. P., and Jansen, R. C. (2007) Sequence polymorphisms cause many false cis eQTLs. *PLoS ONE* **2**, e622
26. Kwan, T., Benovoy, D., Dias, C., Gurd, S., Provencher, C., Beaulieu, P. et al (2008) Genome-wide analysis of transcript isoform variation in humans. *Nat Genet* **40**, 225–231
27. Zhang, W., Duan, S., Kistner, E. O., Bleibel, W. K., Huang, R. S., Clark, T. A. et al (2008) Evaluation of genetic variation contributing to differences in gene expression between populations. *Am J Hum Genet* **82**, 631–640
28. Price, A. L., Patterson, N. J., Plenge, R. M., Weinblatt, M. E., Shadick, N. A., and Reich, D. (2006) Principal components analysis corrects for stratification in genome-wide association studies. *Nat Genet* **38**, 904–909
29. Monks, S. A., Leonardson, A., Zhu, H., Cundiff, P., Pietrusiak, P., Edwards, S. et al (2004) Genetic inheritance of gene expression in human cell lines. *Am J Hum Genet* **75**, 1094–1105
30. Morley, M., Molony, C. M., Weber, T. M., Devlin, J. L., Ewens, K. G., Spielman, R.S. et al (2004) Genetic analysis of genome-wide variation in human gene expression. *Nature* **430**, 743–747
31. Goring, H. H., Curran, J. E., Johnson, M. P., Dyer, T. D., Charlesworth, J., Cole, S. A. et al (2007) Discovery of expression QTLs using large-scale transcriptional profiling in human lymphocytes. *Nat Genet* **39**, 1208–1216
32. Stranger, B. E., Forrest, M. S., Clark, A. G., Minichiello, M. J., Deutsch, S., Lyle, R. (2005) Genome-wide associations of gene expression variation in humans. *PLoS Genet* **1**, e78
33. ENCODE Project Consortium. (2004) The ENCODE (ENCyclopedia Of DNA Elements) Project. *Science* **306**, 636–640
34. Myers AJ, Gibbs JR, Webster JA, Rohrer K, Zhao A, Marlowe L, Kaleem M, Leung D, Bryden L, Nath P, Zismann VL, Joshipura K, Huentelman MJ, Hu-Lince D, Coon KD, Craig DW, Pearson JV, Holmans P, Heward CB, Reiman EM, Stephan D, Hardy J. (2007). A survey of genetic human cortical gene expression. Nature Genetics 39, 1494–1499
35. Spielman RS, Bastone LA, Burdick JT, Morley M, Ewens WJ, Cheung VG. (2007). Common genetic variants account for differences in gene expression among ethnic groups. Nat. Genet 39, 807–808

Chapter 19

Quality Control for Genome-Wide Association Studies

Michael E. Weale

Abstract

This chapter is a comprehensive review of quality control (QC) methods for SNP-based genotyping panels used in genome-wide association studies. These include QC on individuals for missingness, gender checks, duplicates and cryptic relatedness, population outliers, heterozygosity and inbreeding, and QC on SNPs for missingness, minor allele frequency and Hardy-Weinberg equilibrium. The emphasis is on the reasons behind each QC step and on the use of intelligent approaches rather than arbitrary QC thresholds. Scripts and code for performing these QC steps are available at www.kcl.ac.uk/mmg/gwascode/.

Key words: Genome-wide association studies (GWAS), SNP, Genotyping, Quality control, Statistics

1. Introduction

1.1. Scope of this Review

Genome-wide association studies (GWAS) are now commonplace (www.genome.gov/gwastudies; (1)). These studies depend on (relatively) cheap, high-throughput whole-genome genotyping panels that allow the researcher to interrogate hundreds of thousands of single nucleotide polymorphisms (SNPs) across the human genome. Increasingly, these same panels can also be used to infer large-scale copy number variants (CNVs) as well. These panels also have potential applications beyond GWAS, including forensics and population genetics.

Turn to any GWAS report and you will see (or should!) a long section detailing the quality control (QC) steps used on the data, typically resulting in a reduction both in the number of individuals and in the number of SNPs passed on to downstream analysis. A number of reviews on GWAS analysis are already available (2–8). This review is distinguished by its specific focus on QC and on the reasons behind each QC step. My emphasis

Michael R. Barnes and Gerome Breen (eds.), *Genetic Variation: Methods and Protocols*, Methods in Molecular Biology, vol. 628, DOI 10.1007/978-1-60327-367-1_19,

will be on a critical examination of QC in the context of the given dataset, rather than a rigid application of ad hoc QC filtering thresholds. I will focus on QC for case-control GWAS SNP data, but most of the QC steps described here also apply more broadly to whole-genome genotyping data used in other contexts.

1.2. Materials

I will provide solutions based on publically-available open source software. I will use the popular PLINK program for GWAS analysis (9), Unix/Linux text manipulation commands (e.g., awk, grep, head, tail, sort, join) for preparing text files for further analysis, and the R statistical package for additional analysis and graphics (www.r-project.org). These are not the only options available for performing these tasks, but they are commonly used. Scripts and code for performing these QC steps are available at www.kcl.ac.uk/mmg/gwascode/.

2. Methods

2.1. Why QC?

Why this review on SNP-based GWAS QC? Modern GWAS panels have impressive genotyping quality, with claims of call frequency and accuracy in excess of 99.9% not uncommon. These rates are substantially better than for custom-SNP panels (i.e., user-defined SNP content), partly because in predefined GWAS panels any SNPs that do not work well are replaced by others in linkage disequilibrium that do. Thus, many practitioners perceive the real QC challenges to lie elsewhere, for example with calling CNVs.

But there are several reasons why QC for GWAS SNP data is still important. Firstly, genotype call quality is rarely as high in practice as the SNP panel manufacturers claim, because the DNA quality of real GWAS samples is usually less than the carefully prepared samples used to benchmark the panels. Secondly, recent developments such as genome-wide custom SNP panels (e.g., for rare SNPs), separately-genotyped reference control cohorts, and a trend towards searching for smaller and smaller effect sizes via combined analyses of separate studies, have all served to increase the QC problem. Finally, and perhaps most importantly, in the genome-wide context a little bad quality can go a long way. This is because a small amount of systematic error, when applied across hundreds of thousands of SNPs, can still result in hundreds of false positive association signals which swamp the real signals. Increasing the sample size often only makes matters worse – amplifying the false positive signals as much as or more than amplifying the real signals.

2.2. How QC?

QC problems can be characterized in various ways. They can be SNP-specific or individual-specific. They can arise from either misclassified or unclassified data (for SNPs, these correspond to genotype mis-calls and no-calls respectively). Perhaps the most important way to classify QC problems is by their impact, resulting in either false positive or false negative signals of association. False positives (false association hits that pollute the pool of true hits) are the most visible problem, but false negatives (real hits that are not detected) are for their very invisibility perhaps an even bigger problem, as they are much more difficult to correct in downstream analyses. Since QC on SNPs throws away SNPs that don't meet certain thresholds, this type of QC is generally more about controlling false positives than false negatives, although throwing away bad SNPs may still enhance one's power to detect association signals via multiSNP methods such as haplotype analysis, and will also reduce the multiple testing burden if frequentist methods are being used (see also Subheading 10). QC on individuals can in principle fix both false positives and single-SNP false negatives. Examples of how both false positive and false negative signals can arise will be given later.

2.3. Is QC the Only Way?

No. Indeed, ideally one would like to use all the data and have the data quality issues appropriately modelled, rather than crudely hacking out data in order to force the rest of the data into some convenient norm. Methods for dealing with "fuzzy" calls exist ((10), see also the GenABEL package in R (11)), as do methods for accounting for complex relatedness among individuals (12). But there are drawbacks to applying these methods: they don't cover all situations, they are computationally expensive, and they don't extend well.

Thus QC is to some extent a matter of convenience. The above considerations suggest that for best practice we should always perform QC critically rather than impose "off-the-shelf" filters. They also suggest that more than one level of QC – for example a stringent versus a liberal level – should be applied, and the results cross-compared as a form of sensitivity analysis. Thus a 2-tiered QC analysis strategy could consider stringently-QC'd data in Tier 1 and a bigger, less stringently QC'd dataset in Tier 2. Hits from Tier 1 could be accepted with more confidence than those from Tier 2 and prioritized accordingly, while hits from Tier 2 would require even more careful postassociation QC, including referring back to the original allele signal intensity plots. Ultimately, of course, both types of hit would still require replication for validation.

2.4. Is there an Optimal Order for QC Steps?

One can split QC steps into those designed to filter individuals and those designed to filter SNPs. Which order should these be done in? And what about the ordering of steps within these categories? If one filters out "bad" individuals first then one tends to rescue borderline SNPs from being dropped, and vice versa if one filters out "bad" SNPs first. Because one has more SNPs than individuals in a GWAS, it's fairer to think of the problem in terms of proportion of data lost. So which is worse – to lose, say, 5% of your individuals or 5% of your SNPs? Losing 5% of your individuals will result in some loss of power across all your real association signals. The relationship between sample size and power is sigmoidal, meaning that for some signals the loss of power will be negligible, whereas for others it may be substantial. On the other hand, losing 5% of your SNPs may be catastrophic if one of your association signals is most clearly tagged by one these SNPs, but this effect is mitigated by the existence of linkage disequilibrium (LD) among neighbouring SNPs. This allows the lost signal to be recreated – e.g., via imputation techniques (13–15). In short, the relative importance of individual loss versus SNP loss will depend on the specifics of your GWAS: is it well powered; is this a high-LD population; is this a high-density SNP panel?

In what follows, I start with individual QC and move on to SNP QC. This makes sense in terms of the more evidence-based SNP-missingness QC that I propose. In practice, I find that in any case the ordering makes only a marginal difference to which individuals and SNPs get dropped.

2.5. Principal Components Analysis for QC

Principal components analysis (PCA) is a method for dissecting and ranking the correlation structure of multivariate data. In the GWAS context, SNP genotypes are the variables. The technique is most closely associated with the EIGENSTRAT method for correcting population stratification (16, 17), but the method is completely general and can be used as a tool for picking up any types of correlations in the data. Data quality issues often induce such correlations, which is why PCA is a QC tool that is widely used on all types of multivariate data.

There is a conceptually useful geometric interpretation of PCA. Start by considering a GWAS dataset as a very large matrix, with one row for each of n individuals and one column for each of m SNPs. Each cell in the matrix is a genotype call for a given SNP in a given individual, coded $\{0,1,2\}$ to represent a given allele count and then suitably normalised to have zero mean and roughly equal variance. These data can be plotted in m-dimensional "SNP space" by creating one axis for each SNP and plotting the genotype score for each individual along each axis. Individuals that are close to each other in this m-dimensional space have similar genotypes compared to individuals who are far away. PCA can be imagined as a rotation of the axes in this space so as to obey a variance-maximizing algorithm. Consider the variance of all individuals when projected at right-angles onto an axis (call these projections the "scores" for that axis). The first principal component (PC) axis is the one that has the maximum possible variance of its PC scores. The second PC axis has the maximum possible variance of its PC scores, conditional on it being at right-angles to the first PC axis. The jth PC axis has the maximum possible score variance conditional on being at right-angles to all PC axes previously defined. Mathematically, it transpires that this process is equivalent to a singular value decomposition of the original data matrix.

One can also consider plotting the m SNPs in n-dimensional "individual space". SNPs that are close to each other in this space are the ones that are positively correlated among individuals. Local linkage disequilibrium, data quality issues, and population stratification can all induce these correlations, in the same way as they do for individuals in "SNP space". Applying the same variance-maximizing algorithm, one ends up with axes that effectively represent the same PCA solution as do the PC axes in "individual space". Thus the correlation structure of the data is the same, regardless of how it is projected. I shall call the projections of the SNPs onto their axes in "individual space" the SNP "loadings" for that axis. This differs from the terminology used by (16, 17) but is more in line with general PCA terminology given that SNPs rather than individuals are the natural variables in this multivariate analysis.

For GWAS data, correlations in allelic states among individuals could result from shared ancestry, or from local linkage disequilibrium, or from disease causation by many SNPs (a theoretical possibility, although in practice it seems that causal SNPs are too few and of too small effect size to induce PC axes in this way in typical GWAS data), or from laboratory error. Each PC axis from the same PCA may in principle reflect one or possibly more of these underlying causes. Both individual "scores" and SNP "loadings" can be inspected to find valuable clues to the factors influencing a given PC axis. Thus histograms of PC scores (e.g., Fig. 8) can be used to identify individual outliers, indicative of possible data quality issue affecting just those individuals, while Q–Q plots of PC loadings (e.g., Fig. 4) can be used to determine if a given PC axes is being driven by factors acting across the whole genome (e.g., population stratification, cryptic relatedness) or by factors affecting only a subset of SNPs (e.g., local linkage disequilibrium, or SNP-specific data quality issues). For further details, see Subheadings 6 and 7.

[AU1] ***2.6. Quantile–Quantile Plots for QC***

Quantile–quantile plots (Q-Q plots) is another useful tool for QC, particularly in the context of SNP QC. It is often possible to obtain a test statistic, and its related p-value, for every SNP in a GWAS dataset. Examples include tests for departure from Hardy-Weinberg Equilibrium (see Subheading 11) and tests for association (see Subheading 9). P-values generated under the Null hypothesis should, by definition, be drawn from a Uniform distribution between zero and one. It follows that if a set of m p-values are ordered from lowest to highest, then the observed quantile (value) of the jth ordered item should on average be equal to the corresponding expected quantile for m items drawn from a Uniform(0,1) distribution (and this expected value can be shown to be $j/(m+1)$). Thus if the observed quantiles are plotted against the expected ones, one would expect to see a roughly straight line through the origin with a unit slope, albeit with some random variation. It's important to note we still expect this pattern under the Null, even if there is some dependence among the p-values (for example, due to local linkage disequilibrium among SNPs). The overall expectation is the same, although in this case the random variation about the unit slope will be somewhat inflated. "Concentration bands" (see also (18)) can be added to the Q–Q plot, based on the result that the jth ordered value from m independent Uniform(0,1) draws is known to have a Beta(j, $m-j+1$) distribution. Thus Q–Q plots under the Null should broadly fall within the 95% concentration band, although this should be used as a rough guide only given the problem of local dependence expected among GWAS SNPs due to linkage disequilibrium.

If the set of m p-values contains some which are drawn not from the Null but from some alternative hypothesis, then these will skew the distribution of p-values away from a Uniform (0,1) and towards a distribution with a greater preponderance of low p-values. A log scaling will emphasise these low p-values more, and so it is a common practice to plot p-value Q-Q plots on a negative logarithmic scale. SNPs departing from the Null will now appear as points raised above the unit line towards the top right of the plot (see, for example, Fig. 12). It turns out that this scaling is equivalent to converting the expected distribution from a Uniform (0,1) to a chi-square with 2 degrees of freedom, which provides a justification for why Q-Q plots based on raw test statistics are often used when these statistics are expected to have a chi-squared distribution. Q-Q plots can also be generated with respect to any known expected distribution – see Fig. 4 in Subheading 6 for examples using a Normal expectation.

While they are useful tools, I briefly mention here two potential pitfalls that may arise in their use. The first is that p-values derived from Exact Tests do not, as a rule, distribute as Uniform (0,1) even under the Null, so care must be exercised if using them in Q-Q plots. The second is that the point where SNPs start to rise above the unit slope (see, for example, Fig. 12) does not in general indicate that all SNPs above this point are generated under the alternative hypothesis, nor even does it necessarily indicate the point above which a sizeable fraction are generated under the alternative hypothesis, and below which a negligible fraction are. Instead, the starting point at which a dataset becomes "contaminated" with non-Null SNPs may be much higher than the departure point indicated in the Q-Q plot. These points are considered in more detail in Subheading 11.

3. Quality Assurance

3.1. Quality Assurance Versus Quality Control

This review is primarily about quality control – the steps applied between the production of data and its downstream use to ensure it is fit for purpose. Quality assurance addresses those procedures applied further upstream in the production of the data, to increase the chances that most of the data passes QC. Examples of quality assurance steps would be to ensure that good study design and sampling protocols were used; that good quality DNA had been collected; that protocols for DNA extraction and preparation were adequate; that care had been taken to equalize the treatment of cases and controls (for example by assigning samples of both to each plate); that genotyping call rates were routinely monitored to ensure that poorly-manufactured chips were quickly detected and corrected for; and that data on batch allocation, plate allocation, and dates of genotyping were recorded so that these variables can be inspected later for systematic trends (8). A key part of the production of the data is the algorithm used to call the genotypes, and I treat this issue next.

3.2. Genotype Calling and Quality

Figure 1 illustrates the genotype calling problem. GWAS platforms differ in their details but share the common feature that the two possible alleles of a given SNP genotype are assayed, often more than once, by a quantitative measure of signal intensity.

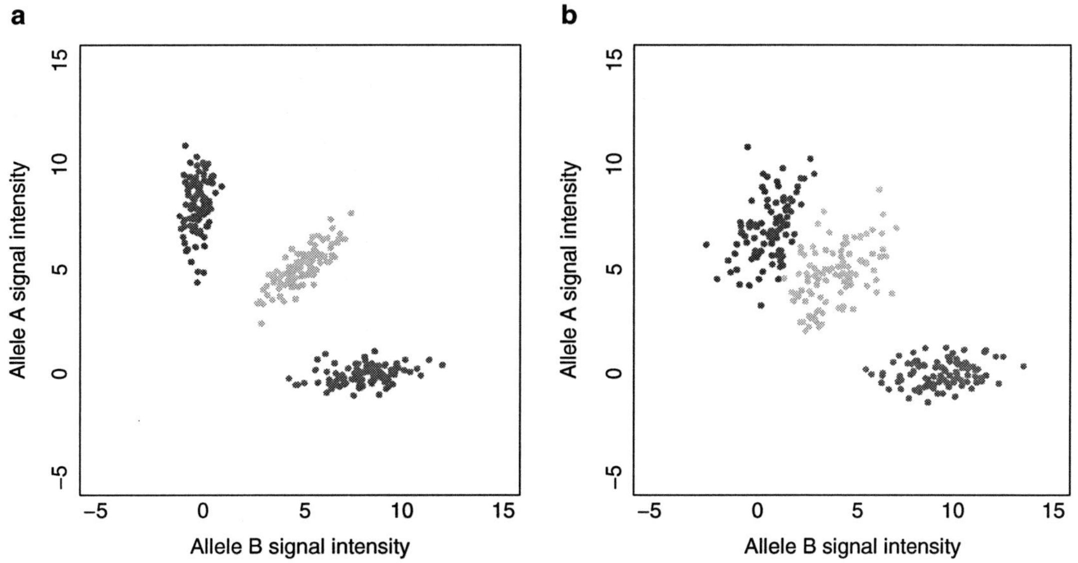

Fig. 1. Examples of allele signal intensity plots. The coloured clusters represent individuals with AA (*to the left*), AB (*in the center*) and BB (*to the right*) genotypes. (**a**) A clean SNP with genotype clusters well separated. (**b**) A problem SNP where the AA and AB genotype clusters overlap. Individuals lying in the contact zone (*grey*) are impossible to classify accurately and so are called as missing by the genotype calling algorithm

The data for a single SNP can therefore be represented, after appropriate normalization, as a scatter-plot of signal intensities for allele A against allele B, with each point representing a different individual. If the underlying chemistry is working well, individuals can be separated into three distinct clusters representing the three possible genotypes, with heterozygotes in the middle (Fig. 1a). Inevitably, problems with some SNPs can arise, for example leading to two of the clusters touching or overlapping with each other (Fig. 1b). This leads to genotype calling algorithms calling individuals along the contact edge as missing, which in turn leads to informative missingness as discussed later.

There is a growing literature on the relative merits of different genotype-calling algorithms (19, 20). Work is still progressing on how best to call SNPs within CNVs, and how best to use Hardy-Weinberg equilibrium to guide correct genotype calling. A general rule is that these algorithms work best when all the samples in a GWAS are analysed together, using these data to perform the clustering rather than relying on external references for where the clusters should lie. Genotype-calling algorithms are often platform-specific, but almost all will result in a quality score for each genotype which reflects one's confidence that the genotype call is correct. Importantly, it is usually a threshold value of this score which determines whether a genotype is called or declared as missing data. Increasing the value of this threshold decreases the proportion of miscalled data, but increases the proportion of missing data and therefore leads to bigger problems of informative missingness (see Note 1). Setting the quality threshold value too high can therefore be counterproductive, so where is the best place to set it?

Given that miscalls are more likely to result in a single-step change in the allele count of the called genotype (i.e., AA→AB rather than AA→BB), it is possible that a miscalled genotype will still add some power to a signal of association provided the underlying genetic model involves a trend on allele counts (rather than a strictly dominant or recessive model, for example). This could imply that a low quality threshold would work well. On the other hand, if the proportion of miscalls differs between cases and controls then false positives or false negatives can arise in a similar way to informative missingness. A high proportion of miscalls would also impact negatively on other downstream analyses such as testing for Hardy-Weinberg Equilibrium or haplotype-calling. One option worth exploring, therefore, is to adopt a two-tier analysis strategy as discussed in Subheading 2.3. A low-stringency (Tier 2) QC analysis could be restricted to simple single-point association analyses only, preferably modelling "fuzzy" genotype calls directly (10), and would therefore include more of the original data by judicious reduction of the quality score threshold.

4. Individual QC: Missingness

4.1. Why Filter?

Informative missingness is one of the biggest problems in GWAS QC. It arises because missingness is not randomly distributed among genotypes but rather is over-represented in, say, AA and AB calls (as in the example in Fig. 1b). False positives will arise if DNA quality differs with phenotype, leading to differences in the frequency of called genotypes (for example, if case data looks like Fig. 1b and control data looks like Fig. 1a). False negatives will arise if the informative missingness signal acts in the opposite direction to the real signal, or more generally by reducing power via a reduced sample size for nonmissing values.

Increased false positives are particularly likely for case-control studies, where case and control samples are likely to have been collected and/or genotyped separately. They are less likely for cohort based studies involving quantitative phenotypes. The problem can also be reduced by ensuring that equal numbers of cases and controls are plated together, so that plate effects can be evenly distributed. Of course, this will not be possible if controls are obtained from a predefined reference database. Experience suggests that some QC for missingness is almost always needed (21).

4.2. How Filter?

The informative missingness issue can be addressed both though individual QC and SNP QC. I will discuss how to set QC thresholds that explicitly address the question of false positives in association testing when dealing with SNP missingness QC in Subheading 9. In this section, I will only consider balancing the cost in terms of the amount of data lost though filtering versus gains in data quality, thus leaving the "fine tuning" for SNP missingness QC to resolve.

The pattern of missingness can be visualized using a plot of one minus missingness against cumulative frequency (see Fig. 2 for an example employing data from (18)). The "elbow" in these figures denotes the point at which one starts to get diminishing returns from increasing QC stringency – i.e. a large number of individuals or SNPs need to be removed in order to effect only a modest change in worst-case missingness for the filtered data. This occurs near 98% call rate for individuals, and 95% coverage for SNPs. We will use this to guide but not determine our SNP missingness QC. For individuals, a threshold between 97 and 98% call rate seems appropriate (97% was used in (18)).

Finally, be aware that we are implicitly assuming missing data does indeed represent failed calls. If multiple genotyping panels have been used, it is possible that some samples have been genotyped on some panels but not others. PLINK has an obligate missingness feature to deal with this problem. Gender mismatches can also be a problem (see Note 2).

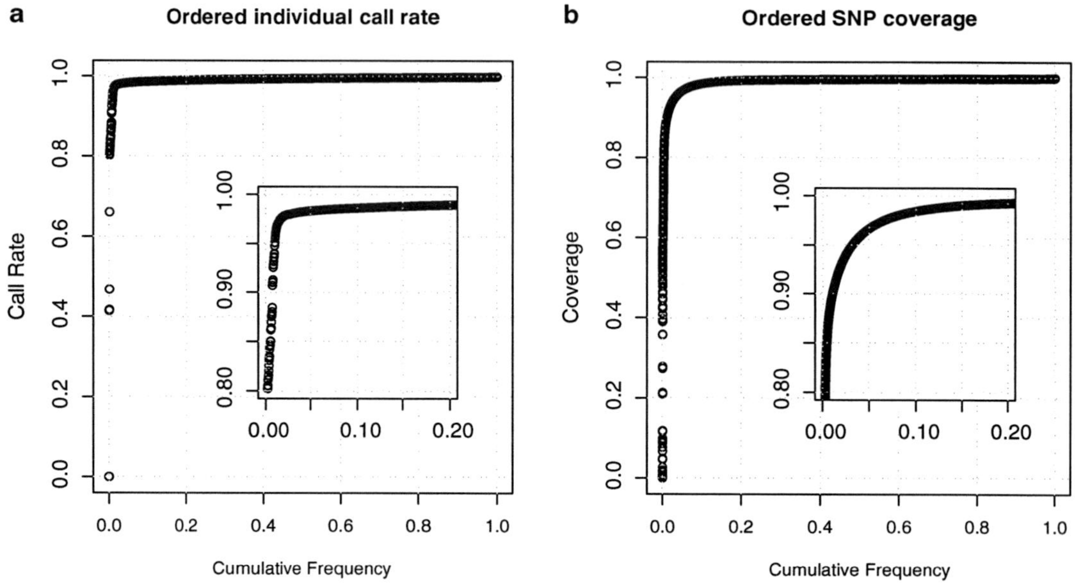

Fig. 2. Plots of one minus missingness ("call rate" for individuals and "coverage" for SNPs) against cumulative frequency (i.e., data points ordered from lowest to highest), using data (58C + NBS + CD cohorts) from (18). Insets show a zoom-in of the top left hand corner. R-code for plotting this figure is available at www.kcl.ac.uk/mmg/gwascode

5. Individual QC: Gender Checks

5.1. Why Filter?

With GWAS data, it is easy to spot individuals who are genetically male but are phenotypically labelled as female, or vice versa, using data from the X and Y chromosomes. One use of this is as a sanity check to make sure that genetic and phenotypic databases are correctly aligned. If around 50% of your samples are mismatched, this indicates a catastrophic randomization of one or other set of sample ID labels. Another use is to identify and assess the small number of mismatches that occur even when the databases are correctly aligned. In many cases, but not always, it will also be a good idea to remove these individuals from analysis.

In some cases, the mismatches may be real, in that they reflect rare medical conditions (see Note 3). Confirmed examples of such medical conditions should usually result in exclusion, on the basis that these individuals are atypical relative to the rest of their cohort.

Gender mismatches sometimes occur with too high a frequency in GWAS datasets, say at ~1% frequency, an order of magnitude greater than what one would expect if these were all due to medical conditions. The implication is that most of these mismatches are due to a labelling error. Usually, it is impossible to tell whether this is a labelling error involving just the gender label (in which case no harm would be done if this individual is retained after correcting the gender label) or whether it is a more serious labelling error linking the wrong DNA sample to the wrong clinical record. What should one do? The conservative approach would be always to assume the worst and exclude the mismatched individuals. However, if the phenotype record is very simple (e.g., just a case-control label plus a gender label), and if cases and controls have been collected separately and genotyped with protocols that prevent DNA label swaps in the lab, then one could argue that the link between DNA and case-control status is secure and thus the individuals can be kept. This is the reasoning adopted by (18), where all gender mismatched individuals were retained.

Sometimes, one also sees an intermediate gender call based on X-chromosome data (i.e., too many heterozygous SNPs to be male, too many homozygous SNPs to be a typical female). The possible reasons for this are discussed in a following section. Safest practice would again be to exclude these individuals.

5.2. How Filter?

Both X-chromosome and Y-chromosome data can be used, but not all GWAS panels include Y-SNPs. A female mislabelled as a male will have many heterozygous X-chromosome calls relative to the other men (who barring miscalls should all have hemizygous genotypes), and all Y-SNPs should be missing data. In contrast, a male mislabelled as a female will have only homozygous X-chromosome calls, and will have called Y-SNP data. An alternative way of describing the X-chromosome pattern is that females will have their X-SNPs more or less in Hardy-Weinberg Equilibrium (HWE), whereas males will depart severely from this by having no heterozygous genotypes. By definition, the inbreeding coefficient F (calculated from X-chromosome data with male hemizygous genotypes coded as though they were homozygous) measures the severity of this departure from HWE – it will be close to zero for genetic females and close to one for genetic males. This is the measure used by PLINK (9) to check to gender mismatches.

Figure 3 illustrates how a histogram plot of F forms, for the most part, two distinct clusters near $F=0$ (females) and $F=1$ (males). There are, however, some individuals who fall part way between these two poles, and some with anomalously low F values. For some, we will see there is evidence for sample contamination, or for membership of a different population. Another possible explanation is X chromosome mosaicism in females, in which some cells have lost part or all of their inactive X-chromosomes while others haven't. This can be seen particular in DNA obtained from old immortalized cell lines. Given the uncertain biological impact of this phenomenon, it is best to exclude all these intermediate-F individuals, who are clearly separated from the "male" or "female" cluster, from further analysis as well.

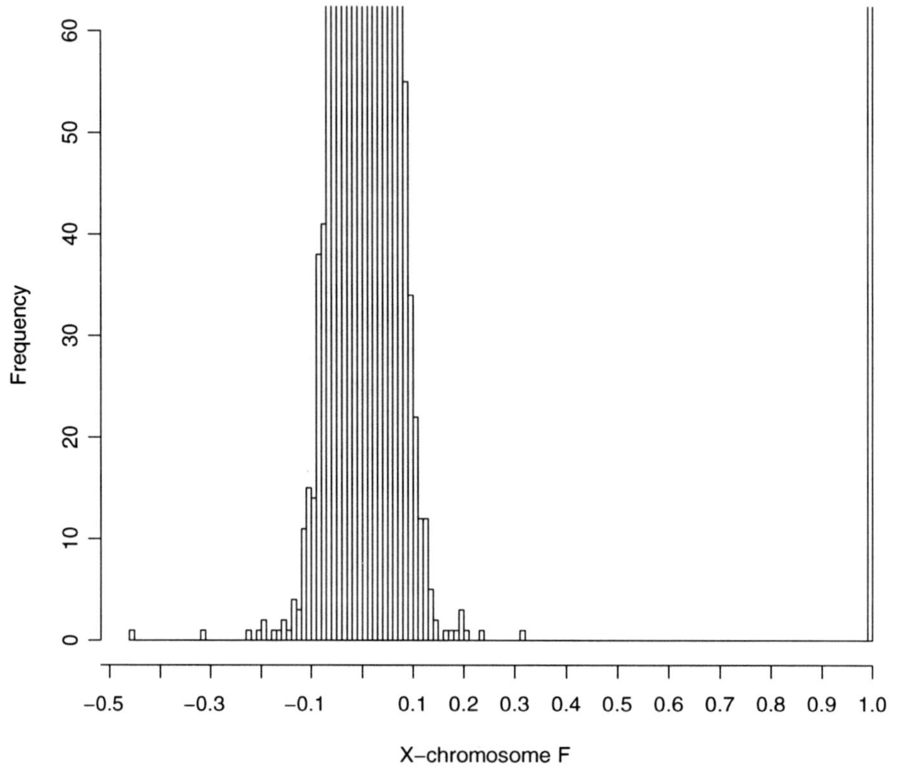

Fig. 3. Histogram of the X-chromosome-specific inbreeding coefficient *F*, a measure of departure from Hardy-Weinberg Equilibrium. Females should have *F* close to 0, and males should have *F* close to 1. Note that natural variability in heterozygosity means that females are more difficult to call with accuracy than males. The *y*-axis has been cropped at 60 counts to allow singleton observations to be seen. Data (58C + NBS + CD cohorts, after individual missingness QC) from (18). R-code for plotting this figure is available at www.kcl.ac.uk/mmg/gwascode

6. Individual QC: Duplicates and Cryptic Relatedness

6.1. Why Filter?

Cryptic relatedness occurs when, for unforeseen reasons, pairs or groups of individuals are more closely related to each other than the population average – thus indicating they are close family members. Sample duplicates can also be treated as an extreme case of cryptic relatedness. Individuals that are closely related induce a correlation structure that will upset downstream association analyses unless properly accounted for. This may introduce false positive and/or false negative results, depending on the situation. This is particularly problematic if the number or degree of cryptic relatedness differs between cases and controls. While it is in principle possible to account for this by appropriate statistical modelling (see (12) for one approach), these methods are complex and do not cover all study designs (importantly, they do not yet cover case-control designs).

Of course, all individuals are related to each other to some extent, and thus efforts at homogenization (both in this section and the next) are ultimately impossible. Pragmatically, however, it seems sensible to apply these filters, as experience shows that certain individuals can be easily spotted as outliers (either too closely or too distantly related) from the rest.

6.2. An LD-Pruned Dataset

For the QC operations defined in this section and the next, it is useful to prepare an LD-pruned dataset containing a smaller set of SNPs than in the total GWAS panel. These SNPs are selected to have minimal linkage disequilibrium (LD) among them. One advantage of doing this is that both cryptic relatedness and population stratification procedures work best under an assumption of no LD among SNPs. A second, pragmatic advantage is that these QC steps can take a long time if performed on the full dataset. Thinning to a set of between 50 and 100 thousand SNPs appears completely adequate for the task of identifying cryptic relatives and population outliers. I recommend a multistage procedure for creating an LD-pruned dataset, one stage for tackling large-scale high-LD regions in the human genome, next for tackling small-scale LD, and a final stage for detecting and correcting for any residual LD effects that may still remain.

Large-scale high-LD regions have recently been detected as the result of applying principal components analysis (PCA) to large GWAS datasets (see Subheading 2.5). Two examples that are big enough to create their own principal component (PC) axes in PCA of GWAS data are a 4 Mb inversion on chromosome 8 (22) and the 8 Mb region spanning the extended MHC region on chromosome 6. The underlying causes of the high LD in these regions is not always clear, but pragmatically it seems best to remove them entirely, as they may not be tackled correctly by methods for removing short-range LD. A list of these high-LD regions as found by PCA in populations of European ancestry is provided in Table 1 of (23) and code for removing these regions from PLINK files is provided at www.kcl.ac.uk/mmg/gwascode.

Small-scale LD can, in the first instance, be pruned using a sliding windows approach implemented in PLINK. Pairwise r^2 directly measures among-SNP correlation and so is the recommended LD measure to use for pruning. A threshold of $r^2 < 0.2$ or $r^2 < 0.3$ should be sufficient to remove high LD. I recommend using large window sizes – say enough SNPs to cover a roughly 2 Mb region, and incrementing by 10% of the window size at each iteration.

In the final stage, one can evaluate how good a job has been done on LD-pruning by examining the distribution of SNP coefficients (here called SNP "loadings" – see Subheading 2.5) from PCA of the LD-pruned dataset. If local LD features have been removed, the top PC axes should reflect features such as

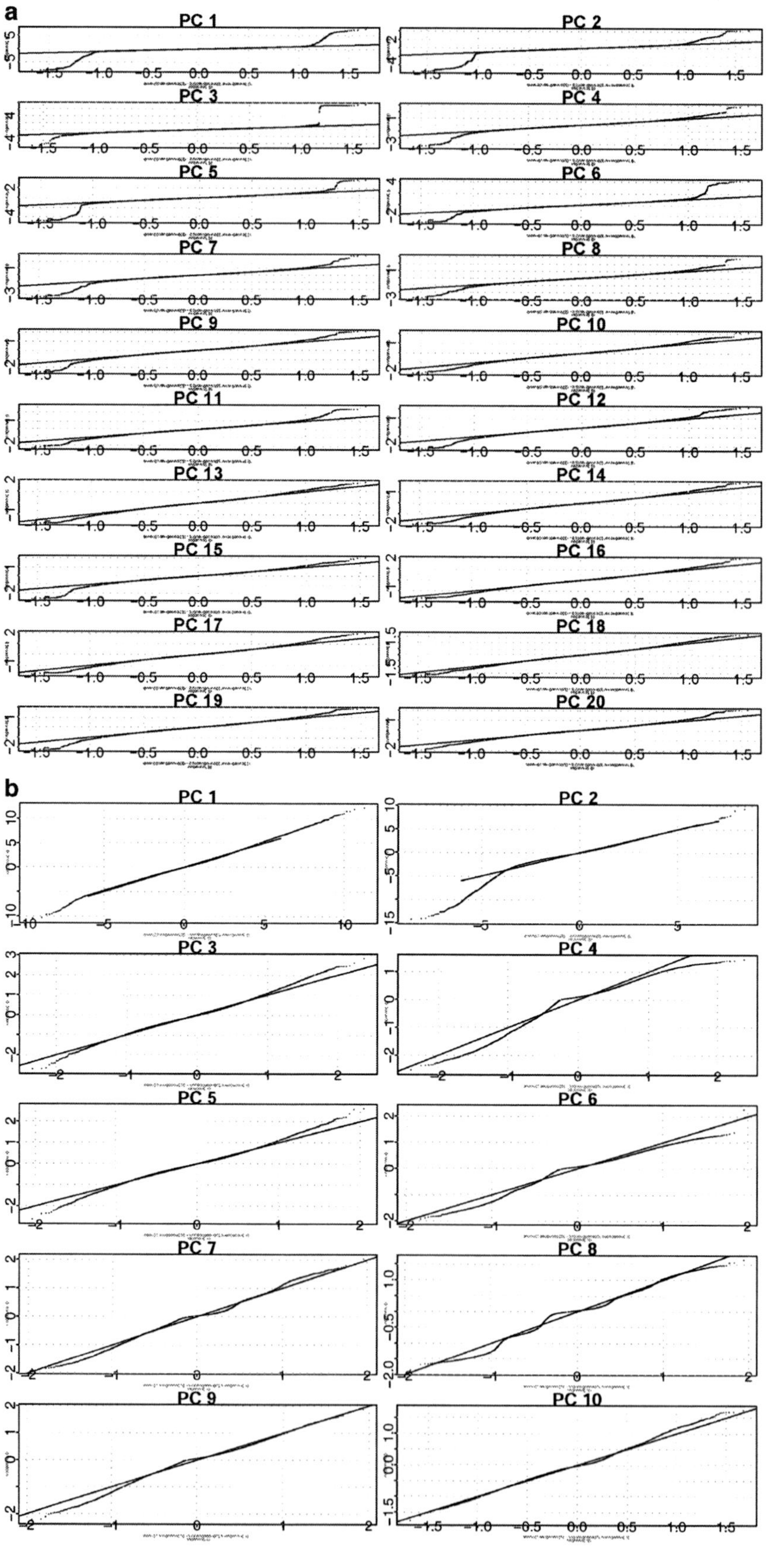
a
PC 1
PC 2
PC 3
PC 4
PC 5
PC 6
PC 7
PC 8
PC 9
PC 10
PC 11
PC 12
PC 13
PC 14
PC 15
PC 16
PC 17
PC 18
PC 19
PC 20
b
PC 1
PC 2
PC 3
PC 4
PC 5
PC 6
PC 7
PC 8
PC 9
PC 10

relatedness and population structure which on average affect all SNPs in the same way. This results in SNP loadings that are approximately normally distributed (Fig. 4b). In contrast, PC axes that are driven by local high-LD features are by definition driven by just those SNPs that belong to these regions, leading to characteristic heavy-tailed quantile–quantile plots (Fig. 4a). If local LD effects are still seen, one can either try more stringent LD-pruning to reduce the dataset still further, or one could apply an alternative method proposed by (24) for correcting local LD, based on regressing each SNP genotype score on the previous *k* SNPs and using the residual to form the correlation matrix for PCA.

6.3. How Filter?

If two individuals come from the same population but are not closely related, one can derive a formula for the probability that they will share 0 (AA vs. BB), 1 (AA or BB vs. AB), or 2 (AA vs. AA, AB vs. AB, BB vs. BB) alleles in common at a given SNP locus. All one requires is an estimate of the frequency of allele A in the population, which can be derived from the GWAS sample itself on the assumption that most individuals do indeed come from this population. Averaging the sharing scores across all autosomal SNP loci in the GWAS panel provides the so-called identity-by-state (IBS) estimate, and again one can derive the expectation of this for two unrelated individuals from the underlying allele frequencies. Two individuals who are closely related will have their observed IBS shifted upwards, because they are more likely to share genotypes than the background chance. This extra degree of sharing is dictated by their identity-by-descent (IBD), which is the proportion of their genomes which are the same due to recent familial transmission. Again formulae can be written for the expected IBS for a given IBD, and this can be used to estimate IBD based on observed IBS (see Note 4).

Figure 5 shows a histogram of IBD estimates using data from (18). The estimates form well-defined clusters around IBD = 1 (duplicates or monozygotic twins), IBD = 0.5 (1st degree relatives: sibs, parent/offspring), and IBD = 0.25 (second degree relative: half-sibs, grand-parent/grand-child, avuncular such as uncle/nephew). This separation provides empirical confidence in calling these close relatives correctly. The large spike at IBD = 1 is due to a large number of duplicate samples in this dataset. The apparent spike at IBD = 0.45 is in fact a single pair of relatives that

Fig. 4. Quantile–quantile (Q–Q) plots of SNP coefficients (here called SNP "loadings") of the top ten principal component axes from principal components analysis as described in (25). (**a**) Data (58C + NBS + CD cohorts, after all individual QC steps carried out) from (18), showing many heavy-tailed Q–Q plots, indicative of principal component axes that are driven by local large-scale high-LD genomic regions. (**b**) Data (58C + NBS + CD cohorts, after LD-pruning), showing approximately normal Q–Q plots, indicative of principal component axes that are driven by pan-genomic correlation effects such as population stratification

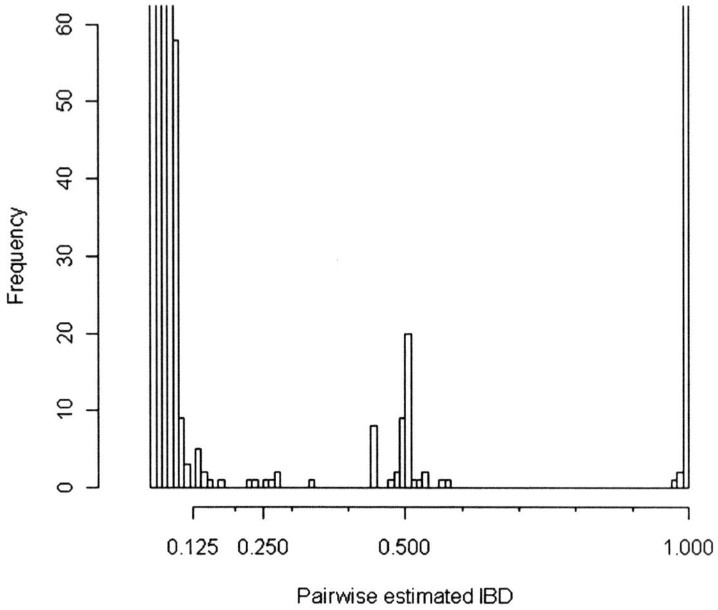

Fig. 5. Histogram of pairwise identity-by-descent (IBD) estimates. Only pairs with IBD > 0.05 are shown. Data (58C + NBS + CD cohorts, after individual missingness and gender QC) from (18)

has become amplified due to the presence of multiple duplicate DNA samples. The single observation at IBD = 0.3325 could stem from a chance departure of the genomic similarity from IBD = 0.25, or could reflect an unusual family situation. For example, if two children share the same father and have mothers who are half-sisters, the children's IBD would be 0.3125, with no inbreeding involved.

Below IBD = 0.25 the calling becomes more problematic, eventually merging with the general background degree of sharing that all individuals in a population have. There is a small hump around IBD = 0.125 (third degree relatives: e.g. full cousins), but it is indistinct. This, together with the small degree of genetic sharing of third degree relatives motivates an empirical threshold for QC set at half-way between second degree and third degree relatives (i.e., at IBD = 0.1875). For related pairs or family groups above this threshold, the usual QC step is to leave one individual in the dataset and drop the other or others, based for example on the one with the least missingness.

Figure 6 shows a network plot which can be useful for visualizing complex family structures involving more than one member. For the data from (18), once duplicate samples have been removed, only two examples of family groups involving more than two individuals are found (Fig. 6a). A more interesting example involving multiple families is provided in Fig. 6b for comparison.

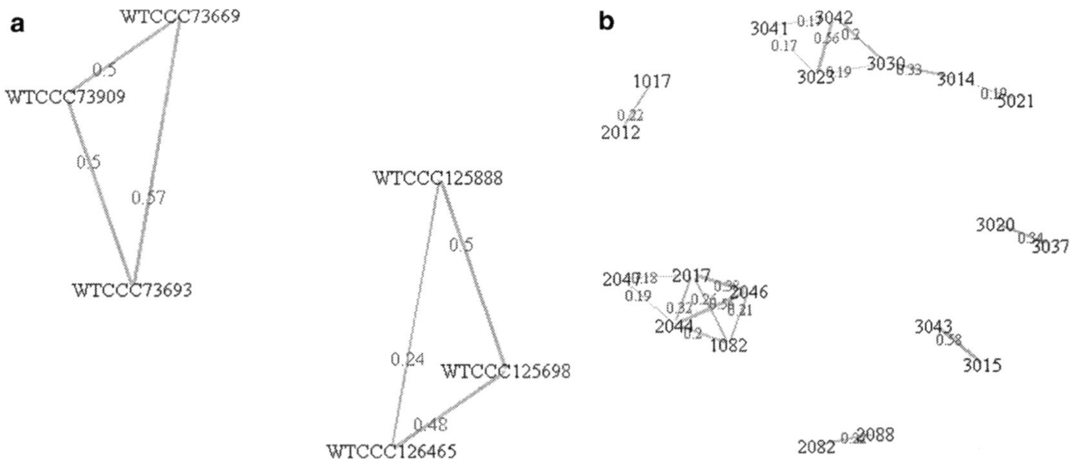

Fig. 6. Networks of pairwise identity-by-descent (IBD) estimates. (**a**) Data (58C + NBS + CD cohorts, after individual missingness and gender QC and after removal of duplicates) from (18). Only networks of size greater than 2, filtered by IBD > 0.1875, are shown. (**b**) Example from a second GWAS dataset, revealing more complex family relationships

7. Individual QC: Population Outliers

7.1. Why Filter?

As with cryptic relatedness, unaccounted-for correlation structure due to unknown population stratification can lead to both false positives and false negatives. Again there are alternatives to filtering, and indeed a very active literature on methods for correcting population stratification (16, 17). EIGENSTRAT (24, 25) has emerged as a popular method for achieving this in GWAS settings, and this is based on using the projection of individuals onto principal component (PC) axes from principal component analysis (PCA) of SNP genotype data as covariates in subsequent association analyses (see Subheading 2.5). In principle, one could use this method to correct for population outliers too, and thus keep them in the analysis. However, unless these outliers lay perfectly along existing PC axes, one would have to introduce one PC axis for each additional outlier, negating any power advantage from keeping the outlier in. It seems safer to remove such outliers, therefore, and keep EIGENSTRAT for correction of subtler and more population stratification trends that may still remain after filtering.

7.2. How Filter?

Two main approaches have arisen for population outlier filtering, one based on the use of reference population data to aid the interpretation of outliers and the other intrinsic to the GWAS data alone. Both have their advantages.

Figure 7 illustrates the first approach, using data from four of the eleven HapMap Phase III populations (www.hapmap.org)

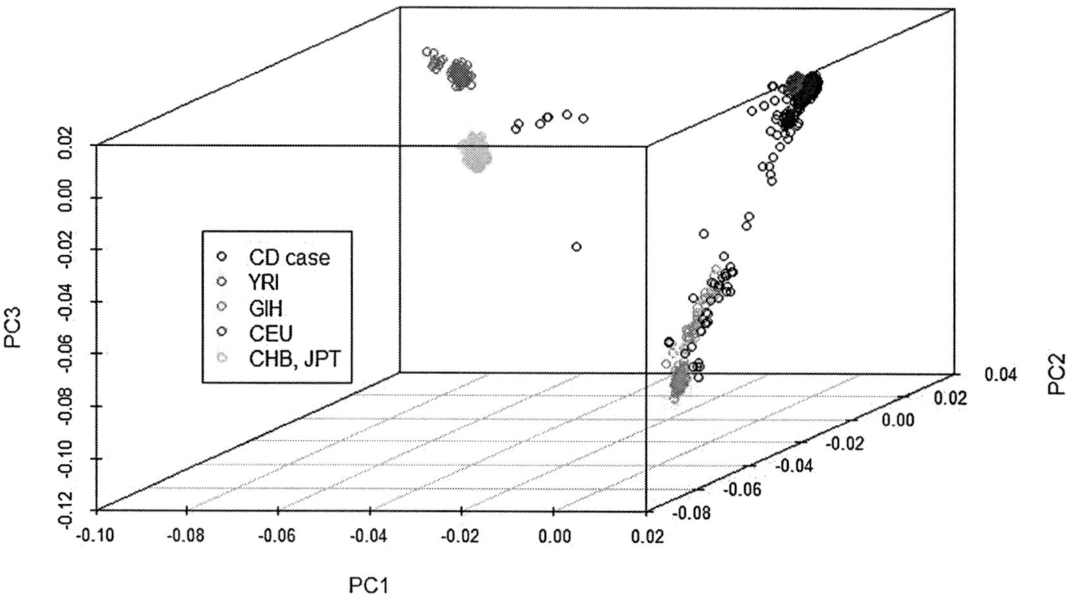

Fig. 7. 3D plot showing the location along three main PC axes of Crohn's Disease samples from (18), when entered into a PCA along with data from unrelated samples taken from four HapMap Phase III populations: CEU (Utah residents with European ancestry), YRI (Yoruba from Ibadan, Nigeria with West African ancestry), CHB + JPT (combined Han Chinese and Japanese with East Asian ancestry) and GIH (Guajarati with sub-continental Indian ancestry). Code for merging reference datasets and plotting this figure is available at www.kcl.ac.uk/mmg/gwascode

combined with Crohn's disease data from (18) and using the EIGENSTRAT-PCA method described in (25). Data have been restricted to an LD-pruned SNP set as described in Subheading 6.2, with some further restrictions to help the merging of different datasets (see Note 5). Figure 7 shows that while most Crohn's samples are clustered with HapMap Phase III European-ancestry data, some are clustered with one of the other HapMap Phase III populations, while others are strung out along a line connection two populations, suggestive of admixed individuals. Until recently (HapMap Phase I and II), only three major HapMap datasets (CEU with European ancestry, YRI with West African ancestry, and CHB + JPT with East Asian ancestry) were available for use as reference populations. Figure 7 shows the added value of adding the Phase III Gujarati dataset as a reference population, as people of Indian ancestry make up a sizable minority of the UK population from which the Crohn's cases were sampled. While there is some ability to pick up this admixture using the two PC axes (PC1 and PC2) that separate CEU from YRI and CHB + JPT, when PC3 (the axis separating out the Gujarati dataset) is added the signal is much clearer. Furthermore, this also allows one to properly distinguish Indian admixture from East Asian admixture.

The advantage of the externally-guided reference-population approach is that one has added reassurance that the outliers one sees are indeed population outliers (i.e., arising fmembership or part-membership of another population). The disadvantage is that an individual may be a population outlier, but the relevant population is not part of the set of reference populations being considered, as with the Indian outliers discussed earlier. With the availability now of eleven HapMap Phase III populations, and of other population samples with whole-genome SNP data such as the Human Genome Diversity Panel (http://hagsc.org/hgdp/index.html), this issue is less likely to cause problems provided one takes the time to enter a larger number of reference populations into the analysis. Nevertheless, there is still value to examining the outliers identified from a purely internally-guided approach, which looks for outliers along any top-ranking PC axis. A useful iterative procedure for doing this is provided in the EIGENSOFT software package (genepath.med.harvard.edu/~reich/Software.htm), which automatically reperforms PCA after every round of outlier removal.

Figure 8 illustrates the internally-guided approach. PC score histograms are presented for the first 20 PC axes of a PCA applied only to GWAS data from (18). Individual outliers are present on almost every axis. A drawback is that one cannot be sure that these are indeed population outliers, as opposed to outliers for some other reason. Cryptically related individuals and poor-quality DNA samples can also appear as outliers in this type of analysis. Thus one typically ends up with many more outliers identified by this approach than by the externally-guided approach. But this can also be an advantage – one can use this approach both as a way of confirming individuals that have also been eliminated from other QC steps and as a way of finding additional outliers, albeit of uncertain provenance.

Finally, note that PCA is not the only method available for this kind of filtering. Direct estimates of admixture among reference populations are possible using programs such as STRUCTURE (26) and FRAPPE (27). Furthermore, both the externally-guided and internally-guided approaches can be performed by other ordination methods. In particular, multidimensional scaling (MDS) has been implemented in PLINK and was used for population outlier QC by (18). The motivating principles behind PCA (a method for maximizing the variance accounted for by each of a ranked set of orthogonal axes) and MDS (a method for minimizing the difference between mapped interindividual distance on k axes and actual distance in a larger n-space) may appear very different, but mathematically they turn out to be identical under certain conditions. These conditions probably don't hold exactly for the situations considered here, but nearly so. The PLINK-MDS method also differs from EIGENSTRAT-PCA

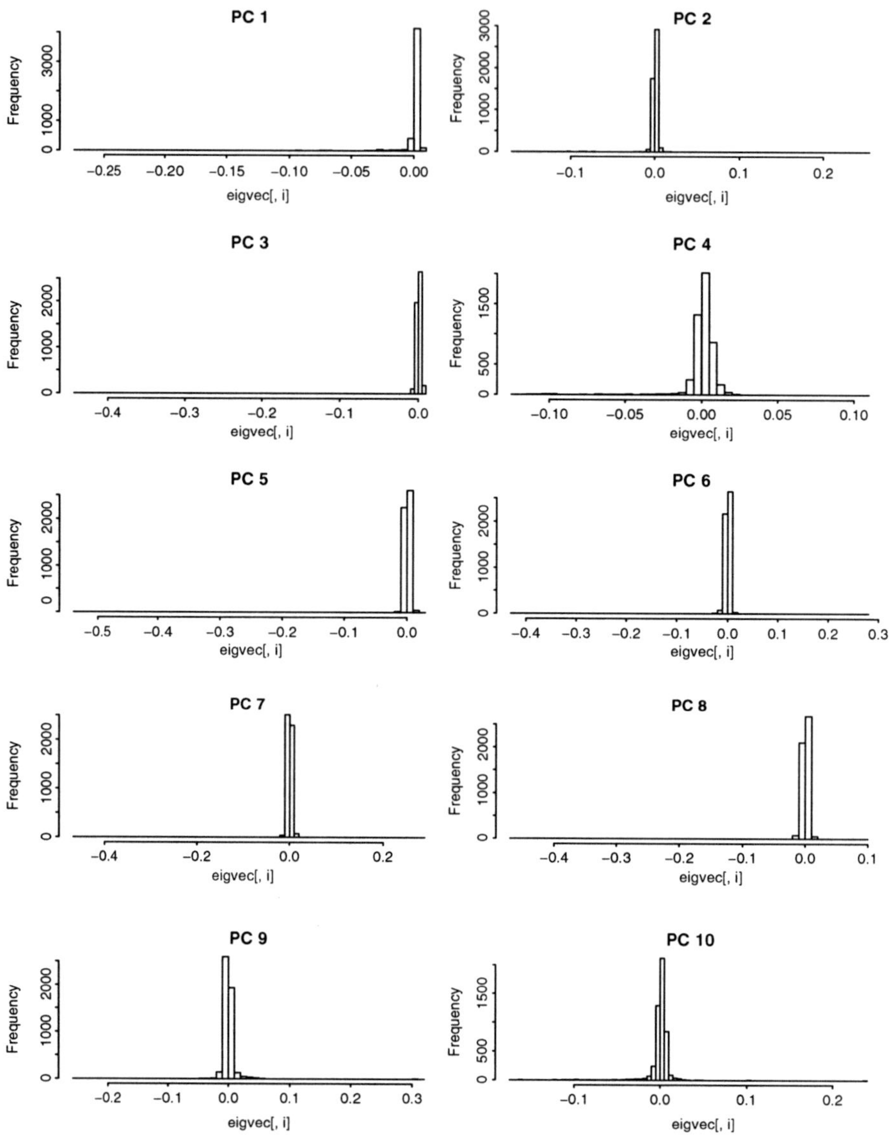

Fig. 8. PC score histograms for the first 20 PC axes of a PCA applied to data (58C + NBS + CD, after individual missingness, gender and cryptic relatedness QC) from (18). The *y*-axis has been cropped at 50 counts to allow singleton observations to be seen. R-code for plotting this figure is available at www.kcl.ac.uk/mmg/gwascode

by working on a different pairwise similarity matrix. While EIGENSTRAT-PCA works on a correlation matrix of scaled genotype scores, PLINK-MDS works on a similarity matrix of IBS scores. But again, the differences between these two similarity matrices are very slight. In practice, therefore, both methods produce very similar ordination plots. Where EIGENSTRAT-PCA gains the advantage, however, is in greater flexibility and more extensive diagnostics. In particular: (a) PC axes can be tested statistically to determine the best number of axes to take forward (24); (b) There is a developed method for correcting local LD by regressing on the previous k SNPs that can be used to adjust the similarity matrix if need be (24); and (c) One can interrogate SNP "loadings" to determine which PC axes are driven by pan-genomic signals (as one would expect for true population structure) or by local genomic high-LD features and/or sets of poor-quality SNPs (see Fig. 4).

8. Individual QC: Heterozygosity and Inbreeding

8.1. Why Filter?

Individuals who are the result of random mating within a single population should have genotypes that are in Hardy-Weinberg equilibrium (HWE). Similarly, under these conditions the proportion of an individual's GWAS panel SNPs which are heterozygous (the individual's panel-specific heterozygosity) is predictable from Hardy-Weinberg expectations and the minor allele frequency at each SNP. Wright's inbreeding coefficient F is directly related to departure from HWE: positive F indicates an excess of homozygotes (low heterozygosity), negative F indicates an excess of heterozygotes (high heterozygosity).

Individuals with anomalously high or low F indicate that the underlying sampling assumptions for that individual have been broken. Anomalously low F (high heterozygosity) can indicate sample contamination (i.e., a mixture of two or more DNAs, leading to more apparent heterozygotes). Anomalously high F (low heterozygosity) can indicate membership of a different population (the Wahlund effect) or indeed could indicate inbreeding. Thus departures from expected heterozygosity indicate either DNA quality problems or problems with the presumed correlation structure of individuals, which provide the justification for their removal.

8.2. How Filter?

The inbreeding coefficient F is somewhat preferable to heterozygosity as the metric of interest, because the latter is dependent on the specific distribution of minor allele frequencies of SNPs in the GWAS panel in question. In practice, however, this makes little odds unless different individuals have been typed on different GWAS panels, and even then the MAF distribution typically does not vary very much and roughly adopts a uniform distribution between 0 and 0.5. Estimation of F proceeds as for the X-chromosome specific F used in gender QC, except now the X-chromosome is excluded and only autosomal SNPs included. Figure 9 illustrates how a histogram plot of F usually forms for the most part a single cluster near $F=0$, but some obvious outliers can also be seen.

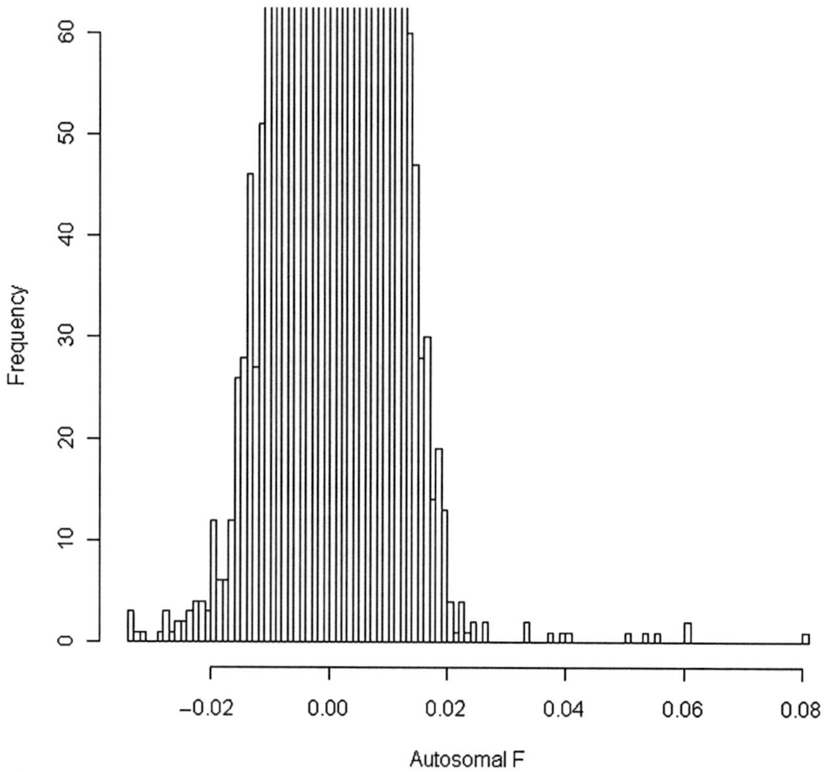

Fig. 9. Histogram of the inbreeding coefficient *F*, a measure of departure from Hardy-Weinberg Equilibrium averaged over all SNPs in a GWAS panel for each individual. The *y*-axis has been cropped at 60 counts to allow singleton observations to be seen. Data (58C + NBS + CD cohorts, after individual missingness, gender, cryptic relatedness and population outlier QC) from (18). R-code for plotting this figure is available at www.kcl.ac.uk/mmg/gwascode

9. SNP QC: Missingness

9.1. Why Filter?

SNP missingness is the complement of individual missingness. It is a must-have QC step due to the strong correlation of missingness with SNP quality and the impact of informative missingness on both false positive and false negative signals of association.

9.2. How Filter?

Figure 10 shows a quantile–quantile (Q–Q) plot of SNP-by-SNP association statistics, after different SNP missingness thresholds have been applied. Q–Q plots are a popular and useful way of visualizing GWAS data (see Subheading 2.6). The rise above the unit diagonal toward the upper end of the plot indicates association "hits" – values that one would not expect under the null hypothesis. Points that rise high enough (indicated by the horizontal line in Fig. 10) are declared "hits".

The pool of hits contains both real and false positive signals. The extent of the latter is indicated by the reduction in departure from the null as QC stringency increases. This either means that

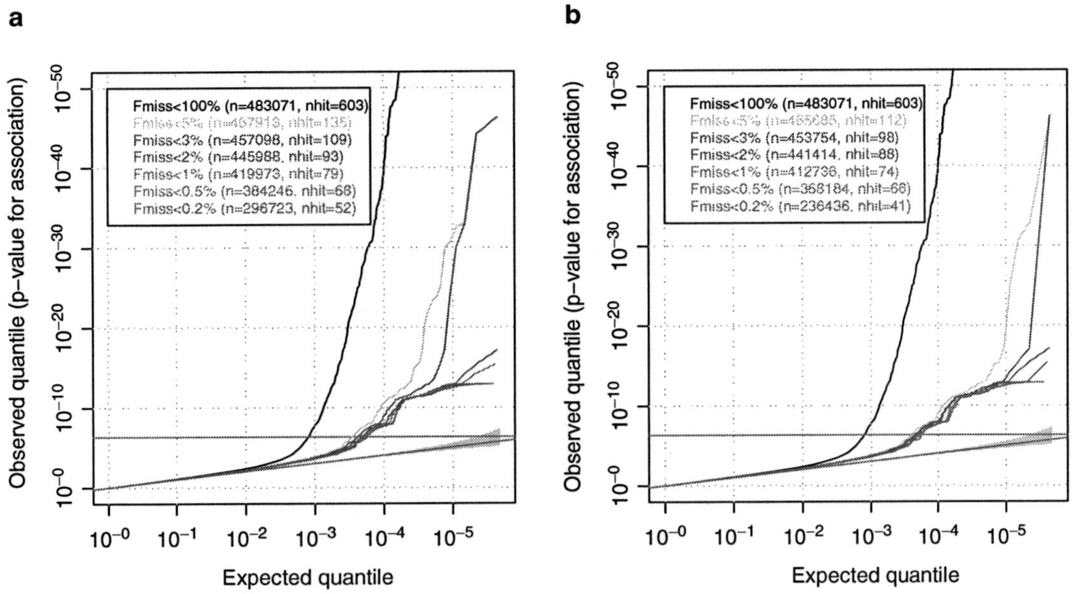

Fig. 10. Quantile–quantile (Q–Q) plot of SNP-by-SNP association statistics, after different SNP missingness thresholds have been applied. Here the association statistics come from the p-value of a logistic regression for trend (genotypes coded 0, 1, 2), with gender added as a cofactor for X-chromosome SNPs. The horizontal line represents the $p = 5 \times 10^{-7}$ threshold. The shaded area indicates the 95% "concentration band" (see Subheading 2.6). Data (58C + NBS + CD cohorts, after individual missingness, gender, cryptic relatedness, population outlier and heterogeneity QC) from (18). (**a**) Missingness thresholds applied to cases and controls combined. (**b**) Missingness thresholds required to be met in cases and controls separately. R-code for plotting this figure is available at www.kcl.ac.uk/mmg/gwascode

true association signals are more prevalent in SNPs with higher missingness (and in general, there is no reason why that should be the case), or that false positive signals of association are being removed (very plausible due to the informative missingness problem – see Subheading 4). Accordingly, this motivates a direct approach to setting an SNP missingness QC threshold, *Fmiss*, based on its effect on association signals. In the case of Fig. 10, setting *Fmiss* < 0.02 appears to have little additional effect on the Q–Q plot, compared to setting *Fmiss* = 0.02, so this setting appears a good choice. It makes little difference whether missingness thresholds are applied to cases and controls combined (Fig. 1a) to cases and controls separately (Fig. 1b), at least for this dataset.

SNP missingness can have an even greater impact on low-MAF SNPs, because rare genotypes can end up with an even smaller ratio of calls to no-calls. It therefore makes sense to explore *Fmiss* for low-MAF SNPs separately. Figure 11 shows the same Q–Q plots as before, but with SNPs split into MAF > 5% and MAF < 5%. Choice of this MAF level is arbitrary, but is empirically motivated in that a sizeable fraction of SNPs (about 20% of the total for this dataset) fall below this level. The plot of MAF > 5% SNPs confirms *Fmiss* = 0.02 as a useful threshold, but suggests a

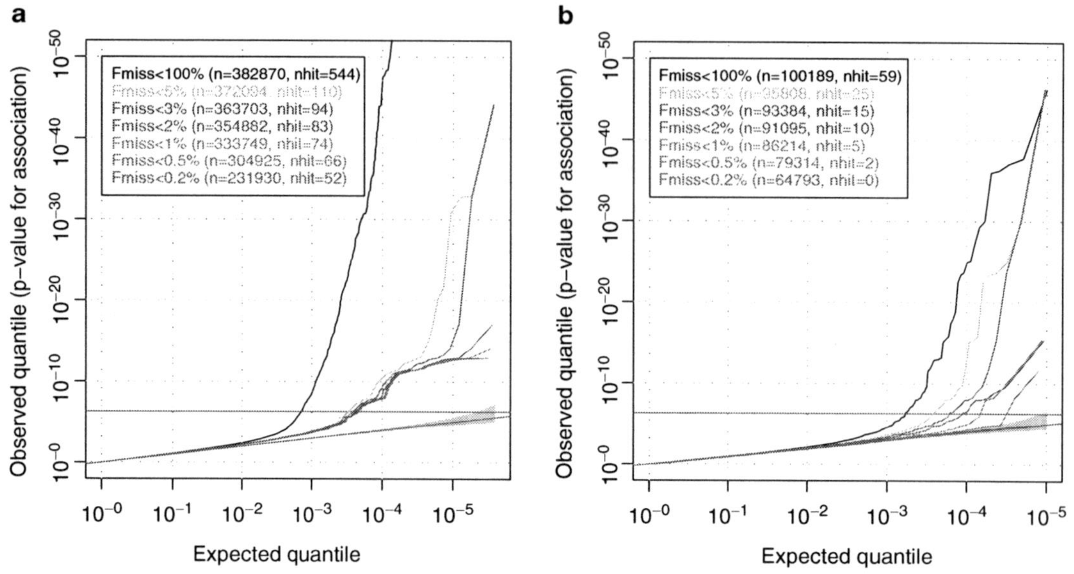

Fig. 11. Quantile–quantile (Q–Q) plots of SNP-by-SNP association statistics (log-transformed *p*-value from logistic trend tests), after different SNP missingness thresholds have been applied. Data and statistics are as for Fig. 10, with missingness thresholds applied to cases and controls combined. (**a**) Data restricted to MAF > 5%. (**b**) Data restricted to MAF < 5%. R-code for plotting this figure is available at www.kcl.ac.uk/mmg/gwascode

more stringent threshold of *Fmiss* = 0.005 would be better for MAF < 5% SNPs. Finally, note that it is important to keep an eye on the number of SNPs being removed as well. Reducing *Fmiss* below 0.005, for example, has the severe effect of removing 50% or more of MAF < 5% SNPs and therefore cannot be recommended.

Finally, note that PLINK also has options to directly test the nonrandom distribution of SNP missingness with phenotype and (through the use of LD at neighbouring SNPs to impute missing genotypes) genotype. These make useful additional checks, although not as direct as the approach outlined earlier, which focuses specifically on the generation of false positive signals of association.

10. SNP QC: Minor Allele Frequency

10.1. Why Filter?

Two main arguments are used to justify filtering by minor allele frequency (MAF). The first is that data quality tends to decrease with decreasing MAF. This is because a low MAF implies rare genotypes which will be seen only a few times in your GWAS dataset. This in turn implies that there is less information that a genotype calling algorithm can use to call this genotype, and so calls are less certain. In addition, as discussed earlier, informative missingness can affect low-MAF SNPs more strongly and thereby increase the chances of a false positive signal. The second argument is that the power to detect an association signal decreases with decreasing MAF. There seems to be little point in including SNPs below a certain MAF in your analysis, because you will never be able to detect an association signal with them. All they will do is increase the number of tests performed and so decrease the power to detect signals in other SNPs by increasing the multiple testing penalty.

Both these arguments have counter-arguments. The data quality issue can be addressed by imposing stricter missingness filters for low-MAF SNPs (see *above*), while the power issue could be addressed via Bayesian approaches in downstream analyses (18, 28). These Bayesian approaches can both correctly account for MAF in evaluating evidence for an association signal and eliminate the need for a multiple testing penalty by instead using a fixed prior for any given SNP being a true hit, depending on the expected number of true hits in the genome rather than the number of tests performed. For these reasons, some GWAS (e.g. (18)) have not used MAF filters at all in their QC. An intermediate approach is to adopt a very light filter as outlined in a following section.

10.2. How Filter?

Two lines of reasoning suggest having MAF > $10/n$, where n is the number of samples, as a useful threshold. Firstly, for large n, an SNP with an observed MAF of $10/n$ is likely to be represented as 20 heterozygotes with the remainder as major homozygotes. Genotype calling algorithms tends to work better when there are at least 20–30 examples of a given genotype (4). Secondly, an SNP with MAF = $10/n$ at least stands some chance in theory of producing a low p-value for association. In a case-control study with equal numbers of cases and controls, if an SNP was represented by 21 heterozygotes, all of them cases, then for large n, the p-value is about $0.5^{21} \approx 5 \times 10^{-7}$, which just about makes the threshold needed for a single-SNP p-value to be considered a "hit".

But again, neither of these arguments is watertight. The problem of <20 examples of a given genotype also arises for SNPs with MAF < $\sqrt{(20/n)}$, a much higher threshold than MAF < $10/n$, this time for the minor homozygote rather than the heterozygote. And just because an SNP cannot achieve a very low p-value on its own does not make it useless. It may still be valuable, for example in haplotype analysis. Furthermore, the achievable single-SNP p-value will be lower for a GWAS with fewer cases than controls, and can be arbitrarily low for a GWAS involving a quantitative trait. The MAF > $10/n$ threshold is therefore suggested merely as a conveniently light threshold with some theoretical justification, to be used perhaps in Tier 1 QC but not in Tier 2 QC (see Subheading 2.3).

11. SNP QC: Hardy-Weinberg Equilibrium

11.1. Why Filter?

Departure from Hardy-Weinberg equilibrium (HWE) can indicate problems with genotype calling. A common scenario is when two of the three clouds representing the three genotypes on an allele intensity plot overlap with one another (as in Fig. 1b), leading the genotype calling algorithm to mistake them for one single genotype. This almost always leads to a very large departure from HWE. A false positive association signal will also be generated if the calling problem is greater in either cases or controls.

In case-control studies, a strong signal of association can itself lead to departure from HWE. Thus while some advocate throwing away such SNPs, others advocate using such departures to add further weight to tests for association (29, 30). However, very extreme departures from HWE are more likely to be caused by failures of genotype calling than by true association signals, and this may be used to guide the QC process (29). Since the definition of "extreme" depends on the underlying null distribution of *p*-values across all SNPs, a sensible approach is to identify bad SNPs by reference to a Q–Q plot (see Section 2.6).

11.2. How Filter?

Typically, a *p*-value for departure from HWE (pHWE) for each SNP is calculated. This can be obtained from a chi-square or from an exact test (31), although *p*-values from exact tests pose problems for Q–Q plots, as I illustrate in a following section.

Proponents of filtering point out that departure from HWE in controls will only occur if the disease prevalence is common, if controls have been selected not to have the disease, and if the genetic model is nonmultiplicative. Thus, filtering by pHWE calculated only from controls seems sensible. However, if cases and controls have been genotyped separately one may also wish to inspect the distribution of pHWE in cases as well.

Figure 12 shows a histogram and Q–Q plot of pHWE values for controls, generated from exact tests. Figure 12a illustrates the potential dangers of using exact tests for this type of analysis. These *p*-values do not come from a Uniform distribution between 0 and 1, and the problem affects both the large-scale and small-scale distribution, as the inset to Fig. 12a shows. This is a problem shared by all exact tests, even in cases where all *p*-values are generated under the Null. A common feature is the large spike at $p = 1$, followed by a noticeable gap between $p = 1$ and the next highest *p*-value. All exact tests are innately conservative, and for example, in this dataset the frequency of *p*-values less than $p = 0.9$ is only 0.78 (in theory it should be 0.9), and this is despite the clear presence of real departures from HWE at the lower end of the *p*-value distribution (see Note 6).

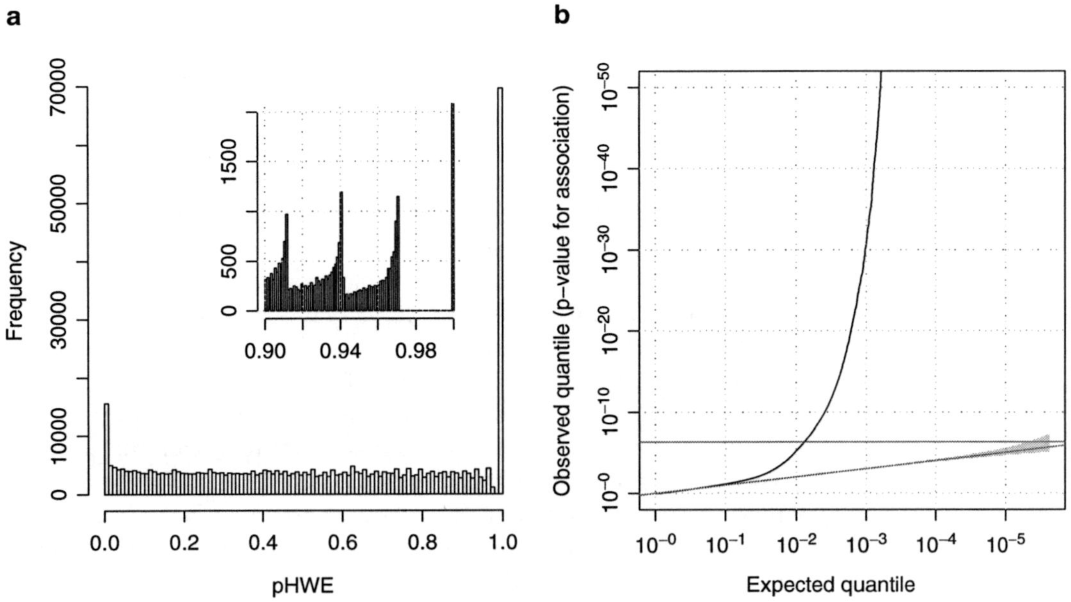

Fig. 12. (**a**) Histogram and (**b**) log-transformed quantile–quantile (Q–Q) plots of *p*-values for departure from Hardy-Weinberg equilibrium (pHWE), calculated from exact tests (31). Data (58C + NBS + CD cohorts, after individual missingness, gender, cryptic relatedness, population outlier, heterogeneity, and SNP missingness QC) from (18). R-code for plotting this figure is available at www.kcl.ac.uk/mmg/gwascode

For datasets with large numbers of SNPs departing very strongly from HWE, the conservative effect of using exact test *p*-values is trivial and does not alter the overall picture (Fig. 12b). Here the Q–Q plot departs sharply above the unit diagonal for values of $-2\log(\text{pHWE}) > 5$ ($\text{pHWE} < 0.082$). Some 46,700 SNPs (10.8% of the total) have *p*-values below this threshold. Considerable care is needed in interpreting Q–Q plots when the errant tail of the distribution is this large. Consider a simple model where N tests are made up of n_0 taken from the Null distribution and n_1 are taken from some alternative distribution such that the top K statistics contain all n_1 alternates. One can show from this model that the resulting Q–Q plot would start to rise above the unit diagonal a long time before one reaches the (N-K)th quantile, provided n_1 is large. Thus in the case of the data in Fig. 12b, it can be shown that a model where $K = 10{,}000$ and $n_1 = 1{,}000$ fits the distribution of *p*-values well (see www.kcl.ac.uk/mmg/gwascode for further details), even though 10,000 is considerably less than 46,700. Of course, this still implies that one would need to throw away 10,000 SNPs (equivalent to setting $\text{pHWE} > 0.002$ as a filter) in order to ensure removing the 1,000 non-Null SNPs, and furthermore this procedure would not address the concern that some of these non-Null SNPs might be due to real association rather than bad clustering.

A better QC approach might proceed as follows, therefore. First, set a genome-wide threshold for pHWE, say at 5×10^{-7}. In this case, 3,334 SNPs fail this threshold. If possible, inspect the cluster plots of all these SNPs to check for bad clustering. Next compute the departure of the observed frequency of heterozygotes from the expected under HWE and compare these against the patterns and magnitudes one might find for real association hits (see (29) for a convenient table). One should find that all these SNPs have excessive departure from HWE that is implausibly high for a real association signal. Next, perform a sensitively analysis for the effect of imposing stronger pHWE thresholds on the propensity for false positive signals of association (Fig. 13). In this case, we find that there is no tendency for SNPs with pHWE $> 5 \times 10^{-7}$ to affect the pool of association hits, which argues against an impact either from false positives or indeed from the loss of strong pHWE signals in true association hits for this dataset. Finally, one should refer back to pHWE information as a part of postassociation QC, when inspecting those SNPs which appear as hits in downstream association analyses.

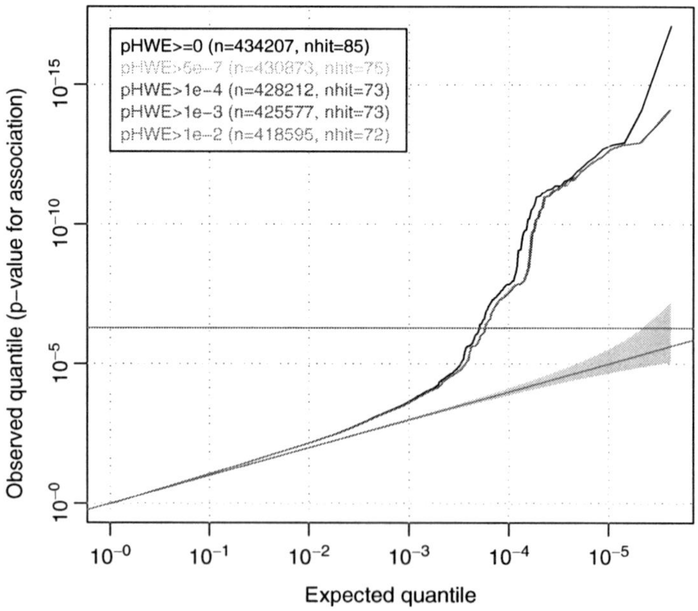

Fig. 13. Quantile–quantile (Q–Q) plots of SNP-by-SNP association statistics (log-transformed *p*-value from logistic trend tests), after different pHWE thresholds have been applied. The horizontal line represents the $p = 5 \times 10^{-7}$ threshold. The shaded area indicates the 95% "concentration band" (see Subheading 2.6). Data are as for Fig. 12. R-code for plotting this figure is available at www.kcl.ac.uk/mmg/gwascode

12. Other QC Filters

12.1. Individuals

Other QC checks for individuals generally revolve around the use of other phenotypic information, if available, that can be checked against the genotype information – for example, blood type and HLA type. Methods for inferring classical HLA types from GWAS SNP data are becoming clearer (32). In family-based GWAS, the same methods as used for uncovering cryptic relatedness can be used for verification of reported relationships.

12.2. SNPs

In family-based GWAS, Mendelian errors (i.e., a child with a genotype incompatible with those of his or her parents) can be used to filter out poor-quality SNPs. In GWAS with sample duplicates, poor concordance among duplicates can be used. Some SNPs (although increasingly few) are still not accurately mapped to the human genome, and so may be filtered out to avoid problems with haplotyping.

12.3. Postassociation SNP QC

Postassociation QC is applied to the pool of hits identified after association analysis. Because this pool of hits is much smaller than the complete set of SNPs in the GWAS panel, these QC checks can be carefully made for each one of these hits. Postassociation QC can therefore be a very useful and efficient method for spotting false positives, but equally it has no value in identifying false negatives.

One highly recommended check at this stage is to refer back to the allele signal intensity plots to inspect by eye for possible genotype calling problems. Another is to check that the association signal does not depend purely on the results from just one SNP. While this can be done simply by looking for some evidence for association at neighbouring SNPs in LD, a more thorough way is to remove the principal hit SNP, impute its genotypes from neighbouring SNPs, and confirm the signal from the imputed genotypes. In general, SNPs that are genomic neighbours are not "neighbours" in any sense on the GWAS panel, reducing the chance that a genotyping error affecting the hit SNP would also affect its genomic neighbours.

13. Concluding Remarks

Despite the increasing quality of genotype calling in GWAS panels, we need to apply quality control. Even subtle differences in DNA quality among individuals, or in their relatedness structure, can lead to a considerable pollution of true hits by false positives.

The affects of such differences only become amplified by increasing sample size and merging different datasets, making QC more rather than less important as "meta" and "mega" analyses of GWAS datasets become more prevalent. While QC is in one sense a clumsy tool in that it results in data being thrown away, the alternative of modelling differences in data quality explicitly becomes increasingly infeasible as datasets become larger, and also requires stronger modelling assumptions. Thus QC remains an essential component of GWAS analysis.

14. Notes

1. The discussion of the Chiamo algorithm in the Supplementary Materials section of (18) also notes the counterintuitive relationship between increased stringency of the genotype calling threshold and increased problems of informative missingness.
2. Note that in PLINK, heterozygous X-chromosome genotypes in male individuals are set as missing data. This can inflate the apparent missingness of females who have been mislabelled as males, since X-chromosome SNPs can make up a large proportion of a GWAS panel. One way to avoid this is to perform individual missingness QC using autosomal SNPs only.
3. Klinefelter syndrome (XXY karyotype with a prevalence of about 0.15% in males) results in a "male" phenotype and a "female" X-chromosome genotype pattern, while Turner syndrome (XO karyotype with a prevalence of about 0.05% in females) results in a "female" phenotype and a "male" X-chromosome genotype pattern. Both these syndromes can be resolved if Y chromosome SNPs form part of the GWAS panel (true for the newer panels). Other conditions, like Androgen Insensitivity Syndrome (genetically male, phenotypically female) and Congenital Adrenal Hyperplasia (genetically female, phenotypically male), are not resolvable with Y chromosome data but are rarer still.
4. For further details of how this is done in PLINK (where estimated IBD is called "PI_HAT"), see (9). Note that IBD estimation works best for large datasets where the majority of individuals are indeed from the same population, which indeed is usually the case for GWAS.
5. When merging datasets, one should ensure that all SNPs have been typed in all datasets and, preferably, that all SNPs are nonsymmetric (meaning that no SNPs are A/T or C/G SNPs). The latter restriction allows unequivocal mapping of all SNPs in all datasets to the same strand. Stripping out

symmetric SNPs is of little consequence for data from Illumina GWAS datasets, as these panels have been designed to have almost no symmetric SNPs in them. Affymetrics GWAS panels, in contrast, have appreciable number of symmetric SNPs (for example, ~15% on their 500 k panel). However, experience suggests that for PCA analyses this still leaves more than enough SNPs for accurate inference of population ancestry patterns, and if LD-pruning is applied after removing symmetric SNPs then the resulting SNP set is unlikely to be that much smaller than with symmetric SNPs included.

6. The general effect of using exact p-values on Q–Q plots is that, as a consequence of the over abundance of $p=1$ values, the plot under the Null still follows a unit slope but this slope is displaced below the main diagonal that goes through the origin. One way to empirically correct for this would be to remove enough $p=1$ values to ensure that 10% of p-values had $p<0.1$, before plotting the Q–Q plot. This is based on the assumption that only p-values generated under the Null have $p<0.1$.

References

1. Johnson, A.D. and O'Donnell, C.J. (2009) An open access database of genome-wide association results. *BMC Med Genet*, **10**, 6.
2. Amos, C.I. (2007) Successful design and conduct of genome-wide association studies. *Hum Mol Genet*, **16 Spec No. 2**, R220–R225.
3. McCarthy, M.I., Abecasis, G.R., Cardon, L.R., Goldstein, D.B., Little, J., Ioannidis, J.P. and Hirschhorn, J.N. (2008) Genome-wide association studies for complex traits: consensus, uncertainty and challenges. *Nat Rev Genet*, **9**, 356–369.
4. Neale, B.M. and Purcell, S. (2008) The positives, protocols, and perils of genome-wide association. *Am J Med Genet B Neuropsychiatr Genet*, **147B**, 1288–1294.
5. Pearson, T.A. and Manolio, T.A. (2008) How to interpret a genome-wide association study. *JAMA*, **299**, 1335–1344.
6. Teo, Y.Y. (2008) Common statistical issues in genome-wide association studies: a review on power, data quality control, genotype calling, and population structure. *Curr Opin Lipidol*, **19**, 133–143.
7. Ziegler, A., Konig, I.R. and Thompson, J.R. (2008) Biostatistical aspects of genome-wide association studies. *Biom J*, **50**, 8–28.
8. Zondervan, K.T. and Cardon, L.R. (2007) Designing candidate gene and genome-wide case-control association studies. *Nat Protoc*, **2**, 2492–2501.
9. Purcell, S., Neale, B., Todd-Brown, K., Thomas, L., Ferreira, M.A., Bender, D., *et al.* (2007) PLINK: a tool set for whole-genome association and population-based linkage analyses. *Am J Hum Genet*, **81**, 559–575.
10. Plagnol, V., Cooper, J.D., Todd, J.A. and Clayton, D.G. (2007) A method to address differential bias in genotyping in large-scale association studies. *PLoS Genet*, **3**, e74.
11. Aulchenko, Y.S., Ripke, S., Isaacs, A. and van Duijn, C.M. (2007) GenABEL: an R library for genome-wide association analysis. *Bioinformatics*, **23**, 1294–1296.
12. Kang, H.M., Zaitlen, N.A., Wade, C.M., Kirby, A., Heckerman, D., Daly, M.J. and Eskin, E. (2008) Efficient control of population structure in model organism association mapping. *Genetics*, **178**, 1709–1723.
13. Anderson, C.A., Pettersson, F.H., Barrett, J.C., Zhuang, J.J., Ragoussis, J., Cardon, L.R. and Morris, A.P. (2008) Evaluating the effects of imputation on the power, coverage, and cost efficiency of genome-wide SNP platforms. *Am J Hum Genet*, **83**, 112–119.
14. Nothnagel, M., Ellinghaus, D., Schreiber, S., Krawczak, M. and Franke, A. (2009) A comprehensive evaluation of SNP genotype imputation. *Hum Genet*, **125**, 163–171.

15. Pei, Y.F., Li, J., Zhang, L., Papasian, C.J. and Deng, H.W. (2008) Analyses and comparison of accuracy of different genotype imputation methods. *PLoS One*, **3**, e3551.
16. Tian, C., Gregersen, P.K. and Seldin, M.F. (2008) Accounting for ancestry: population substructure and genome-wide association studies. *Hum Mol Genet*, **17**, R143–R150.
17. Tiwari, H.K., Barnholtz-Sloan, J., Wineinger, N., Padilla, M.A., Vaughan, L.K. and Allison, D.B. (2008) Review and evaluation of methods correcting for population stratification with a focus on underlying statistical principles. *Hum Hered*, **66**, 67–86.
18. The Wellcome Trust Case Control Consortium. (2007) Genome-wide association study of 14,000 cases of seven common diseases and 3,000 shared controls. *Nature*, **447**, 661–678.
19. Giannoulatou, E., Yau, C., Colella, S., Ragoussis, J. and Holmes, C.C. (2008) GenoSNP: a variational Bayes within-sample SNP genotyping algorithm that does not require a reference population. *Bioinformatics*, **24**, 2209–2214.
20. Lin, Y., Tseng, G.C., Cheong, S.Y., Bean, L.J., Sherman, S.L. and Feingold, E. (2008) Smarter clustering methods for SNP genotype calling. *Bioinformatics*, **24**, 2665–2671.
21. Clayton, D.G., Walker, N.M., Smyth, D.J., Pask, R., Cooper, J.D., Maier, L.M., *et al.* (2005) Population structure, differential bias and genomic control in a large-scale, case-control association study. *Nat Genet*, **37**, 1243–1246.
22. Tian, C., Plenge, R.M., Ransom, M., Lee, A., Villoslada, P., Selmi, C., *et al.* (2008) Analysis and application of European genetic substructure using 300 K SNP information. *PLoS Genet*, **4**, e4.
23. Price, A.L., Weale, M.E., Patterson, N., Myers, S.R., Need, A.C., Shianna, K.V., *et al.* (2008) Long-range LD can confound genome scans in admixed populations. *Am J Hum Genet*, **83**, 132–135; author reply 135–139.
24. Patterson, N., Price, A.L. and Reich, D. (2006) Population structure and eigenanalysis. *PLoS Genet*, **2**, e190.
25. Price, A.L., Patterson, N.J., Plenge, R.M., Weinblatt, M.E., Shadick, N.A. and Reich, D. (2006) Principal components analysis corrects for stratification in genome-wide association studies. *Nat Genet*, **38**, 904–909.
26. Pritchard, J.K., Stephens, M. and Donnelly, P. (2000) Inference of population structure using multilocus genotype data. *Genetics*, **155**, 945–959.
27. Tang, H., Peng, J., Wang, P. and Risch, N.J. (2005) Estimation of individual admixture: analytical and study design considerations. *Genet Epidemiol*, **28**, 289–301.
28. Wakefield, J. (2008) Reporting and interpretation in genome-wide association studies. *Int J Epidemiol*, **37**, 641–653.
29. Wittke-Thompson, J.K., Pluzhnikov, A. and Cox, N.J. (2005) Rational inferences about departures from Hardy-Weinberg equilibrium. *Am J Hum Genet*, **76**, 967–986.
30. Won, S. and Elston, R.C. (2008) The power of independent types of genetic information to detect association in a case-control study design. *Genet Epidemiol*, **32**, 731–756.
31. Wigginton, J.E., Cutler, D.J. and Abecasis, G.R. (2005) A note on exact tests of Hardy-Weinberg equilibrium. *Am J Hum Genet*, **76**, 887–893.
32. Leslie, S., Donnelly, P. and McVean, G. (2008) A statistical method for predicting classical HLA alleles from SNP data. *Am J Hum Genet*, **82**, 48–56.

Chapter 20

Gaining a Pathway Insight into Genetic Association Data

Inti Pedroso

Abstract

The recent application of high throughput genotyping in humans has yielded numerous insights into the genetic basis of human phenotypes and unprecedented amount of genetic variation data. Each genome wide significant finding has explained only a tiny proportion of phenotypic variation, yet genome wide association studies (GWAS) in their entirety can provide unprecedented windows into the molecular genetics of these phenotypes. New methods are emerging to mine modest association signals from these data using information on biological pathways and networks underlying the phenotype variation. These methods promise to enhance the information extracted from GWAS providing grounds for follow up studies of both a genetic and molecular nature.

Key words: Genome, GWAS, Bioinformatics, Variation, Pathway, QTL

1. Introduction

Recent technological advances have consolidated the genetic mapping of human traits as a high throughput and relatively successful approach (1). By the 23rd of April 2009, the Catalog of Published GWAS (http://www.genome.gov/gwastudies/) registered 304 publications and 232 SNPs that reached the widely agreed genome wide significance threshold of 7.5×10^{-8} (2). Although these results do indeed represent a great advance in the field, there are several unresolved problems:

(a) Each of the identified genetic variants explains a tiny proportion of the phenotype variation attributable to genes (3)

(b) After applying the stringent statistical thresholds necessary for multiple testing, there are very few variants (many times none!) representing a very reduced picture of the phenotype's underpinning biology

Michael R. Barnes and Gerome Breen (eds.), *Genetic Variation: Methods and Protocols*, Methods in Molecular Biology, vol. 628, DOI 10.1007/978-1-60327-367-1_20, © Springer Science+Business Media, LLC 2010

(c) We now appreciate that some replicated *loci* do not reach genome-wide significance in individual studies (e.g. PPARG in type 2 diabetes (4)). However, there is no systematic statistical method available to extract information from these numerous true hits immersed in the noise of multiple testing;

(d) There is a lack of formal methods to interpret the findings (e.g. which variants are causative for a given association). Researchers' interpretations can possibly lead to biased conclusions and let aside relevant information.

(e) The coverage of the genotyping platforms may be still low, particularly in regions with high levels of genetic heterogeneity;

(f) Phenotypic heterogeneity may be more pervasive than previously thought. This may be particularly problematic in psychiatric genetics and other fields, where tests for physical symptoms are not available or are poorly defined;

(g) The statistical tests for genotype–phenotype correlation may not capture the real underlying effect of genetic variation, e.g. epistatic effects;

(h) The current approach for GWAS is designed to capture association with common variants, making it less successful for phenotypes with a different genetic architecture (e.g. some variants identified have odds ratios of 3–10, but their frequency tends to be <0.1% in case populations and <0.01% in controls (5)).

In neuropsychiatry, the one notable exception to this is the APOE4 allele (6), which may account for more than half the genetic variance in risk for Alzheimer's disease (AD). Although it was discovered by linkage analysis in the preGWAS era, it is the only replicated locus for Alzheimer's disease in GWASs (as yet), and its discovery has not yet led directly to new treatments for AD, although it is showing the potential to influence clinical practice and treatment on the basis of subclassification of the disease into early and late onset forms (7). Nevertheless, APOE and other examples of replicated *loci* with poor power in diagnostics settings have led to strong criticism of the current GWAS strategy (8).

However, the main strength of GWAS may be to open a window into the molecular mechanism underlying trait variation by providing many thousands of marginal significant results, in which lie the "higher hanging fruit" of real but nongenome-wide significant associations due to either low effect size or a lack of power caused by a low allele frequency and the limited sample sizes utilized in GWAS studies to date. In order to uncover these true associations, we need tools to interpret genetic associations with regard to the genes they may influence and the biological processes these genes act on. These tools aim to identify biological processes or gene networks that are drivers or confer susceptibility to the phenotype under study, improving the interpretation of the results and increasing power to detect genes with small effect sizes that cluster within common biological processes, which may in turn prove a fruitful target for treatment. The need for identification of disease relevant biological processes makes pathway or Gene Set Analysis (GSA) a good candidate to mine information from GWAS.

We give a summary of how to explore GWAS results using biological pathway information to address the questions: (a) if the evidence from a GWAS is overrepresented in some known biological pathways, and (b) how this information can be used to derive information about the biology underlying the experiment results.

2. Materials

All the tools described here are freely available and accessible on the internet. A list of tools and databases mentioned in this review is given in Table 1.

3. Methods

3.1. Mapping Genetic Associations to Genes

The first necessary step is to map genetic variation to genes present in gene-sets (a technical representation of a biological pathway). Gene-sets are genes grouped by a biological criterion, like coexpression or a metabolic pathway, which we will test for enrichment of significant genetic association. This is a sensitive point as it defines which genes and genetic variation will be included in the analysis. At the moment, there is no "gold standard" (see Note 1) and many of the decisions on how to do so rely on what have to be relatively subjectively defined criteria.

There are several public tools that can be used to retrieve or generate SNP-gene mapping (Table 1). The Ensembl project provides an automatic annotation of SNPs to genes that can be accessed using Biomart (9). For large-scale analysis, the Ensembl Perl API can be used allowing for more complex criteria (LD or distance). Similar analyzes can also be performed on a smaller scale using the SNAP interface (Table 1). The Tamal database provides functional annotation for SNPs and SNP to gene association can be retrieved from the web portal (Table 1) or using PLINK's annotation tool. Galaxy (Table 1) provides easy to use tools to perform a wide range of data analyses, such as sorting, pattern matching, filtering and

Table 1
Tools to explore GWAS using GSA

Mapping genetic variation to genes	
Tamal	neoref.ils.unc.edu/tamal/
PLINK	pngu.mgh.harvard.edu/~purcell/plink/
SNAP	www.broad.mit.edu/mpg/snap/
BrainArray	brainarray.mbni.med.umich.edu/Brainarray/
Biomart	www.ensembl.org/biomart/martview/
Ensembl Perl API	www.ensembl.org/info/docs/api/index.html
Galaxy	main.g2.bx.psu.edu/
eQTL resources	
eQTL Web Browser	eqtl.uchicago.edu
GENEVAR	www.sanger.ac.uk/humgen/genevar/
Lab Functional Neurogenomics	labs.med.miami.edu/myers/LFuN/LFuN.html
Pathway definition resources	
BioCarta	www.biocarta.com
KEGG	www.genome.jp/kegg
Gene Ontology	www.geneontology.org
MSigDB	www.broad.mit.edu/gsea/msigdb
Gene set analysis tools	
GSEA	www.broad.mit.edu/gsea
FatiGO	fatigo.bioinfo.cnio.es
Set of tools for Gene Ontology	www.geneontology.org/GO.tools.shtml
GenGen Package	www.openbioinformatics.org/gengen
GSEA-SNP	www.nr.no/pages/samba/area_emr_smbi_gseasnp

operations with genomic coordinates. Video tutorials are available to provide the users with the building bricks to develop their analysis pipeline. Data can be fetched from public databases (like UCSC and Ensembl) or uploaded from users' computers and analysis can be optimized in workflows (10).

Association between genetic variation and mRNA expression levels, so called expression quantitative trait loci (eQTL) can inform on the functional impact of a given associated variant or markers in LD with that variant. Such information has been successfully used to map disease-susceptibility variants affecting mRNA levels, providing a possible biological mechanism to explain the genetic association (11). This functional information can be freely accessed in the public domain (see Table 1 for some eQTL resources) and can be added as a criterion to map genetic variants to genes (see Note 2).

3.2. Gene Set Analysis

Gene set analysis (GSA) or pathway analysis refers to the statistical assessment of enrichment of significant results in a predefined group of genes (gene-set or biological pathway). There are numerous variants of GSA (12), and we have listed some popular software to perform the analysis in Table 1. All listed sites have documentation for different steps of the analysis.

GSA provides evidence for association at the pathway level evaluating the signal across the constituent genes, it seeks for consistent evidence of association across the pathway, and it overcomes the limitations of analyzing just the most significant hits (13, 14). Evidence from pathway analyzes in mRNA microarray expression studies shows that it is possible to detect subtle but consistent association at the pathway level (e.g. average 20% increase in mRNA levels), replication has been suggested to be more frequent at the pathway than at the gene level, and pathways are better predictors of disease status and outcome than single genes (15, 16). In addition, because the functional relationships built into the pathway are known, the findings provide ground for a more biologically comprehensive explanation and help to protect from common biases, such as retrospective hypothesis generation (13) (see Note 3).

Gene-set definition can be obtained by literature review or other high throughput experiments, such as mRNA microarrays, protein–protein or genetic interaction, allowing the incorporation of previous knowledge (e.g. tissue specific expression profiles or manually curated gene–gene interaction information) and the evaluation of convergent evidence towards a specific biological process (13).

Some commonly used sources of gene-sets are listed in Table 1. All provide some level of manual curation of gene-sets. However, it is important to note that manual curation is time consuming and some resources may not be up to date. On the other hand gene-sets extracted from high throughput experiments (e.g. lists of differentially expressed genes) without manual curation may introduce additional noise.

3.3. Gene Set Analysis for GWAS

Among the different variants of GSA listed in Table 1, GenGen and GSEA-SNP are two packages that have been specifically designed to perform GSA on GWAS data. These two packages implement the method introduced by Subramanian et al. (14) with some modification because of differences in data type. A statistically significant GSA result is usually calculated by permuting phenotype labels or genes among gene-set to a derived null distribution (14). In contrast with gene-expression results, in GWAS data, there is a bias toward large genes having more significant findings because more genetic variants are mapped to them leading to a higher rate of false positive associations because of multiple testing. In addition, there are some gene-sets that are enriched in large genes (e.g. gene sets that are rich in large transmembrane proteins, such as those involved in certain neurotransmission pathways). These two factors can potentially produce false positives if one uses the raw *p*-values from GWAS to calculate the significance using gene shuffling (Fig. 1). A possible solution to this problem is to derive a statistic not biased by gene size for each gene (17). This is an interesting alternative approach as it opens the door for multivariate analysis to combine the effect of multiple genetic variants and allows the use of summary statistics from GWAS. Another possibility is to calculate the significance level by permuting the phenotype. However, individual level genotype data are required though these may not be accessible in all cases.

3.4. Interpretation of GSA of GWAS and GWAS Interpretation by GSA

Except for some caveats (see Note 4), statistically significant results point to an enrichment of genetic association in a gene-set. Because the gene-sets are built based on known criteria, the interpretation of the results are easier than analyzing each association independently. Several publications have used GSA methods to interpret GWAS data providing empirical evidence that these can help extract and compile the information in a biologically meaningful manner (18–20).

Because many recent GWAS are somewhat underpowered, it is expected that there are many genes harboring genetic variants with similar effect sizes than can be found at a genome wide significant threshold of association. GSA can help prioritization of these associations by providing statistical support for gene-sets. Genes included in these gene-sets can be screened in silico for putative functional variants. Several tools provide functional annotation for SNPs, including Ensembl, TAMAL and BrainArray (Table 1). The UCSC Table browser (Karolochik, 2004) and Galaxy (Table 1) provide some basic tools which allow users to

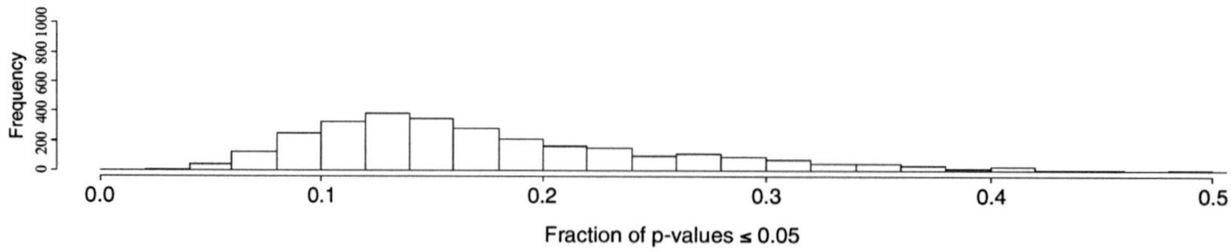

Fig. 1. Type I error of GSA with raw *p*-values from GWAS. A null distribution of association was generated by calculating association *p*-values for 1,000 random phenotype vectors for a sample of 3,000 controls and 2,000 cases genotyped in approx. 360,000 SNPs. GSA was performed on each GWAS result mapping SNPs to genes within 20 kb, using the minimum *p*-value of each gene and shuffling genes among pathway to calculate the gene-set significance. The figure shows the distribution of the fraction of times a *p*-value ≤ 0.05 was obtained for each pathway. Using raw *p*-values from GWAS and calculating the gene-set significance by gene shuffling produce in average ~15% false positives, with some gene-sets reaching *p*-values ≤ 0.05 as often as 50% of the times

develop custom annotations, which may be focussed on unpublished experimental data (allowing analysis of user-submitted data), or it may be more specific for some functional information (e.g. epigenetic regulation).

Perhaps, the main advantage of GSA is to provide a biological hypothesis which can be interrogated to test the effect of putatively functional genetic variations found in association studies. For example, if GSA identifies a pathway involved in the regulation of hormone secretion, it may be a sensible follow up to study the effect of genetic variation in genes connected with hormone secretion.

There is theoretical background as well as some empirical evidence supporting the use of gene-set methods for GWAS (21–23). The main argument being that the underlying cause of phenotypic variability are changes in biological processes at the cellular and organismal level, meaning that perturbations can arise from any element acting on these process. This would predict that *loci* influencing phenotype variation could be found in different genes belonging to the same biological process. Given the power characteristics of GWAS, it is expected that different studies may show associations in different *loci* which are connected within the same pathway and hence lead to the same or similar phenotypes. At first glance, using conventional genetic association analysis, association at different loci in the same pathway would be seen as lack of replication, but GSA represents a framework to interpret these apparent discrepancies and provide replication at the gene-set level. Wang et al. (19) applied GSA to several Crohn's disease datasets showing significant enrichment in the IL12 pathway in all data sets. Examination of the association of each gene across the different studies showed not all genes reach significance in all studies. This strongly suggests that GSA can aid both the interpretation of association under the stringent statistical threshold of GWAS and overcome genetic heterogeneity by providing a broader, yet meaningful, framework for replication of GWAS findings.

3.5. Conclusions

We have reviewed the main necessary steps to perform GSA on GWAS results. Some of the steps may require additional data analyses, but in general public resources provide the necessary tools to perform the analysis. Application of GSA on GWAS has revealed interesting insights into the molecular pathway underlying trait variation. As its application becomes more common, replication at the gene-set level will provide stronger support for the application of this analyses. Current evidence suggests that application of GSA on GWAS data can help prioritizing genes and molecular phenotypes for additional studies, extracting information from hundreds, or perhaps thousands, of weak but possibly true associations and overcoming genetic heterogeneity by providing a biologically meaningful framework for replication of GWAS findings.

4. Notes

1. SNP to gene Mapping: The first question to ask is what kind of functional relationship one wants to map. For example, one may want to map genetic variation affecting protein function (e.g. nonsynonymous variants) or include variation affecting gene transcript expression. Both are not mutually exclusive, but the results can be quite different. In the first case, it may be sufficient to take associated variants that map directly to a gene; however, a more sensitive approach should also map variants in Linkage Disequilibrium (LD) with associated variants to genes. "Mapping to a gene" should include the coding region but also introns (impact on splicing can affect protein function as well) and untranslated region (possibly impacting microRNA regulation). In the case of variants impacting transcript expression, it is necessary to define a distance limit because including genetic variation affecting possible enhancers (which can be hundreds of kb away) may introduce excessive noise in downstream analysis. A sensible approach may define the limits based on evidence from eQTL studies. These studies have shown that the 95% of genetic variation affecting transcript levels is within 20 kb of the transcription start and end sites (24). Examples of both approaches can be found in the literature (25), but no systematic comparison of the effect on downstream analysis has been done.
2. It should be noted that most eQTL studies (as with most current GWAS) are underpowered to detect subtle effects on transcription and in the case of *trans*-eQTLs (eQTLs caused by variants remote to the gene locus), they are highly prone to type I error. The error rate of *cis*-eQTLS (effects caused by variants in the gene locus) type I error should be much lower, however, there is still a possibility that the observed effect is caused by variants in the expression probe binding site.
3. There are several variants of GSA and network analysis, but all of them test a list of genes for enrichment in specific gene-sets grouped using biological criteria, such as membership of a metabolic or signalling pathway, or protein complex, etc. (26). A drawback of GSA is its strong dependence on the availability and quality of gene-sets. Better annotation is expected for well studied processes, such as cancer related pathways, when compared with less well characterized biological processes (e.g. Psychiatric disease pathways).
4. GSA was originally designed and tested on gene-expression data. Gene transcript expression can be highly correlated, making the identification of significant gene-sets somehow easier because

expression of genes have been shown to change in a concerted manner. Furthermore, gene-expression analyses capture changes that are both cause and consequence of the phenotype, providing a *double signal* to pick up. In addition, changes in gene expression can be of several orders of magnitude making detection relatively easy. In contrast, genotype-phenotype correlations of common variants with complex traits (the focus of current GWAS) have subtle effect sizes, they are not correlated with each other (although epistasis is expected to play a role it has been challenging to find in GWAS) and the magnitude of change is much smaller than in gene-expression studies. These have important practical consequences. Whereas in gene-expression, a robust gene-set finding is supported by the several genes, providing evidence that a significant portion of a biological process is deregulated, in a GWAS one significant association in the same gene-set may not be statistically significant in the gene-set analysis, but its functional consequences may indeed perturb the entire biological process. Although this does not invalidate the application of these methods, it must be taken into account when interpreting the results. On the other hand, statistical support from GSA or network analysis provide strong evidence for the biological process contributing to phenotype variation because each of the signals is independent from each other (as far the genetic variants are not in LD), each represents additional evidence and they are not a consequence of the phenotype.

References

1. Altshuler, D., Daly, M.J. and Lander, E.S. (2008) Genetic mapping in human disease. *Science*, **322**, 881–888.
2. Dudbridge, F. and Gusnanto, A. (2008) Estimation of significance thresholds for genomewide association scans. *Genet. Epidemiol.*, **32**, 227–234.
3. McCarthy, M.I. and Hirschhorn, J.N. (2008) Genome-wide association studies: potential next steps on a genetic journey. *Hum. Mol. Genet.*, **17**, R156–R165.
4. Altshuler, D., Hirschhorn, J.N., Klannemark, M., Lindgren, C.M., Vohl, M.C., Nemesh, J., *et al.* (2000) The common PPARgamma Pro12Ala polymorphism is associated with decreased risk of type 2 diabetes. *Nat. Genet.*, **26**, 76–80.
5. Stefansson, H., Rujescu, D., Cichon, S., Pietilainen, O.P., Ingason, A., Steinberg, S., *et al.* (2008) Large recurrent microdeletions associated with schizophrenia. *Nature*, **455**, 232–236.
6. Pericak-Vance, M.A., Bebout, J.L., Gaskell, P.C., Jr., Yamaoka, L.H., Hung, W.Y., Alberts, M.J., *et al.* (1991) Linkage studies in familial Alzheimer disease: evidence for chromosome 19 linkage. *Am. J. Hum. Genet.*, **48**, 1034–1050.
7. Roses, A.D., Saunders, A.M., Huang, Y., Strum, J., Weisgraber, K.H. and Mahley, R.W. (2007) Complex disease-associated pharmacogenetics: drug efficacy, drug safety, and confirmation of a pathogenetic hypothesis (Alzheimer's disease). *Pharmacogenomics J.*, **7**, 10–28.
8. Goldstein, D.B. (2009) Common genetic variation and human traits. *N. Engl. J. Med.*, **360**, 1696–1698.
9. Smedley, D., Haider, S., Ballester, B., Holland, R., London, D., Thorisson, G. and Kasprzyk, A. (2009) BioMart – biological queries made easy. *BMC Genomics*, **10**, 22.
10. Taylor, J., Schenck, I., Blankenberg, D. and Nekrutenko, A. (2007) Using galaxy to perform large-scale interactivve data

analyses. *Curr. Protoc. Bioinformatics*, **Chapter 10**, Unit.

11. Webster, J.A., Gibbs, J.R., Clarke, J., Ray, M., Zhang, W., Holmans, P., *et al.* (2009) Genetic control of human brain transcript expression in Alzheimer disease. *Am. J. Hum. Genet.*, **84**, 445–458.
12. Liu, Q., Dinu, I., Adewale, A.J., Potter, J.D. and Yasui, Y. (2007) Comparative evaluation of gene-set analysis methods. *BMC Bioinformatics*, **8**, 431.
13. Lehner, B. and Lee, I. (2008) Network-guided genetic screening: building, testing and using gene networks to predict gene function. *Brief. Funct. Genomic. Proteomic.*, 7, 217–227.
14. Subramanian, A., Tamayo, P., Mootha, V.K., Mukherjee, S., Ebert, B.L., Gillette, M.A., *et al.* (2005) Gene set enrichment analysis: a knowledge-based approach for interpreting genome-wide expression profiles. *Proc. Natl. Acad. Sci. U.S.A.*, **102**, 15545–15550.
15. Chuang, H.Y., Lee, E., Liu, Y.T., Lee, D. and Ideker, T. (2007) Network-based classification of breast cancer metastasis. *Mol. Syst. Biol.*, **3**, 140.
16. Jiang, Z. and Gentleman, R. (2007) Extensions to gene set enrichment. *Bioinformatics*, **23**, 306–313.
17. Li, J. and Ji, L. (2005) Adjusting multiple testing in multilocus analyses using the eigenvalues of a correlation matrix. *Heredity*, **95**, 221–227.
18. Bergholdt, R., Storling, Z.M., Lage, K., Karlberg, E.O., Olason, P.I., Aalund, M., *et al.* (2007) Integrative analysis for finding genes and networks involved in diabetes and other complex diseases. *Genome Biol.*, **8**, R253.
19. Wang, K., Zhang, H., Kugathasan, S., Annese, V., Bradfield, J.P., Russell, R.K., *et al.* (2009) Diverse genome-wide association studies associate the IL12/IL23 pathway with Crohn Disease. *Am. J. Hum. Genet.*, **84**, 399–405.
20. Baranzini, S.E., Galwey, N.W., Wang, J., Khankhanian, P., Lindberg, R., Pelletier, D., *et al.* (2009) Pathway and network-based analysis of genome-wide association studies in multiple sclerosis. *Hum. Mol. Genet.*
21. Gibson, G. (2009) Decanalization and the origin of complex disease. *Nat. Rev. Genet.*, **10**, 134–140.
22. Wu, X., Jiang, R., Zhang, M.Q. and Li, S. (2008) Network-based global inference of human disease genes. *Mol. Syst. Biol.*, **4**, 189.
23. Loscalzo, J., Kohane, I. and Barabasi, A.L. (2007) Human disease classification in the postgenomic era: a complex systems approach to human pathobiology. *Mol. Syst. Biol.*, **3**, 124.
24. Veyrieras, J.B., Kudaravalli, S., Kim, S.Y., Dermitzakis, E.T., Gilad, Y., Stephens, M. and Pritchard, J.K. (2008) High-resolution mapping of expression-QTLs yields insight into human gene regulation. *PLoS Genet.*, **4**, e1000214.
25. Moskvina, V., Craddock, N., Holmans, P., Nikolov, I., Pahwa, J.S., Green, E., Owen, M.J. and O'Donovan, M.C. (2009) Gene-wide analyses of genome-wide association data sets: evidence for multiple common risk alleles for schizophrenia and bipolar disorder and for overlap in genetic risk. *Mol. Psychiatry*, **14**, 252–260.
26. Nam, D. and Kim, S.Y. (2008) Gene-set approach for expression pattern analysis. *Brief. Bioinformatics*, **9**, 189–197.

INDEX

Michael R. Barnes and Gerome Breen (eds.), *Genetic Variation: Methods and Protocols*, Methods in Molecular Biology, vol. 628, DOI 10.1007/978-1-60327-367-1, © Springer Science+Business Media, LLC 2010

R

S

T

U

V

W

Y

LaVergne, TN USA
20 September 2010
197683LV00003B/3/P

9 781603 273664